原版影印说明

1.《聚合物百科词典》(5册) 是 Springer Reference *Encyclopedic Dictionary of Polymers*(2nd Edition) 的影印版。为使用方便，由原版2卷改为5册：

第1册 收录 A–C 开头的词组；

第2册 收录 D–I 开头的词组；

第3册 收录 J–Q 开头的词组；

第4册 收录 R–Z 开头的词组；

第5册 为原书的附录部分及参考文献。

2. 缩写及符号、数学符号、字母对照表、元素符号等查阅说明各册均完整给出。

由 Jan W. Gooch 主编的《聚合物百科词典》是关于高分子科学与工程领域的参考书，2007年出版第一版，2011年再版。本书收录了7 500多个高分子材料方面的术语，涉及高分子材料的各个方面，如粘合剂、涂料、油墨、弹性体、塑料、纤维等，还包括生物化学和微生物学方面的术语，以及与新材料、新工艺相关的术语；并且不仅包括其物理、电子和磁学性能方面的术语，还增加了数据处理的统计和数值分析以及实验设计方面的术语。每个词条方便查找，并给出了简洁的定义，以及相互参照的相关术语。为了说明得更清晰，全书给出1 160个图、73个表。有的词条还给出方程式、化学结构等。

材料科学与工程图书工作室

联系电话 0451–86412421

0451–86414559

邮　　箱 yh_bj@aliyun.com

xuyaying81823@gmail.com

zhxh6414559@aliyun.com

聚合物百科词典

Jan W. Gooch

Encyclopedic Dictionary of Polymers
2nd Edition

VOLUME 2
D–I

黑版贸审字08-2014-010号

Reprint from English language edition:
Encyclopedic Dictionary of Polymers
by Jan W.Gooch
Copyright © 2011 Springer New York
Springer New York is a part of Springer Science+Business Media
All Rights Reserved

This reprint has been authorized by Springer Science & Business Media for distribution in China Mainland only and not for export therefrom.

图书在版编目（CIP）数据

聚合物百科词典.2, D~I:英文/（美）古驰（Gooch, J. W.）主编.—哈尔滨:哈尔滨工业大学出版社，2014.3

（Springer词典精选原版系列）

ISBN 978-7-5603-4443-0

Ⅰ.①聚… Ⅱ.①古… Ⅲ.①聚合物–词典–英文 Ⅳ.①O63-61

中国版本图书馆CIP数据核字（2013）第292198号

责任编辑　许雅莹　张秀华　杨　桦
出版发行　哈尔滨工业大学出版社
社　　址　哈尔滨市南岗区复华四道街10号　邮编150006
传　　真　0451-86414749
网　　址　http://hitpress.hit.edu.cn
印　　刷　哈尔滨市石桥印务有限公司
开　　本　787mm×1092mm 1/16　印张　15
版　　次　2014年3月第1版　2014年3月第1次印刷
书　　号　ISBN 978-7-5603-4443-0
定　　价　118.00元

（如因印刷质量问题影响阅读，我社负责调换）

Acknowledgements

The editor wishes to express his gratitude to all individuals who made available their time and resources for the preparation of this book: James W. Larsen (Georgia Institute of Technology), for his innovations, scientific knowledge and computer programming expertise that were invaluable for the preparation of the Interactive Polymer Technology Programs that accompany this book; Judith Wiesman (graphics artist), for the many graphical presentations that assist the reader for interpreting the many complex entries in this publication; Kenneth Howell (Springer, New York), for his continued support for polymer science and engineering publications; and Daniel Quinones and Lydia Mueller (Springer, Heidelberg) for supporting the printed book and making available the electronic version and accompanying electronic interactive programs that are important to the scientific and engineering readers.

Preface

The second edition of Encyclopedic Dictionary of Polymers provides 40% more entries and information for the reader. A Polymers Properties section has been added to provide quick reference for thermal properties, crystallinity, density, solubility parameters, infrared and nuclear magnetic spectra. Interactive Polymer Technology is available in the electronic version, and provides templates for the user to insert values and instantly calculate unknowns for equations and hundreds of other polymer science and engineering relationships. The editor offers scientists, engineers, academia and others interested in adhesives, coatings, elastomers, inks, plastics and textiles a valuable communication tool within this book. In addition, the more recent innovations and biocompatible polymers and adhesives products have necessitated inclusion into any lexicon that addresses polymeric materials. Communication among scientific and engineering personnel has always been of critical importance, and as in any technical field, the terms and descriptions of materials and processes lag the availability of a manual or handbook that would benefit individuals working and studying in scientific and engineering disciplines. There is often a challenge when conveying an idea from one individual to another due to its complexity, and sometimes even the pronunciation of a word is different not only in different countries, but in industries. Colloquialisms and trivial terms that find their way into technical language for materials and products tend to create a communications fog, thus unacceptable in today's global markets and technical communities.

The editor wishes to make a distinction between this book and traditional dictionaries, which provide a word and definition. The present book provides for each term a complete expression, chemical structures and mathematic expression where applicable, phonetic pronunciation, etymology, translations into German, French and Spanish, and related figures if appropriate. This is a complete book of terminology never before attempted or published.

The information for each chemical entry is given as it is relevant to polymeric materials. Individual chemical species (e.g., ethanol) were taken from he *CRC Handbook of Chemistry and Physics*, 2004 Version, the Merck Index and other reference materials. The reader may refer to these references for additional physical properties and written chemical formulae. Extensive use was made of ChemDraw®, CambridgeSoft Corporation, for naming and drawing chemical structures (conversion of structure to name and vice versa) which are included with each chemical entry where possible. Special attention was given to the IUPAC name that is often given with the common name for the convenience of the reader.

The editor assembled notes over a combined career in the chemical industries and academic institutions regarding technical communication among numerous colleagues and helpful acquaintances concerning expressions and associated anomalies. Presently, multiple methods of nomenclature are employed to describe identical chemical compounds by common and IUPAC names (eg. acetone and 2-propanone) because the old systems (19th century European and trivial) methods of nomenclature exists with the modern International Union of Pure and Applied Chemistry, and the conflicts between them are not likely to relent in the near future including the weights and measures systems because some nations are reluctant to convert from English to metric and, and more recently, the International Systems of Units (SI). Conversion tables for converting other systems to the SI units are included in this book for this purpose. In addition, there are always the differences in verbal pronunciation, but the reasons not acceptable to prevent cogent communication between people sharing common interests.

In consideration of the many challenges confronting the reader who must economize time investment, the structure of this book is optimized with regard the convenience of the reader as follows:

- Comprehensive table of contents
- Abbreviations and symbols
- Mathematics signs
- English, Greek, Latin and Russian alphabets
- Pronunciation/phonetic symbols
- Main body of terms with entry term in English, French German and Italian
- Conversion factors

- Microbiology nomenclature and terminology
- References

The editor acknowledges the utilization of many international sources of information including journals, books, dictionaries, communications, and conversations with people experienced in materials, polymer science and engineering. A comprehensive reference section contains all of the sources of information used in this publication. Pronunciation, etymological, cross-reference and related information is presented in the style of the 11th Edition of the Meriam-Webster Dictionary, where known, for each term. The spelling for each term is presented in German, French, and Spanish where translation is possible. Each term in this book includes the following useful information:

- Spelling (in **bold** face) of each term and alternative spellings where more than one derivation is commonly used
- Phonetic spelling \-\ using internationally published phonetic symbols, and this is the first book that includes phonetic pronunciation information missing in technical dictionaries that allows the reader to pronounce the term
- Parts of speech in English following each phonetic spelling, eg. *n.*, *adj.*
- Cross-references in CAPITALS letters
- Also called *example* in italics
- Etymological information [-] for old and new terms that provides the reader the national origins of terms including root words, prefixes and suffixes; historical information is critical to the appreciation of a term and its true meaning
- French, German, Italian and Spanish spellings of the term { - }
- A comprehensive explanation of the term
- Mathematical expressions where applicable
- Figures and tables where applicable
- A comprehensive reference section is included for further research

References are included for individual entries where a publication(s) is directly attributable to a definition or description. Not all of the references listed in the Reference section are directly attributable to entries, but they were reviewed for information and listed for the reader's information. Published dictionaries and glossaries of materials were very helpful for collecting information in the many diverse and smaller technologies of the huge field of polymers. The editor is grateful that so much work has been done by other people interested in polymers.

The editor has attempted to utilize all relevant methods to convey the meaning of terms to the reader, because a term often requires more information than a standard entry in a textbook dictionary, so this book is dedicated to a complete expression. Terminology and correct pronunciation of technical terms is continuously evolving in scientific and industrial fields and too often undocumented or published, and therefore, not shared with others sometimes leading to misunderstandings. Engineering and scientific terms describe a material, procedure, test, theory or process, and communication between technical people must involve similar jargon or much will be lost in the translation as often has been the editor's experience. The editor has made an attempt to provide the reader who has an interested in the industries that have evolved from adhesives, coatings, inks, elastomers, plastics and textiles with the proper terminology to communicate with other parties whether or not directly involved in the industries. This publication is a single volume in the form of a desk-handbook that is hoped will be an invaluable tool for communicating in the spoken and written media.

Physics, electronic and magnetic terms because they are related to materials and processes (e.g., *ampere*).

Biomolecular materials and processes have in the recent decade overlapped with polymer science and engineering. Advancements in polymeric materials research for biomolecular and medical applications are rapidly becoming commercialized, examples include biocompatible adhesives for sutureless tissue bonding, liquid dressings for wounds and many other materials used for *in vitro* and *in vivo* medical applications. To keep pace with these advancements, the editor has included useful terms in the main body that are commonly used in the material sciences for these new industries.

A microbiology section has been included to assist the reader in becoming familiar with the proper nomenclature of bacteria, fungi, mildew, and yeasts – organisms that affect materials and processes because they are ubiquitous in our environment. Corrosion of materials by microorganisms is commonplace, and identification of a specific organism is critical to prevent its occurrence. Engineers and materials scientists will appreciate the extensive sections on different types of microorganisms together with a section dedicated to microbiology terminology that is useful for communicating in the jargon of biologists instead of referring to all organisms as "bugs."

New materials and processes, and therefore new terms, are constantly evolving with research, development and global commercialization. The editor will periodically update this publication for the convenience of the reader.

Statistics, numerical analysis other data processing and experimental design terms are addressed as individual terms and as a separate section in the appendix, but only as probability and statistics relate to polymer technology and not the broad field of this mathematical science. The interactive equations are listed in the Statistics section of the Interactive Polymer Technology program.

Interactive Polymer Technology Programs

Along with this book we are happy to provide a collection of unique and useful tools and interactive programs along with this Springer Reference. You will find short descriptions of the different functions below. Please download the software at the following website: http://extras.springer.com/2011/978-1-4419-6247-8

Please note that the file is more than 200 MB. Download the ZIP file and unzip it. It is strongly recommended to read the **ReadMe.txt** before installing. The software is started by opening the file InPolyTech.pdf and following the instructions. Detailed instructions can be found under 'Help Instructions'.

The software consists of 15 programs and tools that are briefly described in the appendix.

Abbreviations and Symbols

Abbreviations	Symbols
An	absorption (formerly extinction) (= log t_i^{-1})
A	Area
A	surface
A	Helmholtz energy ($A = U - TS$)
A	preexponential constant [in $k = A \exp(-E^{\ddagger}/RT)$]
A_2	second virial coefficient
a	exponent in the property/molecular weight relationship ($E^{\ddagger} = KM^a$); always with an index, e.g., a_η, a_s, etc.
a	linear absorption coefficient, $a = l^{-1}$
absolute	abs
acre	spell out
acre-foot	acre-ft
air horsepower	air hp
alternating-current (as adjective)	a-c
A^m	molar Helmholtz energy
American Society for Testing and Materials	ASTM
amount of a substance (mole)	n
ampere	A or amp
ampere-hour	amp-hr
amplitude, an elliptic function	am.
angle	β
angle, especially angle of rotation in optical activity	∝
Angstrom unit	Å
antilogarithm	antilog
a_o	constant in the Moffit–Yang equation
Area	A
Atactic	at
atomic weight	at. wt
Association	Assn.
atmosphere	atm

Abbreviations	Symbols
average	avg
Avogadro number	N_L
avoirdupois	avdp
azimuth	az or α
barometer	bar.
barrel	bbl
Baumé	Bé
b_o	constant in the Mofit–Yang equation
board fee (feet board measure)	fbm
boiler pressure	spell out
boiling point	bp
Boltzmann constant	k
brake horsepower	bhp
brake horsepower-hour	bhp-hr
Brinell hardness number	Bhn
British Standards Institute	BSI
British thermal unit[1]	Btu or B
bushel	bu
C	heat capacity
c	specific heat capacity (formerly; specific heat); c_p = specific isobaric heat capacity, c_v = specific isochore heat capacity
c	"weight" concentration (= weight of solute divided by volume of solvent); IUPAC suggests the symbol ρ for this quantity, which could lead to confusion with the same IUPAC symbol for density
c	speed of light in a vacuum
c	speed of sound
calorie	cal
candle	c
candle-hour	c-hr
candlepower	cp
ceiling temperature of polymerization, °C	T_c

Abbreviations	Symbols
cent	c or ¢
center to center	c to c
centigram	cg
centiliter	cl
centimeter or centimeter	cm
centimeter-gram-second (system)	cgs
centipoise	cP
centistokes	cSt
characteristic temperature	Θ
chemical	chem.
chemical potential	μ
chemical shift	δ
chemically pure	cp
circa, about, approximate	ca.
circular	cir
circular mils	cir mils
cis-tactic	ct
C^m	molar heat capacity
coefficient	coef
cologarithm	colog
compare	cf.
concentrate	conc
conductivity	cond, λ
constant	const
continental housepower	cont hp
cord	cd
cosecant	csc
cosine	cos
cosine of the amplitude, an elliptic function	cn
cost, insurance, and freight	cif
cotangent	cot
coulomb	spell out
counter electromotive force	cemf
C_{tr}	transfer constant ($C_{tr} = k_{tr}/k_p$)
cubic	cu
cubic centimeter (liquid, meaning milliliter. ml)	cu, cm, cm^3
cubic centimeter	cm^3 cubic expansion coefficient ∝
cubic foot	cu ft
cubic feet per minute	cfm
cubic feet per second	cfs

Abbreviations	Symbols
cubic inch	cu in.
cubic meter	cu m or m^3
cubic micron	cu μ or cu mu or μ^3
cubic millimeter	cu mm or mm^3
cubic yard	cu yd
current density	spell out
cycles per second	spell out or c
cylinder	cyl
D	diffusion coefficient
D_{rot}	rotational diffusion coefficient
day	spell out
decibel	db
decigram	d.g.
decomposition, °C	T_{dc}
degree	deg or °
degree Celsius	°C
degree centigrade	C
degree Fahrenheit	F or °
degree Kelvin	K or none
degree of crystallinity	∝
degree of polymerization	X
degree Réaumur	R
delta amplitude, an elliptic function	dn
depolymerization temperature	T_{dp}
density	ρ
diameter	diam
Dictionary of Architecture and Construction	DAC
diffusion coefficient	D
dipole moment	p
direct-current (as adjective)	d-c
dollar	$
dozen	doz
dram	dr
dynamic viscosity	η
E	energy (E_k = kinetic energy, E_p = potential energy, $E^{\ddagger}$ = energy of activation)
E	electronegativity
E	modulus of elasticity, Young's modulus ($E = \sigma_{ij}/\varepsilon_{ij}$)
E	general property

Abbreviations	Symbols
E	electrical field strength
e	elementary charge
e	parameter in the Q-e copolymerize-tion theory
e	cohesive energy density (always with an index)
edition	Ed.
Editor, edited	ed.
efficiency	eff
electric	elec
electric polarizability of a molecule	α
electrical current strength	I
electrical potential	V
electrical resistance	R or X
electromotive force	emf
electronegativity	E
elevation	el
energy	E
enthalpy	H
entropy	S
equation	eq
equivalent weight	equiv wt
et alii (and others)	et al.
et cetera	etc.
excluded volume	u
excluded volume cluster integral	β
exempli gratia (for example)	e.g.
expansion coefficient	α
external	ext
F	force
f	fraction (excluding molar fraction, mass fraction, volume fraction)
f	molecular coefficient of friction (e.g., f_s, f_D, f_{rot})
f	functionality
farad	spell out or f
Federal	Fed.
feet board measure (board feet)	fbm
feet per minute	fpm
feet per second	fps
flash point	flp

Abbreviations	Symbols
fluid	fl
foot	ft
foot-candle	ft-c
foot-Lambert	ft-L
foot-pound	ft-lb
foot-pound-second (system)	fps
foot-second (see cubic feet per second)	
fraction	$\int$
franc	fr
free aboard ship	spell out
free alongside ship	spell out
free on board	fob
freezing point	fp
frequency	spell out
fusion point	fnp
G	Gibbs energy (formerly free energy or free enthalpy) ($G = H - TS$)
G	shear modulus ($G = \sigma_{ij}$/angle of shear)
G	statistical weight fraction ($G_i = g_i/\Sigma_i g_i$)
g	gravitational acceleration
g	statistical weight
g	*gauche* conformation
g	parameter for the dimensions of branched macromolecules
G^m	molar Gibbs energy
gallon	gal
gallons per minute	gpm
gallons per second	gps
gauche conformation	g
Gibbs energy	G
grain	spell out
gram	g
gram-calorie	g-cal
greatest common divisor	gcd
H	enthalpy
H^m	molar enthalpy
h	height
h	Plank constant
haversine	hav

Abbreviations	Symbols
heat	Q
heat capacity	C
hectare	ha
henry	H
high pressure (adjective)	h-p
hogshead	hhd
horsepower	hp
horsepower-hour	hp-hr
hour	h or hr
hundred	C
hundredweight (112 lb)	cwt
hydrogen ion concentration, negative logarithm of	pH
hyperbolic cosine	cosh
hyperbolic sine	sinh
hyperbolic tangent	tanh
I	electrical current strength
I	radiation intensity of a system
i	radiation intensity of a molecule
ibidem (in the same place)	ibid.
id est (that is)	i.e.
inch	in.
inch-pound	in-lb
inches per second	ips
indicated horsepower	ihp
indicated horsepower-hour	ihp-hr
infrared	IR
inside diameter	ID
intermediate-pressure (adjective)	i-p
internal	int
International Union of Pure and Applied Chemistry	IUPAC
isotactic	it
J	flow (of mass, volume, energy, etc.), always with a corresponding index
joule	J
K	general constant
K	equilibrium constant
K	compression modulus ($p = -K \Delta V/V_o$)
k	Boltzmann constant

Abbreviations	Symbols
k	rate constant for chemical reactions (always with an index)
Kelvin	K (Not °K)
kilocalorie	kcal
kilocycles per second	kc
kilogram	kg
kilogram-calorie	kg-al
kilogram-meter	kg-m
kilograms per cubic meter	kg per cu m or kg/m^3
kilograms per second	kgps
kiloliter	Kl
kilometer or kilometer	km
kilometers per second	kmps
kilovolt	kv
kilovolt-ampere	kva
kilowatt	kw
kilowatthour	kwhr
Knoop hardness number	KHN
L	chain end-to-end distance
L	phenomenological coefficient
l	length
lambert	L
latitude	lat or ϕ
least common multiple	lcm
length	l
linear expansion coefficient	Y
linear foot	lin ft
liquid	liq
lira	spell out
liter	l
logarithm (common)	log
logarithm (natural)	log. or ln
kibgutyde	kibg. or λ
loss angle	δ
low-pressure (as adjuective)	l-p
lumen	1*
lumen-hour	1-hr*
luments per watt	lpw
M	"molecular weight" (IUPAC molar mass)
m	mass
mass	spell out or m
mass fraction	w

Abbreviations	Symbols
mathematics (ical)	math
maximum	max
mean effective pressure	mep
mean horizontal candlepower	mhcp
meacycle	mHz
megohm	MΩ
melting point, -temperature	mp, T_m
meter	m
meter-kilogram	m-kg
metre	m
mho	spell out
microsmpere	μa or mu a
microfarad	μf
microinch	μin.
micrometer (formerly micron)	μm
micromicrofarad	μμf
micromicron	μμ
micron	μ
microvolt	μv
microwatt	μw or mu w
mile	spell out
miles per hour	mph
miles per hour per second	mphps
milli	m
milliampere	ma
milliequivalent	meq
milligram	mg
millihenry	mh
millilambert	mL
milliliter or milliliter	ml
millimeter	mm
millimeter or mercury (pressure)	mm Hg
millimicron	mμ or m mu
million	spell out
million gallons per day	mgd
millivolt	mv
minimum	min
minute	min
minute (angular measure)	′

Abbreviations	Symbols
minute (time) (in astronomical tables)	m
mile	spell out
modal	m
modulus of elasticity	E
molar	M
molar enthalpy	H_m
molar Gibbs Energy	G_m
molar heat capacity	C_m
mole	mol
mole fraction	x
molecular weight	mol wt or M
month	spell out
N	number of elementary particles (e.g., molecules, groups, atoms, electrons)
N_L	Avogadro number (Loschmidt's number)
n	amount of a substance (mole)
n	refractive index
nanometer (formerly millimicron)	nm
National Association of Corrosion Engineers	NACE
National Electrical Code	NEC
newton	N
normal	N
number of elementary particles	N
Occupational Safety and Health Administration	OSHA
ohm	Ω
ohm-centimeter	ohm-cm
oil absorption	O.A.
ounce	oz
once-foot	oz-ft
ounce-inch	oz-in.
outside diameter	OD
osomotic pressure	Π
P	permeability of membranes
p	probability
p	dipole moment
$\mathbf{p}_i$	induced dipolar moment
p	pressure

Abbreviations	Symbols
p	extent of reaction
Paint Testing Manual	PTM
parameter	Q
partition function (system)	Q
parts per billion	ppb
parts per million	ppm
pascal	Pa
peck	pk
penny (pency – new British)	p.
pennyweight	dwt
per	diagonal line in expressions with unit symbols or (see Fundamental Rules)
percent	%
permeability of membranes	P
peso	spell out
pint	pt.
Planck's constant (in $E = hv$) (6.62517 +/− 0.00023 x 10^{-27} erg sec)	h
polymolecularity index	Q
potential	spell out
potential difference	spell out
pound	lb
pound-foot	lb-ft
pound-inch	lb-in.
pound sterling	£
pounds-force per square inch	psi
pounds per brake horsepower-hour	lb per bhp-hr
pounds per cubi foot	lb per cut ft
pounds per square foot	psf
pounds per square inch	psi
pounds per square inch absolute	psia
power factor	spell out or pf
pressure	p
probability	p
Q	quantity of electricity, charge
Q	heat
Q	partition function (system)
Q	parameter in the Q–e copolymerize-tion equation

Abbreviations	Symbols
Q, Q	polydispersity, polymolecularity in-dex ($Q = \overline{M_w}/\overline{M_n}$)
q	partition function (particles)
quantity of electricity, charge	Q
quart	qt
quod vide (which see)	q.v.
R	molar gas constant
R	electrical resistance
R_G	radius of gyration
R_n	run number
R_ϑ	Rayleigh ratio
r	radius
r_o	initial molar ratio of reactive groups in polycondensations
radian	spell out
radius	r
radius of gyration	R_G
rate constant	k
Rayleigh ratio	R_ϑ
reactive kilovolt-ampere	kvar
reactive volt-ampere	var
reference(s)	ref
refractive index	n
relaxation time	τ
resistivity	ρ
revolutions per minute	rpm
revolutions per second	rps
rod	spell out
root mean square	rms
S	entropy
S^m	molar entropy
S	solubility coefficient
s	sedimentation coefficient
s	selectivity coefficient in osmotic measurements)
Saybolt Universal seconds	SUS
secant	sec
second	s or sec
second (angular measure)	″
second-foot (see cubic feet per second)	

Abbreviations	Symbols
second (time) (in astronomical tables)	s
Second virial coefficient	A_2
shaft horsepower	shp
shilling	s
sine	sin
sine of the amplitude, an elliptic function	sn
society	Soc.
Soluble	sol
solubility coefficient	S
solubility parameter	δ
solution	soln
specific gravity	sp gr
specific heat	sp ht
specific heat capacity (formerly: specific heat)	c
specific optical rotation	$[\alpha]$
specific volume	sp vol
spherical candle power	scp
square	sq
square centimeter	sq cm or cm^2
square foot	sq ft
square inch	sq in.
square kilometer	sq km or km^2
square meter	sq m or m^2
square micron	sq μ or μ^2
square root of mean square	rms
standard	std
Standard	Stnd.
Standard deviation	σ
Staudinger index	$[\eta]$
stere	s
syndiotactic	st
T	temperature
t	time
t	trans conformation
tangent	tan
temperature	T or temp
tensile strength	ts
threodiisotactic	tit
thousand	M
thousand foot-pounds	kip-ft
thousand pound	kip

Abbreviations	Symbols
ton	spell out
ton-mile	spell out
trans conformation	t
trans-tactic	tt
U	voltage
U	internal energy
U^m	molar internal energy
u	excluded volume
ultraviolet	UV
United States	U.S.
V	volume
V	electrical potential
v	rate, rate of reaction
v	specific volume always with an in-dex
vapor pressure	vp
versed sine	vers
versus	vs
volt	v or V
volt-ampere	va
volt-coulomb	spell out
voltage	U
volume	V or vol.
Volume (of a publication)	Vol
W	weight
W	work
w	mass function
watt	w or W
watthour	whr
watts per candle	wpc
week	spell out
weight	W or w
weight concentration*	c
work	y yield
X	degree of polymerization
X	electrical resistance
x	mole fractio y yield
yard	yd
year	yr
Young's	E
Z	collision number
Z	z fraction
z	ionic charge

Abbreviations	Symbols
z	coordination number
z	dissymmetry (light scattering)
z	parameter in excluded volume theory
α	angle, especially angle of rotation in optical activity
α	cubic expandion coefficient [$\alpha = V^{-1} (\partial V/\partial T)_p$]
α	expansion coefficient (as reduced length, e.g., α_L in the chain end-to-end distance or α_R for the radius of gyration)
α	degree of crystallinity (always with an index)
α	electric polarizability of a molecule
[α]	"specific" optical rotation
β	angle
β	coefficient of pressure
β	excluded volume cluster integral
Γ	preferential solvation
γ	angle
γ	surface tension
γ	linear expansion coefficient
δ	loss angle
δ	solubility parameter
δ	chemical shift
ε	linear expansion ($\varepsilon = \Delta l/l_o$)
ε	expectation
ε_r	relative permittivity (dielectric number)
η	dynamic viscosity
[η]	Staudinger index (called J_o in DIN 1342)
Θ	characteristic temperature, especial-ly theta temperature
θ	angle, especially angle of rotation
ϑ	angle, especially valence angle
κ	isothermal compressibility [$\kappa = V^{-1} (\partial V/\partial p)_T$]
κ	enthalpic interaction parameter in solution theory

Abbreviations	Symbols
λ	wavelength
λ	heat conductivity
λ	degree of coupling
μ	chemical potential
μ	moment
μ	permanent dipole moment
ν	mement, with respect to a reference value
ν	frequency
ν	kinetic chain length
ξ	shielding ratio in the theory of random coils
Ξ	partition function
Π	osmotic pressure
ρ	density
σ	mechanical stress (σ_{ii} = normal stress, σ_{ij} = shear stress)
σ	standard deviation
σ	hindrance parameter
τ	relaxation time
τ_i	internal transmittance (transmission factor) (represents the ratio of transmitted to absorbed light)
φ	volume fraction
φ(r)	potential between two segments separated by a distance r
Φ	constant in the viscosity-molecular-weight relationship
[Φ]	"molar" optical rotation
χ	interaction parameter in solution theory
ψ	entropic interaction parameter in solution theory
ω	angular frequency, angular velocity
Ω	angle
Ω	probability
Ω	skewness of a distribution

*(= weight of solute divided by volume of solvent); IUPAC suggests the symbol ρ for this quantity, which could lead to confusion with the same IUPAC symbol for density.

Notations

The abbreviations for chemicals and polymer were taken from the "Manual of Symbols and Terminology for Physicochemical Quantities and Units," *Pure and Applied Chemistry* **21***1) (1970), but some were added because of generally accepted use.

The ISO (International Standardization Organization) has suggested that all extensive quantities should be described by capital letters and all intensive quantities by lower-case letters. IUPAC doe not follow this recommendation, however, but uses lower-case letters for specific quantities.

The following symbols are used above or after a letter.

Symbols Above Letters

— signifies an average, e.g., $\overline{M}$ is the average molecular weight; more complicated averages are often indicated by $\langle \rangle$, e.g., $\langle R_G^2 \rangle$ is another way of writing $\overline{(R_G^2)}_z$

— stands for a partial quantity, e.g., $\tilde{v}_A$ is the partial specific volume of the compound A; V_A is the volume of A, whereas $\tilde{V}_A^m$xxx is the partial molar volume of A.

Superscripts

°	pure substance or standard state
∞	infinite dilution or infinitely high molecular weight
m	molar quantity (in cases where subscript letters are impractical)
(q)	the *q* order of a moment (always in parentheses)
‡	activated complex

Subscripts

Initial	State
1	solvent
2	solute
3	additional components (e.g., precipitant, salt, etc.)
am	amorphous
B	brittleness
bd	bond
cr	crystalline
crit	critical
cryst	crystallization
e	equilibrium

Initial	State
E	end group
G	glassy state
i	run number
i	initiation
i	isotactic diads
ii	isotactic triads
Is	heterotactic triads
j	run number
k	run number
m	molar
M	melting process
mon	monomer
n	number average
p	polymerization, especially propagation
pol	polymer
r	general for average
s	syndiotactic diads
ss	syndiotactic triads
st	start reaction
t	termination
tr	transfer
u	monomeric unit
w	weight average
z	z average
Prefixes	
at	atactic
ct	*cis*-tactic
eit	erythrodiisotactic
it	isotactic
st	syndiotactic
tit	threodiisotactic
tt	*trans*-tactic

Square brackets around a letter signify molar concentrations. (IUPAC prescribes the symbol *c* for molar councentrations, but to date this has consistently been used for the mass/volume unit.)

Angles are always given by °.

Apart from some exceptions, the meter is not used as a unit of length; the units cm and mm derived from it are used. Use of the meter in macromolecular science leads to very impractical units.

Mathematical Signs

Sign	Definition
Operations	
+	Addition
−	Subtraction
×	Multiplication
·	Multiplication
÷	Division
/	Division
∘	Composition
∪	Union
∩	Intersection
±	Plus or minus
∓	Minus or plus
Convolution	
⊕	Direct sum, variation
⊖	Various
⊗	Various
⊙	Various
:	Ratio
⨿	Amalgamation
Relations	
=	Equal to
≠	Not equal to
≈	Nearly equal to
≅	Equals approximately, isomorphic
<	Less than
<<	Much less than
>	Greater than
>>	Much greater than
≤	Less than or equal to
≦	Les than or equal to
≦	Less than or equal to
≥	Greater than or equal to
≥	Grean than or equalt o
≧	Greater than or equal to
≡	Equivalent to, congruent to
≢	Not equivalent to, not congruent to
|	Divides, divisible by
~	Similar to, asymptotically equal to
:=	Assignment

Sign	Definition
∈	A member of
⊂	Subset of
⊆	Subset of or equal to
⊃	Superset of
⊇	Superset of or equal to
∝	Varies as, proportional to
≐	Approaches a limit, definition
→	Tends to, maps to
←	Maps from
↦	Maps to
↪ or ↩	Maps into
□	d'Alembertian operator
Σ	Summation
Π	Product
∫	Integral
∮	Contour integral
Logic	
∧	And, conjunction
∨	Or, distunction
¬	Negation
⇒	Implies
→	Implies
⇔	If and only if
↔	If and only if
∃	Existential quantifier
∀	Universal quantifier
∈	A member o
∉	Not a member of
⊢	Assertion
∴	Hence, therefore
∵	Because
Radial units	
′	Minute
″	Second
°	Degree
Constants	
π	pi (≈ 3.14159265)
e	Base of natural logarithms (≈ 2.71828183)

Sign	Definition
Geometry	
⊥	Perpendicular
∥	Parallel
∦	Not parallel
∠	Angle
∢	Spherical angle
$\stackrel{v}{=}$	Equal angles
Miscellaneous	
i	Square root of -1
′	Prime
″	Double prime
‴	Triple prime
√	Square root, radical
$\sqrt[3]{}$	Cube root
$\sqrt[n]{}$	nth root
!	Factorial
!!	Double factorial
∅	Empty set, null set
∞	Infinity

Sign	Definition
∂	Partial differential
Δ	Delta
∇	Nabla, del
∇^2, Δ	Laplacian operator

English–Greek–Latin Numerical Prefixes

English	Greek	Latin
2	bis	di
3	tris	tri
4	tetrakis	tetra
5	pentakis	penta
6	hexakis	hexa
7	heptakis	hepta
8	octakis	octa
9	nonakis	nona
10	decakis	deca

Greek-Russian-English Alphabets

Greek letter		Greek name	English equivalent	Russian letter		English equivalent
A	α	Alpha	(ä)	А	а	(ä)
B	β	Beta	(b)	Б	б	(b)
				В	в	(v)
Γ	γ	Gamma	(g)	Г	г	(g)
Δ	δ	Delta	(d)	Д	д	(d)
E	ε	Epsilon	(e)	Е	е	(ye)
Z	ζ	Zeta	(z)	Ж	ж	(zh)
				З	з	(z)
H	η	Eta	(ā)	И	и	(i, ē)
Θ	θ	Theta	(th)	Й	й	(ē)
I	ι	Iota	(ē)	К	к	(k)
				Л	л	(l)
K	k	Kappa	(k)	М	м	(m)
Λ	λ	Lambda	(l)	Н	н	(n)
				О	о	(ô, o)
M	μ	Mu	(m)	О	о	(ô, o)
				П	п	(p)
N	ν	Nu	(n)	Р	р	(r)
Ξ	ξ	Xi	(ks)	С	с	(s)
				Т	т	(t)
O	o	Omicron	α	У	у	ōō
Π	π	Pi	(P)	Ф	ф	(f)
				Х	х	(kh)
P	ρ	Rho	(r)	Х	х	(kh)
				Ц	ц	(t_s)
Σ	σ	Sigma	(s)	Ч	ч	(ch)
T	τ	Tau	(t)	Ш	ш	(sh)
Υ	υ	Upsilon	(ü, ōō)	Щ	щ	(shch)
				Ъ	ъ	8
Φ	ø	Phi	(f)	Ы	ы	(ë)
X	χ	Chi	(H)	ь	ь	(ë)
Ψ	ψ	Psi	(ps)	Э	э	(e)
				Ю	ю	(ū)
Ω	ω	Omega	(ō)	Я	я	(yä)

English-Greek-Latin Numbers

English	Greek	Latin
1	mono	uni
2	bis	di
3	tris	tri
4	tetrakis	tetra
5	pentakis	penta
6	hexakis	hexa
7	heptakis	hepta
8	octakis	octa
9	nonakis	nona
10	decakis	deca

International Union of Pure and Applied Chemistry: Rules Concerning Numerical Terms Used in Organic Chemical Nomenclature (specifically as prefixes for hydrocarbons)

1	mono- or hen-	10 deca-	100 hecta-	1000 kilia-
2	di- or do-	20 icosa-	200 dicta-	2000 dilia-
3	tri-	30 triaconta-	300 tricta-	3000 trilia-
4	tetra-	40 tetraconta-	400 tetracta-	4000 tetralia-
5	penta-	50 pentaconta-	500 pentactra	5000 pentalia-
6	hexa-	60 hexaconta-	600 hexacta	6000 hexalia-
7	hepta-	70 hepaconta-	700 heptacta-	7000 hepalia-
8	octa-	80 octaconta-	800 ocacta-	8000 ocatlia-
9	nona-	90 nonaconta-	900 nonactta-	9000 nonalia-

Source: IUPAC, Commission on Nomenclature of Organic Chemistry (N. Lorzac'h and published in *Pure and Appl. Chem* 58: 1693–1696 (1986))

Elemental Symbols and Atomic Weights

Source: International Union of Pure and Applied Chemistry (IUPAC) 2001Values from the 2001 table *Pure Appl. Chem.*, **75**, 1107–1122 (2003). The values of zinc, krypton, molybdenum and dysprosium have been modified. The *approved name* for element 110 is included, see *Pure Appl. Chem.*, **75**, 1613–1615 (2003). The *proposed name* for element 111 is also included.

A number in parentheses indicates the uncertainty in the last digit of the atomic weight.

List of Elements in Atomic Number Order

At No	Symbol	Name	Atomic Wt	Notes
1	H	Hydrogen	1.00794(7)	1, 2, 3
2	He	Helium	4.002602(2)	1, 2
3	Li	Lithium	[6.941(2)]	1, 2, 3, 4
4	Be	Beryllium	9.012182(3)	
5	B	Boron	10.811(7)	1, 2, 3
6	C	Carbon	12.0107(8)	1, 2
7	N	Nitrogen	14.0067(2)	1, 2
8	O	Oxygen	15.9994(3)	1, 2
9	F	Fluorine	18.9984032(5)	
10	Ne	Neon	20.1797(6)	1, 3
11	Na	Sodium	22.989770(2)	
12	Mg	Magnesium	24.3050(6)	
13	Al	Aluminium	26.981538(2)	
14	Si	Silicon	28.0855(3)	2
15	P	Phosphorus	30.973761(2)	
16	S	Sulfur	32.065(5)	1, 2
17	Cl	Chlorine	35.453(2)	3
18	Ar	Argon	39.948(1)	1, 2
19	K	Potassium	39.0983(1)	1
20	Ca	Calcium	40.078(4)	1
21	Sc	Scandium	44.955910(8)	
22	Ti	Titanium	47.867(1)	
23	V	Vanadium	50.9415(1)	
24	Cr	Chromium	51.9961(6)	
25	Mn	Manganese	54.938049(9)	
26	Fe	Iron	55.845(2)	
27	Co	Cobalt	58.933200(9)	
28	Ni	Nickel	58.6934(2)	
29	Cu	Copper	63.546(3)	2
30	Zn	Zinc	65.409(4)	
31	Ga	Gallium	69.723(1)	
32	Ge	Germanium	72.64(1)	
33	As	Arsenic	74.92160(2)	
34	Se	Selenium	78.96(3)	
35	Br	Bromine	79.904(1)	
36	Kr	Krypton	83.798(2)	1, 3
37	Rb	Rubidium	85.4678(3)	1
38	Sr	Strontium	87.62(1)	1, 2
39	Y	Yttrium	88.90585(2)	
40	Zr	Zirconium	91.224(2)	1
41	Nb	Niobium	92.90638(2)	
42	Mo	Molybdenum	95.94(2)	1
43	Tc	Technetium	[98]	5
44	Ru	Ruthenium	101.07(2)	1
45	Rh	Rhodium	102.90550(2)	
46	Pd	Palladium	106.42(1)	1
47	Ag	Silver	107.8682(2)	1
48	Cd	Cadmium	112.411(8)	1
49	In	Indium	114.818(3)	
50	Sn	Tin	118.710(7)	1
51	Sb	Antimony	121.760(1)	1
52	Te	Tellurium	127.60(3)	1
53	I	Iodine	126.90447(3)	
54	Xe	Xenon	131.293(6)	1, 3
55	Cs	Caesium	132.90545(2)	
56	Ba	Barium	137.327(7)	
57	La	Lanthanum	138.9055(2)	1
58	Ce	Cerium	140.116(1)	1
59	Pr	Praseodymium	140.90765(2)	
60	Nd	Neodymium	144.24(3)	1
61	Pm	Promethium	[145]	5
62	Sm	Samarium	150.36(3)	1
63	Eu	Europium	151.964(1)	1
64	Gd	Gadolinium	157.25(3)	1
65	Tb	Terbium	158.92534(2)	
66	Dy	Dysprosium	162.500(1)	1
67	Ho	Holmium	164.93032(2)	
68	Er	Erbium	167.259(3)	1

At No	Symbol	Name	Atomic Wt	Notes
69	Tm	Thulium	168.93421(2)	
70	Yb	Ytterbium	173.04(3)	1
71	Lu	Lutetium	174.967(1)	1
72	Hf	Hafnium	178.49(2)	
73	Ta	Tantalum	180.9479(1)	
74	W	Tungsten	183.84(1)	
75	Re	Rhenium	186.207(1)	
76	Os	Osmium	190.23(3)	1
77	Ir	Iridium	192.217(3)	
78	Pt	Platinum	195.078(2)	
79	Au	Gold	196.96655(2)	
80	Hg	Mercury	200.59(2)	
81	Tl	Thallium	204.3833(2)	
82	Pb	Lead	207.2(1)	1, 2
83	Bi	Bismuth	208.98038(2)	
84	Po	Polonium	[209]	5
85	At	Astatine	[210]	5
86	Rn	Radon	[222]	5
87	Fr	Francium	[223]	5
88	Ra	Radium	[226]	5
89	Ac	Actinium	[227]	5
90	Th	Thorium	232.0381(1)	1, 5
91	Pa	Protactinium	231.03588(2)	5
92	U	Uranium	238.02891(3)	1, 3, 5
93	Np	Neptunium	[237]	5
94	Pu	Plutonium	[244]	5
95	Am	Americium	[243]	5
96	Cm	Curium	[247]	5
97	Bk	Berkelium	[247]	5
98	Cf	Californium	[251]	5
99	Es	Einsteinium	[252]	5
100	Fm	Fermium	[257]	5
101	Md	Mendelevium	[258]	5
102	No	Nobelium	[259]	5
103	Lr	Lawrencium	[262]	5
104	Rf	Rutherfordium	[261]	5, 6
105	Db	Dubnium	[262]	5, 6
106	Sg	Seaborgium	[266]	5, 6
107	Bh	Bohrium	[264]	5, 6
108	Hs	Hassium	[277]	5, 6
109	Mt	Meitnerium	[268]	5, 6
110	Ds	Darmstadtium	[281]	5, 6
111	Rg	Roentgenium	[272]	5, 6
112	Uub	Ununbium	[285]	5, 6
114	Uuq	Ununquadium	[289]	5, 6
116	Uuh	Ununhexium		see Note above
118	Uuo	Ununoctium		see Note above

1. Geological specimens are known in which the element has an isotopic composition outside the limits for normal material. The difference between the atomic weight of the element in such specimens and that given in the Table may exceed the stated uncertainty.
2. Range in isotopic composition of normal terrestrial material prevents a more precise value being given; the tabulated value should be applicable to any normal material.
3. Modified isotopic compositions may be found in commercially available material because it has been subject to an undisclosed or inadvertent isotopic fractionation. Substantial deviations in atomic weight of the element from that given in the Table can occur.
4. Commercially available Li materials have atomic weights that range between 6.939 and 6.996; if a more accurate value is required, it must be determined for the specific material [range quoted for 1995 table 6.94 and 6.99].
5. Element has no stable nuclides. The value enclosed in brackets, e.g. [209], indicates the mass number of the longest-lived isotope of the element. However three such elements (Th, Pa, and U) do have a characteristic terrestrial isotopic composition, and for these an atomic weight is tabulated.
6. The names and symbols for elements 112-118 are under review. The temporary system recommended by J Chatt, *Pure Appl. Chem.*, **51**, 381–384 (1979) is used above. The names of elements 101-109 were agreed in 1997 (See *Pure Appl. Chem.*, 1997, **69**, 2471–2473) and for element 110 in 2003 (see *Pure Appl. Chem.*, 2003, **75**, 1613–1615). The proposed name for element 111 is also included.

List of Elements in Name Order

At No	Symbol	Name	Atomic Wt	Notes
89	Ac	Actinium	[227]	5
13	Al	Aluminium	26.981538(2)	
95	Am	Americium	[243]	5
51	Sb	Antimony	121.760(1)	1

At No	Symbol	Name	Atomic Wt	Notes
18	Ar	Argon	39.948(1)	1, 2
33	As	Arsenic	74.92160(2)	
85	At	Astatine	[210]	5
56	Ba	Barium	137.327(7)	
97	Bk	Berkelium	[247]	5
4	Be	Beryllium	9.012182(3)	
83	Bi	Bismuth	208.98038(2)	
107	Bh	Bohrium	[264]	5, 6
5	B	Boron	10.811(7)	1, 2, 3
35	Br	Bromine	79.904(1)	
48	Cd	Cadmium	112.411(8)	1
55	Cs	Caesium	132.90545(2)	
20	Ca	Calcium	40.078(4)	1
98	Cf	Californium	[251]	5
6	C	Carbon	12.0107(8)	1, 2
58	Ce	Cerium	140.116(1)	1
17	Cl	Chlorine	35.453(2)	3
24	Cr	Chromium	51.9961(6)	
27	Co	Cobalt	58.933200(9)	
29	Cu	Copper	63.546(3)	2
96	Cm	Curium	[247]	5
110	Ds	Darmstadtium	[281]	5, 6
105	Db	Dubnium	[262]	5, 6
66	Dy	Dysprosium	162.500(1)	1
99	Es	Einsteinium	[252]	5
68	Er	Erbium	167.259(3)	1
63	Eu	Europium	151.964(1)	1
100	Fm	Fermium	[257]	5
9	F	Fluorine	18.9984032(5)	
87	Fr	Francium	[223]	5
64	Gd	Gadolinium	157.25(3)	1
31	Ga	Gallium	69.723(1)	
32	Ge	Germanium	72.64(1)	
79	Au	Gold	196.96655(2)	
72	Hf	Hafnium	178.49(2)	
108	Hs	Hassium	[277]	5, 6
2	He	Helium	4.002602(2)	1, 2
67	Ho	Holmium	164.93032(2)	
1	H	Hydrogen	1.00794(7)	1, 2, 3
49	In	Indium	114.818(3)	
53	I	Iodine	126.90447(3)	
77	Ir	Iridium	192.217(3)	
26	Fe	Iron	55.845(2)	
36	Kr	Krypton	83.798(2)	1, 3
57	La	Lanthanum	138.9055(2)	1
103	Lr	Lawrencium	[262]	5
82	Pb	Lead	207.2(1)	1, 2
3	Li	Lithium	[6.941(2)]	1, 2, 3, 4
71	Lu	Lutetium	174.967(1)	1
12	Mg	Magnesium	24.3050(6)	
25	Mn	Manganese	54.938049(9)	
109	Mt	Meitnerium	[268]	5, 6
101	Md	Mendelevium	[258]	5
80	Hg	Mercury	200.59(2)	
42	Mo	Molybdenum	95.94(2)	1
60	Nd	Neodymium	144.24(3)	1
10	Ne	Neon	20.1797(6)	1, 3
93	Np	Neptunium	[237]	5
28	Ni	Nickel	58.6934(2)	
41	Nb	Niobium	92.90638(2)	
7	N	Nitrogen	14.0067(2)	1, 2
102	No	Nobelium	[259]	5
76	Os	Osmium	190.23(3)	1
8	O	Oxygen	15.9994(3)	1, 2
46	Pd	Palladium	106.42(1)	1
15	P	Phosphorus	30.973761(2)	
78	Pt	Platinum	195.078(2)	
94	Pu	Plutonium	[244]	5
84	Po	Polonium	[209]	5
19	K	Potassium	39.0983(1)	1
59	Pr	Praseodymium	140.90765(2)	
61	Pm	Promethium	[145]	5
91	Pa	Protactinium	231.03588(2)	5
88	Ra	Radium	[226]	5
86	Rn	Radon	[222]	5
75	Re	Rhenium	186.207(1)	
45	Rh	Rhodium	102.90550(2)	
111	Rg	Roentgenium	[272]	5, 6
37	Rb	Rubidium	85.4678(3)	1
44	Ru	Ruthenium	101.07(2)	1
104	Rf	Rutherfordium	[261]	5, 6
62	Sm	Samarium	150.36(3)	1
21	Sc	Scandium	44.955910(8)	
106	Sg	Seaborgium	[266]	5, 6
34	Se	Selenium	78.96(3)	
14	Si	Silicon	28.0855(3)	2

At No	Symbol	Name	Atomic Wt	Notes
47	Ag	Silver	107.8682(2)	1
11	Na	Sodium	22.989770(2)	
38	Sr	Strontium	87.62(1)	1, 2
16	S	Sulfur	32.065(5)	1, 2
73	Ta	Tantalum	180.9479(1)	
43	Tc	Technetium	[98]	5
52	Te	Tellurium	127.60(3)	1
65	Tb	Terbium	158.92534(2)	
81	Tl	Thallium	204.3833(2)	
90	Th	Thorium	232.0381(1)	1, 5
69	Tm	Thulium	168.93421(2)	
50	Sn	Tin	118.710(7)	1
22	Ti	Titanium	47.867(1)	
74	W	Tungsten	183.84(1)	

At No	Symbol	Name	Atomic Wt	Notes
112	Uub	Ununbium	[285]	5, 6
116	Uuh	Ununhexium		see Note above
118	Uuo	Ununoctium		see Note above
114	Uuq	Ununquadium	[289]	5, 6
92	U	Uranium	238.02891(3)	1, 3, 5
23	V	Vanadium	50.9415(1)	
54	Xe	Xenon	131.293(6)	1, 3
70	Yb	Ytterbium	173.04(3)	1
39	Y	Yttrium	88.90585(2)	
30	Zn	Zinc	65.409(4)	
40	Zr	Zirconium	91.224(2)	1

Pronounciation Symbols and Abbreviations

ə	Banana, collide, abut
ˈə, ˌə	Humdrum, abut
ᵊ	Immediately preceding \l\, \n\, \m\, \ŋ\, as in battle, mitten, eaten, and sometimes open \ˈō-pᵊm\, lock and key \-ᵊ ŋ-\; immediately following \l\, \m\, \r\, as often in French table, prisme, titre
ər	further, merger, bird
ˈə-, ˈə-r	As in two different pronunciations of hurry \ ˈhər-ē, \ ˈhə-rē\
a	mat, map, mad, gag, snap, patch
ā	day, fade, date, aorta, drape, cape
ä	bother, cot, and, with most American speakers, father, cart
á	father as pronounced by speakers who do not rhyme it with bother; French patte
aú	now, loud, out
b	baby, rib
ch	chin, nature \ˈnā-chər\
d	did, adder
e	bet, bed, peck
ˈē, ˌē	beat, nosebleed, evenly, easy
ē	easy, mealy
f	fifty, cuff
g	go, big, gift
h	hat, ahead
hw	whale as pronounced by those who do not have the same pronunciation for both whale and wail
i	tip, banish, active
ī	site, side, buy, tripe
j	job, gem, edge, join, judge
k	kin, cook, ache
ḵ	German ich, Buch; one pronunciation of loch
l	lily, pool
m	murmur, dim, nymph
n	no, own
ⁿ	Indicates that a preceeding vowel or diphthong is pronounced with the nasal passages open, as in French un bon vin blanc \œⁿ-bōⁿvaⁿ-bläⁿ\
ŋ	sing \ˈsiŋ\, singer \ˈsiŋ-ər\, finger \ˈfiŋ-gər\, ink \ˈiŋk\
ō	bone, know, beau
ȯ	saw, all, gnaw, caught
ü	fool
u̇	took
œ	French coeuf, German Hölle
œ̅	French feu, German Höhle
ȯi	coin, destroy
p	pepper, lip
r	red, car, rarity
s	source, less
sh	as in shy, mission, machine, special (actually, this is a single sound, not two); with a hyphen between, two sounds as in grasshopper \ˈgras-ˌhä-pər\
t	tie, attack, late, later, latter
th	as in thin, ether (actually, this is a single sound, not two); with a hyphen between, two sounds as in knighthood \ˈnīt-ˌh----d\
th	then, either, this (actually, this is a single sound, not two)
ü	rule, youth, union \ˈyün-yən\, few \ˈfyü\
u̇	pull, wood, book, curable \ˈky ú r-ə-bəl\, fury \ˈfy----r-ē\
ue	German füllen, hübsch
ue	French rue, German fühlen
v	vivid, give
w	we, away
y	yard, young, cue \ˈkyü\, mute \ˈmyüt\, union \ˈyün-yən\
ʸ	indicates that during the articulation of the sound represented by the preceding character the front of the tongue has substantially the position it has for the articulation of the first sound of yard, as in French digne \dēnʸ\
z	zone, raise
zh	as in vision, azure \ˈa-zhər\ (actually this is a single sound, not two).
\	reversed virgule used in pairs to mark the beginning and end of a transcription: \ˈpen\
ˈ	mark preceding a syllable with primary (strongest) stress: \ˈpen-mən-ˌship\
ˌ	mark preceding a syllable with secondary (medium) stress: \ˈpen-mən-ˌship\
-	mark of syllable division

( )	indicate that what is symbolized between is present in some utterances but not in others: *factory* \ **❙**fak-t(ə-)rē
÷	indicates that many regard as unacceptable the pronunciation variant immediately following: *cupola* \ **❙**kyü-pə-lə, ÷- **❙**lō\

Explanatory Notes and Abbreviations

(date)	date that word was first recorded as having been used
[...]	etomology and origin(s) of word
{...}	usage and/or languages, including French, German, Italian and Spanish
adj	adjective
adv	adverb
B.C.	before Christ
Brit.	Britain, British
C	centigrade, Celsius
c	century
E	English
Eng.	England
F	French, Fahrenheit
Fr.	France
fr.	from
G	German
Gr.	Germany
L	Latin
ME	middle English

n	noun
neut.	neuter
NL	new Latin
OE	old English
OL	old Latin
pl	plural
prp.	present participle
R	Russian
sing.	singular
S	Spanish
U.K.	United Kingdom
v	verb

Source: From *Merriam-Webster's Collegiate© Dictionary*, Eleventh Editioh, ©2004 by Merriam-Webster, incorporated, (www.Merriam-Webster.com). With permission.

Languages

French, German and Spanish translations are enclosed in {--} and preceded by F, G, I and S, respectively; and gender is designated by f-feminine, m-masculine, n-neuter. For example: **Polymer**--{F polymere m} represents the French translation "polymere" of the English word polymer and it is in the masculine case. These translations were obtain from multi-language dictionaries including: *A Glossary of Plastics Terminology in 5 Languages*, 5th Ed., Glenz, W., (ed) Hanser Gardner Publications, Inc., Cinicinnati, 2001. By permission).

D

d *n* \ˈdē\ (1) SI abbreviation for prefix ▶ Deci-. (2) Abbreviation in use with SIU system for time interval of 1 day (= 86,400 s).

D (1) Symbol for diameter. (2) Chemical symbol for the hydrogen isotope of atomic weight 2, deuterium.

D_{6500} or D_{65} *adj* Refers to the daylight special power distribution curve with a correlated color temperature of 6,500 K. See ▶ Daylight Illuminates, ▶ C.I.E, and ▶ Correlated Color Temperature.

da *n* SI abbreviation for prefix ▶ Deca-.

Dabber *n* \ˈdab-\ *n* [ME *dabbe*] (14c) Dome-shaped brush of soft hair for applying spirit varnish or for polishing and finishing gilding.

DABCO *n* (former trade name, now generic). Abbreviation for 1,4-Diazabicyclo-2,2,2-Octane. See ▶ Triethylenediamine.

DAC Abbreviation for ▶ Diallyl Chlorendate.

Dacron \ˈdā-ˌkrän\ {*trademark*} Fiber from poly(ethylene terephthalate). Manufactured by DuPont, U.S.

Dado \ˈdā-(ˌ)dō\ *n* [It, die, plinth] (1664) (1) Lower part of an interior wall when paneled or decorated, customarily of chair rail height. See ▶ Wainscot. (2) Paper that provides this architectural division.

DAF Abbreviation for Diallyl Fumarate, a polymerizable monomer.

Dagincoleic Acid *n* $C_{22}H_{44}O_4$. Acid constituent of Borneo dammer. MP, 170°C.

Dagingenoleic Acid *n* $C_{13}H_{26}O_3$. Acidic constituent of Borneo dammer. MP 126°C.

DAIP *n* Abbreviation for Diallyl Isophthalate Resin. See ▶ Allyl Resin.

Dalamar Yellow *n* Pigment Yellow 74 and 65 (11741). An azo coupling of m-nitro-o-anisidine with acetoacet-o-anisidine; it is one of the growing group of close chemical relatives of Hansa Yellow G (an azo coupling of m-nitro-p-toluidine with acetoacetanilide). The high strength toner is superior in tinting strength and fastness properties to Hansa Yellow G.

Dalton's Law of Partial Pressures *n* The pressure exerted by a mixture of gases is equal to the sum of the separate pressures which each gas would exert if it alone occupied the whole volume. This fact is expressed in the following formula:

$$PV = V(p_1 + p_2 + p_3, \text{etc}).$$

(Giambattista A, Richardson R, Richardson RC, Richardson B (2003) College physics. McGraw Hill Science/Engineering/Math, New York)

Dam *n* A ridge around the perimeter of a mold that retains excess resin during pressing and curing.

DAM *n* Abbreviation for ▶ Diallyl Maleate. See ▶ Allyl Resin.

Damaged Selvage See ▶ Cut Selvage.

Damask \ˈda-məsk\ *n* [ME *damaske*, fr. ML *damascus*, fr *Damascus*] (14c) A firm, glossy, Jacquard-patterned fabric that may be made from linen, cotton, rayon, silk, or a combination of these with various manufactured fibers. Similar to brocade, but flatter and reversible, damask is used for napkins, tablecloths, draperies, and upholstery.

Dammar (Damar) \ˈda-mər\ *n* [Malay] (1698) Natural resinous exudation from trees of the *Dipterocarpaceae* family, which grow chiefly in the East Indies (now Indonesia) and Malaya. Very pale in color (practically colorless to deep yellow). Soluble in hydrocarbons without the necessity of running. Average acid value of 30. A fossil resin used as an ingredient in printing ink varnishes.

Dammarolic Acid *n* $C_{56}H_{80}O_8$. Acidic constituent of dammar.

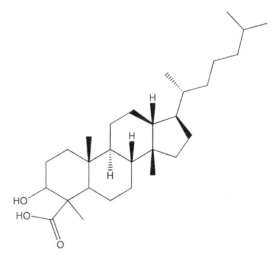

Dammar Wax See ▶ Resene, β.

Dammarylic Acid *n* $C_{38}H_{60}O_3$. Acidic constituent of dammar.

Dampener In lithography, cloth-covered, parchment paper, or rubber rollers that distribute the dampening solution to the press plate.

Dampending *n* (in Tire Cord) The relative ability to absorb energy and deaden oscillation after excitation.

Dampening Solution See ▶ Fountain Solution.

Damping *n* Reduction of vibration amplitude due to viscous or frictional resistance within a material or to drag of its environment upon a structure. The vibration may be mechanical, sonic, or electronic. Energy dissipates per cycle is called ▶ Hysteresis loss. See also ▶ Critical Damping.

Dancer Roll *n* A roller mounted on an axis which is movable with respect to axes of other rollers in an apparatus, used to control or measure tension of a continuous web or strand as it passes through a series of rollers. Dancer rolls are used as tension-sensing devices in the extrusion coating of wire, and as tension-maintaining devices in film winding.

DAP Abbreviation for ▶ Diallyl Phthalate.

Darkfield Illumination *n* Incident or transmitted illumination of the specimen by indirect light whereby no direct light is admitted directly to the objective.

Dart Impact Test See ▶ Free-Falling-Dart Test.

Darvic *n* Poly(vinyl chloride), manufactured by ICI, Great Britain.

Dashpot \ █dash- ▌pät\ *n* (1861) (1) A device used for damping vibration and cushioning shock in hydraulic systems. Typically, a dashpot is a liquid- or gas-filled cylinder with a piston that is attached to a moving machine part. (2) A modeling concept useful in visualizing the mechanical behavior of viscoelastic materials, a purely viscous element that may operate alone or connected in series and/or parallel with springs and sliders.

Datiric Acid \sə- █tir-ik\ *n* $CH_3(CH_2)_{15}COOH$. Another name for margaric acid. A straight chain aliphatic acid. MP, 59.5°C.

Daub \ █dób, █däb\ *v* [ME, fr. MF *dauber*] (14c) To apply a coating by crude unskillful strokes.

Daylight \ █dā- ▌līt\ *n* (13c) Natural illumination that is the result of various mixtures of direct sunlight covering a wide range of correlated color temperatures. For precision in definitions, see ▶ North Light, ▶ Artificial Daylight, and ▶ Daylight Illuminants.

Daylight Fluorescent Pigment See ▶ Fluorescent Pigment.

Daylight Illuminants *n*, CIE Series of illuminant spectral power distribution curves based on measurements of natural daylight and recommended by the CIE in 1965. Values are defined for the wavelength region 300–830 nm. They are described in terms of the correlated color temperature. The most important is D_{6500} because of the closeness of its correlated color temperature to that of Illuminant C, 6,774 K. D_{7500} bluer than D_{6500} and D_{5500} yellower than D_{6500} are also used.

Daylight Opening *n* The clearance between two platens of a molding press when in the open position.

dB *n* Abbreviation for ▶ Decibel.

DBEP *n* Abbreviation for ▶ Dibutoxyethyl Phthalate.

DBP *n* Abbreviation for ▶ Dibutyl Phthalate.

DBPC *n* Abbreviation for Di-*tert*-Butyl-*p*-Cresol.

DBS *n* Abbreviation for ▶ Dibutyl Sebacate.

DBTDL *n* Abbreviation for ▶ Dibutyltin Dilaurate.

DC Drive *n* (direct-current drive) A machine drive, particularly that of an extruder, powered by a direct-current motor. The availability of economical solid-state rectifiers, and the good torque-versus-speed characteristic and tight speed regulation of DC drives have made them the most popular choice today for variable-speed service.

DCHP *n* Abbreviation for ▶ Dicyclohexyl Phthalate.

DCO *n* Nondrying castor oil which has been converted into a drying oil by the catalytic removal of water from its principal fatty acid (80% ricinoleic acid), forming approximately equal quantities with 9–11 and 9–12 unsaturation. The term DCO rather than "dehydrated castor oil" has been suggested and is being used for this oil in order to reduce the tendency to confuse castor oil with DCO, which is entirely different in its properties from the oil from which it is derived.

DCP *n* Abbreviation for ▶ Dicapryl Phthalate.

DDM *n* Abbreviation for ▶ 4,4′-Diaminodiphenyl Methane.

DDP *n* Abbreviation for ▶ Didecyl Phthalate.

DE *n* Abbreviation for ▶ Diatomaceous Earth. See ▶ Diatomite.

Dead Burned Calcium Sulfate See ▶ Calcium Sulfate, ▶ Anhydrous.

Deadcoloring See ▶ Abbozzo.

Dead-End Polymerization *n* A free radical polymerization in which a very active initiator is used so that the rate of polymerization continuously decreases as initiator is consumed and a limiting conversion of monomer to polymer occurs.

Dead Flat *n* Of a finish, the quality of having no luster or gloss. Syn: ▶ Lusterless.

Dead Fold *n* A fold that does not spontaneously unfold; a crease.

Dead Oil See Creosote.

Dead Zone *n* In control work, a range of the quantity sensed in which a small change in the quantity causes no change in the indication of it, nor any control action. Compare ▶ Induction Period.

Deaerate \dē-ar-āt\ *vt* (1791) To remove air from a substance. Deaeration is an important step in the production of vinyl plastisols and most casting operations accomplished by subjecting the liquid to a moderate vacuum with or without agitation, to remove air that would cause objectionable bubbles or blisters in finished products.

Debloomed Oil See ▶ Oil, Debloomed.

Debonding *n* In a bonded joint, separation of the bonded surfaces. In a laminate, separation of layers or fibers from the matrix.

Deborah Number \de-b(ə-)rə nəm-bər\ (De, N_{De}) The dimensionless ratio of the relaxation time of a viscoelastic fluid to a characteristic time for the flow process being considered. Relaxation time is often taken as the ratio of viscosity to modulus, while the characteristic time is a significant length divided by average velocity. The Deborah number has been useful in studying such phenomena as die-exit swell in extrusion. If *De* is high, elastic effects probably dominate; if *De* is low, the flow is essentially viscous.

Deburr \də-bər\ To remove from a machined part rough edges or corners left by the machining operations. Compare ▶ Deflashing.

Debye Equation \də-bī i-kwā-zhən\ A relationship for the relative permittivity (ε) of a dielectric material as a function of the electronics and atomic polarizability (α_a) and the orientational polarizability terms.

$$(\varepsilon - 1)/(\varepsilon + 2) = (4\pi N_1/3)(\alpha_a + \mu^2/3kT)$$

N_1 = number of molecules present per cubic centimeter, μ = dipole moment of the molecule concerned, k = Boltzmann's constant and T = temperature.

Deybe–Hückle Theory of Strong Electrolytes *n* This theory assumes that strong electrolytes are completely dissociated in solution, but that the hydrated ions have enough residual attraction for each other that they are not truly independent of each other. The theory assumes that each ion has surrounding it on the average more ions of opposite charge, counterions, than of the same charge, and that the number of counterions near a given ion increases with increasing concentration. The presence of this diffuse ionic atmosphere relatively rich in counterions restricts the movement of each ion in the solution. The theory permits a quantitative prediction of how the conductivity should change with concentration, a prediction that agrees with experimental results. The theory also permits a calculation of the ionic-atmosphere effect on colligative properties. (Russell JB (1980) General chemistry. McGraw-Hill, New York)

Deca- {*combining form*} [ME, fr. L. fr. Gk *deka-*, *dek-*, fr. *deka*] (da) SI-acceptable prefix meaning × 10.

Decabromodiphenyl Oxide \-brō-( )mō-( )dī- fe-nºl äk- sīd\ *n* (deca-bromodiphenyl either) Diphenyl ether in which all ten phenyl hydrogens have been replaced by bromine, containing 83% bromine. The commercial product, a free-flowing white powder, is much used as a flame retardant in High-Impact Polystyrene and Structural Foam, usually with synergistic Antimony Oxide. Typical would be 12% decabromodiphenyl oxide and 4% antimony oxide in the resin.

Decahydronaphthalene \-hī-( )drō-naf-thə-lēn\ *n* (Decalin®) A saturated bicyclic hydrocarbon, $C_{10}H_{18}$, essentially two ▶ Cyclohexane rings fused together, sharing two hydrogen atoms. The commercial product is a mixture of *cis-* and *trans-* isomers. It is a colorless liquid with an aromatic odor, derived by treating molten naphthalene with hydrogen in the presence of a catalyst. It is a solvent for many resins. Hydrocarbon solvent with a boiling range for the commercial product (175–205°C), slightly higher than that for mineral spirit. Sp gr, 0.894/20°C; flp, 60°C (140°F); refractive index, 1.467; and vp, 2 mm Hg/30°C. In Europe, *Decaline.*

Decalcomania \di-ˌkal-kə-ˈmā-nē-ə\ *n* [F *décalcomanie*, fr. *décalquer* to copy by tracing (fr. *dé-* de- + *calquer* to trace, fr. It *calcare*, literally, to tread, fr. L) + *manie* mania, from LL *mania*] (1884) (decal) A printed design on a temporary carrier such as paper or film, and subsequently transferred to the item to be decorated. Deals are widely used for decorating many materials including plastics, paper, china, glass, furniture, etc. The imprint is adhered to plastic surfaces by means of a pressure-sensitive adhesive, solvent welding, or heat and pressure. (Skeist I (ed) (1990) Handbook of adhesives. Van Nostrand Reinhold, New York)

Decals *n* An abbreviation for ▶ Decalcomania. A printed design on a temporary carrier.

Decanedioic Acid Syn: ▶ Sebacic Acid.

Decantation \ˌdē-ˌkan-ˈtā-shən\ *n* [NL *decantare*, fr. L *de-* + ML *cantus* edge, fr. L, iron ring round a wheel] (1633) or siphoning off the liquid from a precipitate or sediment or the upper layer of two immiscible liquids as a partial means of separating the phases.

Decarboxylate \-ˈbäk-sə-ˌlāt\ *vt* (1922) To remove from an organic acid its carboxylic-acid groups.

Decarboxylation *n* Removal of carboxyl groups from organic acids, usually as the result of heating.

Decating See ▶ Decatizing.

Decating Mark *n* A crease mark or impression extending fillingwise across the fabric near the beginning or end of the piece.

Decatizing *n* A finishing process in which fabric, wound tightly on a perforated roller, either has hot water circulated through it (wet decatizing), or has steam blown through it (dry decatizing). The process is aimed chiefly at improving the hand and removing wrinkles.

Decay \di-ˈkā\ *n* [ME, fr. ONF *decaïr*, fr. LL *decadere* to fall, sink, fr. L *de-* + *cadere* to fall] (15c) Diminution of a radioactive substance due to nuclear emission of alpha or beta particles, gamma rays or positrons. (Giambattista A, Richardson R, Richardson RC, Richardson B (2003) College physics. McGraw Hill Science/Engineering/Math, New York)

Deci- {*combining form*} [F *déci-*, fr. L *decimus* tenth, fr. *decem* ten] (d) The SI-approved prefix meaning × 0.1.

Decibel \ˈde-sə-ˌbel\ *n* [ISV *deci-* + *bel*] (1928) (dB) In acoustics and electronic circuits, a change in sound intensity or circuit power by the factor $10^{0.1} = 1.259$. These changes correspond to changes in sound *pressure* or circuit *voltage* by the square root of the above factor, 1.122. In acoustics most measurements are referred to a sound-power level of 10^{-12} W, corresponding to a sound-pressure level of 0.00002 Pa, the presumed threshold of human audibility. 1 dB is the least change in intensity detectable by normal ears. Thus, because of the exponential definition of the decibel, an 80-dB sound is ten times as intense as a 70-dB sound. OSHA regulations prescribe maximum sound levels permitted for employee exposure. In many plants, scrap grinders, which used to be major noise generators, now run much more quietly or are isolated by enclosures, or both. (Giambattista A, Richardson R, Richardson RC, Richardson B (2003) College physics. McGraw Hill Science/Engineering/Math, New York)

Decitex *n* One-tenth of a tex.

Deckle \ˈde-kəl\ *n* [Gr *Deckel*, literally, cover, fr. *decken* to cover, fr. OHGr *decchen*] (1816) (deckle rod, cut-off plate) In extrusion of film or sheet, or extrusion coating, a small rod or plate attached to each end of the die that symmetrically shortens the length of the die opening, thus facilitating the production of a web of le than the maximum width.

Deck Paint *n* A paint having a high degree of resistance to mechanical wear; especially used on surfaces such as porch floors and ships' decks.

Declination \ˌde-klə-ˈnā-shən\ *n* [ME *declinacioun*, fr. MF *declination*, fr. L *declination-*, *declinatio* angle of the heavens, turning aside] (14c) The angle between the vertical plane containing the direction of the earth's field at any point and a plane containing the geographic north and south meridian.

Decomposition \(ˌ)dē-ˌkäm-pə-ˈzi-shən\ *n* [F *décomposer*, fr. *dé-* de + *composer* to compose] (ca. 1751) Breakdown of molecular structure by chemical or thermal action. With polymers, depending on the severity of conditions, decomposition products can range from subpolymers and oligomers down to monomers and even atoms and ions. In similar terms it is the chemical separation of a substance into two or more substances, which may differ from each other and from the original substances (Zaiko GE (ed) (1995) Degradation and stabilization of polymers. Nova Science Publishers, New York). See also ▶ Degradation.

Decomposition Temperature n (T_{dc}) The temperature at which a polymer or other material (usually organic compound) decomposes by breaking chemical bonds, symbolized by (no symbol is available, but T_{dc} will be used for this purpose). This property should not be confused with the temperature of depolymerization, T_{dp} (Groenewoud WM (2001) Characterization of polymers by thermal analysis. Elsevier Science and Technology Books, New York; Lenz RW (1967) Organic chemistry of synthetic high polymers. Interscience, New York). {G Zersetzungstemperatur f, F température de décomposition, température f, S temperatura de descomposición, temperatura f, I temperatura di decomposizione, temperatura f}.

Decomposition Potential \-pə-❘ten(t)-shəl\ n The minimum voltage which must be applied across a pair of inert electrodes immersed in a medium in order to electrolyze the medium.

Decompression Zone \❘dē-kəm-❘pre-shən ❘zōn\ In a vented extruder, the zone of deep flights immediately forward of the first metering zone and beneath the vent, where unwanted gases and vapors are released from the (typically) foaming melt. In some vented single-screw machines and most twin-screw machines, the two zones are separated by a dynamic value, whose resistance to flow helps to increase the superheating of the melt prior to decompression.

Décor \dā-❘kór, di-❘; ❘de-❘kór, ❘dā-❘\ n [F *décor*, fr. *décorer* to decorate, fr. L *decorare*] (1897) The combination of materials, furnishing, and objects used in interior decorating to create an atmosphere or style.

Decorated \❘de-kə-❘rāt\ *vt* [L *decoratus*, pp of *decorare*, fr. *decor-*, decus ornament, honor] (1530) Adorned, embellished, or made more attractive by means of color or surface detail.

Decorating n The processes used for decorating plastics are defined under the following headings:
Airless Spraying in-mold Decorating
Ashing Letterpress Printing
Decalcomania Metallizing
Double-shot Molding Offset Printing
Electroless Plating Painting of Plastics
Electrophoretic Deposition Printing on Plastics
Electroplating on Plastics Roller Coating
Electrostatic Printing Rubber-plate Printing
Electrostatic Spray Coating Screen Printing
Embedment Decorating Second-Surface
Embossing Decorating
Fill-and-Wipe Spray-and-Wipe Painting
Flexographic Printing Spray Coating
Flocking Thermographic-Transfer
Flow Coating Process
Gravure Coating Vacuum Metalizing
Gravure Printing Valley Printing
Hot Stamping
(Carley JF (ed) (1993) Whittington's dictionary of plastics. Technomic, Lancaster)

Decorative Board n (decorative laminate) A special term for laminates used in the furniture and cabinetry industries, which are defined by the Decorative Board Section of NEMA as "…a product resulting from the impregnation or coating of a decorative web of paper, cloth, or other carrying medium with a thermosetting type of resin and consolidation of one or more of these webs with a cellulosic substrate under heat and pressure of less than 500 pounds per square inch". This includes all boards that were formerly called low-pressure melamine and polyester laminates, but does not include vinyls. "Cellulosic" here means impregnated paper, wood, or plywood.

Decorative Painting n Architectural or aesthetical painting.

Decorticating \(❘)dē-❘kór-tə-❘kā-iŋ\ n [L *decortication-*, *decorticatio*, fr. *decorticare* to remove the bark from, fr. *de-* + *cortic-*, *cortex* bark] (ca. 1623) A mechanical process for separating the woody matter from the bast fiber of such plants as ramie and hemp.

Decoupage \❘dā-(❘)kü-❘päzh\ n [F *découpage*, literally, act of cutting out, fr MF, fr. *decouper* to cut out, fr. *de-* + *couper* to cut] (1946) Modern technique of "burying" a picture design, which has been affixed to a plain surface, in multiple coats of a clear, varnish-type coating. Many coats are applied until the surface no longer reveals the thickness of the design. This art-form is said to have developed from an old French peasant craft consisting of the decoration of a plain surface with cut-out designs of cloth or paper, and then coating it with shellac, varnish or clear lacquer. Another version is that the art of decoupage was originated by craftsmen of eighteenth century Europe in an effort to duplicate Chinese and Japanese lacquer-ware. (Merriam-Webster's Collegiate Dictionary (2004), 11th edn. Merriam-Webster, Springfield)

Decyl Butyl Phthalate \-❘byü-t⁹l ❘tha-❘lāt\ See ▶ Butyl Isodecyl Phthalate.

Decyl-Octyl Methacrylate \- me- tha-krə- lāt\ $H_2C=C(CH_3)COO(CH_2)_nCH_3$, with n = 7, 8 and 9. A polymerizable mixed monomer for acrylic plastics.

Decyl Tridecyl Phthalate n $C_{10}H_{21}COOC_6H_4OOCC_{13}H_{27}$. A plasticizer for vinyls, cellulosics, and polystyrene.

Deep \ dēp\ *adj* [ME *dep*, fr. OE *dēop*; akin to OHGr *tiof* deep, OE *dyppan* to dip] (before 12c) Having depth (in color); often used improperly for dark.

Deep Drawing *n* The process of forming a thermoplastics sheet in a mold in which the ratio of depth to the shortest lateral opening is 1:1 or greater.

Deep-Dyeing Variants \- ver-ē-ənt\ *n* Polymers that have been chemically modified to increase their dyeability. Fibers and fabrics made therefrom can be dyed to very heavy depth.

Defect \ dē- fekt\ *n* [ME, fr. MF, fr. L *defectus* lack, fr. *deficere* to desert, fail, fr. *de-* + *facere* to do] (15c) An internal irregularity or flaw in the structure of a crystal.

Defects *n* A general term that refers to some flaw in a textile product that detracts from either performance or appearance properties.

Definite Proportions, Law of *n* In every sample of each compound substance the proportions by weight of the constituent elements are always the same.

Deflashing \di- flash-eŋ\ *n* The removal of flash or rind left on molded plastic articles by spaces between the mold-cavity edges. Methods include ▶ Tumbling, ▶ Blast Finishing, use of dry or wet abrasive belts, and hand methods using knives, scrapers, broaching tools, and files. Soft thermoplastic parts are sometimes deflashed by tumbling them while immersed in a severe coolant such as liquid nitrogen. See also ▶ Abrasive Finishing, ▶ Airless Blast Deflashing.

Deflashing Equipment *n* Equipment used to remove excess, unwanted material from a processed plastic.

Deflection Temperature \di- flek-shən tem-pə(r)- chúr\ *n* According to ASTM D 648, the stress-labeled temperature at which a standard test bar, centrally loaded as a simple beam to develop a (theoretical) maximum stress of 455 or 1,820 kPa (66 or 264 psi), and warmed at 2°C/min, deflects 0.25 mm. For many years this temperature was called the "heat-distortion point," a term now deprecated and fading from use {*deflection* G Durchbiegung f, F déflexion f, S deflección f, I deflessione f}.

Deflocculant \de- flä-kyə- lāt\ An additive that prevents pigments in suspension from coalescing to form flocs. See ▶ Dispersant (▶ Dispersing Agent). *Also known as Deflocculating Agent*.

Deflocculating Agent *n* A substance that breaks down agglomerates into primary particles or prevents the latter from combining into agglomerates (ISO).

Deflocculation *n* State or condition of a dispersion of a solid in a liquid in which each solid particle remains independent and unassociated with adjacent particles. A deflocculated suspension shows zero or very low yield value.

Defoamer \de- fōm-\ *n* (defoaming agent) A substance that, when added in small percentages to a liquid containing gas bubbles, causes the small bubbles to coalesce into larger ones that rise to the surface and break. Additive used to reduce or eliminate foam formed in a coating or a coating constituent.

Deformation *n* (1) Change in dimension or shape of a plastic product, particularly a test specimen. (2) In a tensile or compression test, the change in length of the specimen in the direction of applied force. Deformation divided by original length equals *elongation*, often expressed as a percent. (3) In a shear mode, the angle of shear in radians. (Shah V (1998) Handbook of plastics testing technology. Wiley, New York) {G Deformation f; Verformung f, F déformation f, S deformación f, I deformazione f}.

Degassing *n* (breathing) In injection or transfer molding, the momentary opening and closing of a mold during the early stages of the cycle to permit the escape of air or gas from the heated compound. See also ▶ Deaerate.

Degating *n* The removal of material left on a plastic part formed by the passage between the runner and the cavity, i.e., the gate. The operation is sometimes performed automatically by a mold element. Otherwise, the gate may be removed by manual breaking or cutting, sometimes followed by sanding or burnishing.

Degating Equipment *n* Equipment used to separate the molded part, automatically or manually, from an injection molded solid runner system.

Degradable Plastic See ▶ Biodegradation.

Degradation \ˌde-grə-ˈdā-shən\ *n* (ca. 1535) A deleterious change in chemical structure, physical properties, or appearance of a plastic caused by exposure to heat (*thermal degradation*), light (*photodegradation*), oxygen (*oxidative degradation*), or weathering. The ability of plastics to withstand such degradation is called *stability* (Zaiko GE (ed) (1995) Degradation and stabilization of polymers. Nova Science, New York; Dissado LA, Fothergill CJ (eds) (1992) Electrical degradation and breakdown of polymers. Institution of Electrical Engineering (IEE), London). See also ▶ Artificial Weathering, ▶ Autocatalytic Degradation, ▶ Biodegradation, ▶ Corrosion Resistance, ▶ Decomposition, ▶ Deterioration, ▶ Pink Staining, ▶ Xenon-Arc Aging.

Degras *n* Natural dark wool grease from which the various grades of lanolin are prepared.

Degreaser \(ˌ)dē-ˈgrē-sər\ *n* Solvent or compounded material for removing oils, fats, or grease from a substrate. Also the apparatus in which this operation is carried out.

Degreasing *n* Removal of grease, oil and other fatty matter (from metals, fabrics, etc.) by the use of solvents or chemical cleaners, electro, or heat processes.

Degree of Crosslinking \di-ˈgrē əv ˈkrós-ˌlink-eŋ\ *n* The fraction of mer units of a polymer that are crosslinked, equal to the quotient of the mer weight and the average molecular weight of segments between crosslinks.

Degree of Crystallinity \ˈkris-tə-ˈli-nə-tē\ *n* The amount of crystalline material in a partially crystalline polymer, usually expressed as a percentage. For a particular polymer the value depends on the crystallization conditions, especially the rate of cooling and the crystallization temperature. (Elias H-G (2003) An introduction to plastics. Wiley, New York)

Degree of Cure \ˈkyúr\ *n* The extent to which curing or hardening of a thermosetting resin has progressed. See also ▶ A-Stage and ▶ C-Stage.

Degree of Efflorescence \ˌe-flə-ˈre-sᵊn(t)s\ *n* (1) *Flatting and Glossing* – A nonuniform decrease or increase in gloss noticed when the surface is illuminated and viewed at near grazing angles. There is no apparent change in color when viewed perpendicular to the surface. (2) *Fine Efflorescence* – A barely discernible whitening of the surface when viewed perpendicularly. (3) *Medium Efflorescence* – A readily noticeable whitening of the surface without a marked masking of the color. (4) *Heavy Efflorescence* – A white deposit sufficient to mask the color.

Degree of Esterification \-e-ˌster-ə-fə-ˈkā-shən\ *n* The extent to which the acid groups of terephthalic and/or other acids have reacted with diols to form ester groups in polyester polymer production.

Degree of Freedom \-ˈfrē-dəm\ *n* (1867) The number of the variables determining the state of a system (usually pressure, temperature, and concentrations of the components) to which arbitrary values can be assigned.

Degree of Metamerism \-mə-ˈta-mə-ˌri-zəm\ *n* See ▶ Metamerism, Degree of.

Degree of Orientation \-ˌōr-ē-ən-ˈtā-shən\ *n* The extent of crystalline deformation after application of stress to a polymer.

Degree of Polymerization \-pə-ˌli-mə-rə-ˈzā-shən\ *n* (DP, chain length) The average number of monomer units per polymer molecule, $DP = M_n/M_{mer}$, where M_n = the polymer's ▶ Number-Average Molecular Weight, and M_{mer} = the molecular weight of the mer (which may differ from monomer molecular weight). In a polymer having worthwhile mechanical properties, *DP* exceeds 500, and many commercial thermoplastics have *DP*s between 1,000 and 10,000. Specifically, the average number of anhydroglucose units (or derivative units) per molecule of cellulose (or cellulose derivative). The type of average obtained depends upon the method used for the determination. Hence, the method must always be specified. See also ▶ Molecular Weight.

Degumming \(ˌ)dē-ˈgəm-eŋ\ *vt* (1887) The removal of gum from silk by boiling in a mildly alkaline solution. Usually accomplished on the knit or woven fabric.

DEHP *n* (DOP) Abbreviation for Di(2-Ethylehexyl) Phthalate.

Dehydrated Castor Oil \(ˌ)dē-ˈhī-ˌdrāt-ed ˈkas-tər ˈói(ə)l\ *n* A castor oil chemically treated to improve its drying qualities. See ▶ DCO.

Dehydrated Gypsum \ˈjip-səm\ *n* See ▶ Calcium Sulfate, ▶ Anhydrous.

Dehydration *n* The removal of water from a substance either by ordinary drying or heating, or by absorption, chemical action, condensation of water vapor, or by centrifugal force or filtration. The term is generally applied to the removal of combined water as, for instance, from hydrates.

Dehydroacetic Acid \(ˌ)dē-ˌhī-drə-ə-ˈsē-tik\ *n* Abbreviated is DHA. It is a rosin derivative present in hy-drogenated rosin. A heterocyclic compound used as a plasticizer, fungicide, and bactericide.

Dehydrogenated Oils *n* Semidrying oils which have been converted into drying oils by first treating them with a halogen (usually chlorine), then eliminating hydrogen halide. This results in conjugated double bonds in the fatty acids chain.

Dehydrogenation \ˌdē-(ˌ)hī-ˌdrä-jə-ˈnā-shən\ *n* (1866) The removal of hydrogen from a compound by chemical means. See also ▶ Cracking (2).

Dehydrohalogenation \-ˌha-lə-jə-ˈnā-shən\ *n* The splitting of hydrogen chloride or other hydrohalide from polymers such as PVC, by action of excessive heat or light.

Deionization \(ˌ)dē-ˈī-ə-nə-ˈsā-shən\ *n* The removal of ions from a solution by ion exchange.

Deka- (da) A prefix permitted by SI and meaning × 10.

Delaminate \(ˌ)dē-ˈla-mə-ˌnāt\ *vt* To separate existing layers.

Delaminated Clay \ˈklā\ *n* Pigment produced by mechanically separating the platelets of natural kaolin so that particles of approximately ¼ micrometer thickness and up to 10 μm in diameter are produced. This pigment is used to increase the opacifying efficiency of titanium dioxide by keeping the titanium dioxide particles separated by the thickness of the calcined clay particle. This spacing allows each titanium dioxide particle to intercept a wavelength of visible light and thereby results in maximum light-scattering efficiency.

Delamination *n* Separation of the wood surfaces at the bonded joints caused by a cohesive failure in the adhesive or a failure of the adhesive at the surface. When the separation occurs in the wood, even though very close to the bonded joint, the separation is termed *wood failure* or *checking* rather than delamination. Magnification is often necessary to determine whether the failure is in the bond or in the wood. The term also applies to failure occurring between successive coatings.

Delated Deformation \di-ˈlāt-ˌdē-ˌfór-ˈmā-shən\ *vt* Deformation that is time-dependent and is exhibited by material subjected to a continuing load; creep. Delayed deformation may be recoverable following removal of the applied load.

Deleafing *n* Loss of leafing of metallic pigments in paints, giving rise to reduced metallic luster. See ▶ Leafing.

Deliquescence \ˌde-liˈkwe-sənt\ *adj* [L *deliquescent-, deliquescens*, pp of *deliquescere*] (1791) Property possessed by some materials of absorbing moisture from the air and forming a solution in the water so absorbed.

Deliquescent *n* Said of some salts that are so strongly attractive of water that they can absorb enough from room air to liquefy.

Delocalized Molecular Orbital *n* An orbital which extends over more than two atoms in a molecule, ion, or larger aggregate.

Delrin *n* Poly(oxymethylene) (from formaldehyde). Manufactured by DuPont, U.S.

Delta E, ΔE \ˈdel-tə\ *n* Total color difference computed by use of a color difference equation. It is generally calculated as the square root of the sum of the squares of the chromaticity difference, ΔC, and the lightness difference, ΔL: $\Delta E = [(\Delta C)^2 + (\Delta L)^2]^{1/2}$. See ▶ Color Difference Equations.

Delustering *n* Subduing or dulling the natural luster of a textile material by chemical or physical means. The term often refers to the use of titanium dioxide or other white pigments as delustrants in textile materials.

Delustrant *n* A chemical agent used either before or after spinning to produce dull surfaces on synthetic fibers to obtain a more natural, silk like appearance.

Denaturation \(ˌ)dē-ˌnā-chə-ˈrā-shən\ *n* (1685) Addition of unpleasant or toxic substances to a product to make it unfit for human consumption.

Denatured Alcohol *n* Ethyl alcohol to which denaturant (odorous and/or nauseating substances) has been added in small quantities to make it unpotable without diminishing its solvent power. Typical denaturants are benzene, acetaldehyde, and pyridine.

Dendrite \ˈden-ˌdrīt\ *n* (1751) (dendritic) A branched, treelike crystal habit usually associated with rapid crystal growth.

Denier \di-ˈnī(-ə)r\ *n* [ME *denere*, fr. MF *denier*, fr. L *denarius*, coin worth ten asses, fr. *denarius* containing ten, fr. *deni* ten each, fr. *decem* ten] (15c) A unit of weight expressing the size or coarseness but particularly the fineness of a relatively continuous fiber or yarn (Joseph ML (1986) Textile science, 5th edn. CBS College Publishing, New York). The weight in grams of 9,000 m of a fiber in the form of continuous filament. Although deprecated in SI, this is still the most widely used unit of lineal density in the textile industry to indicate fineness of natural or synthetic fibers. 1 denier = 1.1111×10^{-7} kg/m. See also ▶ CUT, ▶ Grex Number, and ▶ TEX.

Denier per Filament (dpf) \fi-lə-mənt\ *n* The denier of an individual continuous filament or an individual staple fiber if it were continuous. In filament yarns, it is the yarn denier divided by the number of filaments. (Joespeh ML (1986) Textile science, 5th edn. CBS College Publishing, New York)

Denier Variation *n* Usually variation in diameter, or other cross-sectional dimension, along the length of a filament or bundle of filaments. It is caused by malfunction or lack of process control in fiber manufacturing and degrades resulting fabric appearance or performance (Joespeh ML (1986) Textile science, 5th edn. CBS College Publishing, New York)

Denim \de-nəm\ *n* [F (*serge*) *de Nîmes* serge of *Nîmes*, France] (1695) A firm 2 × 1 or 3 × 1 twill-weave fabric, often having a whitish tinge, obtained by using white filling yarns with colored warp yarns. Heavier weight denims, usually blue or brown, are used for dungarees, work clothes, and men's and women's sportswear. Lighter weight denims with softer finish are made in a variety of colors and patterns and are used for sportswear and draperies.

Densitometer \den(t)-sə-tä-mə-tər\ *n* (1901) Instrument to measure the optical density of a transmitting material, or an instrument to measure the negative log of the reflectance of a reflecting material. Such instruments do not measure color. They are widely used in the graphic arts and photographic industries for process control. In spectrographic and dispersive X-ray analysis utilizing photography, they are adapted for scanning and plotting the density of the photographed transparency as a function of the wavelength.

Density \den(t)-sə-tē\ *n* (1603) (absolute density) Mass per unit volume of a substance. The SI unit is kg/m³, but others in common use are g/cm³, lb/ft³, and lb/gallon. It is an important criterion in specifying some plastics, e.g., polyethylene, which can vary in density from 0.92 to 0.98 g/cm³, with significant associated variation in properties such as modulus. Since density decreases with rising temperature, the temperature of measurement should be stated. See www.astm.org for current methods of determining density. See also ▶ Apparent Density, ▶ Bulk Density, ▶ Bulking Value, ▶ Gradient-Tube Density Method, ▶ Relative Density, and ▶ Specific Gravity.

Density, Apparent See ▶ Apparent Density.

Density Gradient \grā-dē-ənt\ *n* A gradient of density such as in a gradient column for measuring density of plastic materials.

Density, Optical Negative log of the transmittance. See ▶ Beer-Bouguer Law and ▶ Absorption Coefficient.

Densothene *n* Poly(ethylene), manufactured by Metal Box, Great Britain.

Dent *n* On a loom, the space between the wires of a reed.

Deoxy- (desoxy-) A prefix denoting replacement of a hydroxyl group with hydrogen.

Deoxyribonucleic Acid *n* [*deoxyribose* + *nucleic acid*] (1938) See ▶ DNA.

DEP Abbreviation for ▶ Diethyl Phthalate.

Depolymerization *n* The reversion of a polymer to its monomer, or to a polymer of lower molecular weight. Such reversion occurs in most plastics when they are exposed to very high temperatures in the absence of air. Splitting of polymers into molecules or polymers of lower molecular weight. The temperature of depolymerization is symbolized as T_d. (Odian GC (2004) Principles of polymerization. Wiley, New York; Lenz RW (1967) Organic chemistry of synthetic high polymers. Interscience, New York)

Depolymerized Rubber *n* Rubber which has been extensively milled or otherwise treated so that it has sufficient solubility in aromatic hydrocarbons for the resulting solution to be used as a varnish or varnish constituent.

Deposit Attack *n* Excessive corrosion that may occur under solid substances that may be deposited on a metal surface and thus shield it from ready access to oxygen or ions in the solution. This is a form of concentration cell corrosion.

Deposition *n* (1) Usually referred to in metal treatment as the amount of treatment by weight (mg/sq ft). (2) Process of applying a material to a base by means of vacuum, evaporation, sputtering, electrolysis, chemical reaction, vapor methods, etc.

Depropagation *n* The sequential chain scission step during depolymerization responsible for the formation of monomer. Has a lower activation energy than propagation and hence is favored at high temperatures.

Depth of Draw *n* In thermoforming, the depth of the lowest point in the formed object relative to the clamped edge of the sheet. Also see ▶ Draw Ratio (2).

Depth of Field *n* (1911) The thickness of the preparation brought into good focus at a single setting of the focusing adjustment. It lies in the plane of the specimen and decreases with an increase in numerical aperture.

Depth of Finish *n* The appearance phenomenon of a coating having an "apparent" depth. Also referred to as "seeing into the film."

Depth of Focus *n* The thickness of the image of the preparation appearing in good focus at one setting of the focusing adjustment.

Depth of Screw Syn: ▶ Flight Depth.

Depth of Shade Concept which embodies the subjective or visual equality of concentration of colorants, dyes, or pigments. See ▶ Depth of Shade, ▶ Standard.

Depth of Shade, Standard *n* An arbitrarily chosen depth of shade for all hues, from which a uniform depth of shade may be determined for purposes of comparison. Depth of shade can then be described as a multiple or fraction of standard depth. (Developed primarily, and used most widely, in Europe.)

Derby Red \ˈdər-bē, *esp British* ˈdär-\ *n* Basic lead chromate. See ▶ Chrome Orange, Light and Deep.

Deregistering (Crimp) Process of disordering or disaligning the crimp in a tow band to produce bulk. Also see ▶ Threaded-Roll Process.

Derived High Polymer A polymer which has been produced by chemical alteration of a primary high polymer or a natural high polymer. See also ▶ Primary High Polymer.

Dermatitis \ˌdər-mə-ˈtī-təs\ *n* (1876) Inflammation or irritation of the skin. Industrial dermatitis is an occupational skin disease. There are two general types of skin reaction: primary irritation dermatitis and sensitization dermatitis.

Descaling *n* Removal of mill sale or caked rust from steel by chemical and/or mechanical means, sometimes assisted by flame cleaning.

Desensitize \(ˌ)dē-ˈsen(t)-sə-ˌtīz\ *vt* (1898) To treat non-image areas of a water lithographic plate to make them water-receptive and ink-repellant.

Desiccant \ˈde-se-kənt\ *n* (1676) A substance capable of absorbing water vapor from air and other materials enclosed with it; a drying agent. Typically used to maintain low humidity in a storage or test vessel or package.

Desiccant Dryers *n* Materials which absorb water from air (e.g., silica gel).

Desiccator \de-si-ˈkā-tər\ *n* A plastic or heavy-glass bowl with a tight-fitting lid, containing a drying agent, in which a sample can be kept in a controlled, dry atmosphere, typically under a mild vacuum.

Designers Color *n* Opaque watercolor paints that dry flat and are not necessarily permanent.

Design Life *n* The period through which a part or product is expected to perform satisfactorily, remaining within preset performance tolerances.

Desmodur *n* Isocyanate grades for polyurethane. Manufactured by Bayer, Germany.

Desmophen *n* Polyester for polyurethanes. Manufactured by Bayer, Germany.

Desorption \-ˈsórp-shən\ *n* (1924) The escape or loss of a substance previously absorbed into, or adsorbed onto, another.

Destaticization *n* The treating of a plastic to minimize its tendency to accumulate static electric charge, or removal of charge. See also ▶ Antistatic Agent, ▶ Static Eliminator, and ▶ Soot-Chamber Test.

Destructive Distillation \di-ˈstrek-tiv ˌdis-tə-ˈlā-shən\ *n* (ca. 1831) Decomposition of an organic material by heating in the absence of air. Destructive distillation of wood, which produces charcoal and several small organic molecules, was once the major source of methanol.

Destructively Distilled Pine Oil See ▶ Pine Oil.

Destructively Distilled Wood Turpentine See ▶ Turpentine.

Desulfurizing \(ˌ)dē-ˈsəl-fər-ˌīz-eŋ\ *vt* An after treatment to remove sulfur from newly spun viscose rayon by passing the yarn through a sodium sulfide solution.

DETA *n* Abbreviation for ▶ Diethylenetriamine.

Detection Limit *n* Of chemical quantitative analyses, the least quantity of concentration of the substance being sought that can be detected with a stated level of confidence. This widely used term may be giving way to ▶ Minimum Detectable Amount.

Detergent \-jənt\ *adj* (1616) A surface-active agent that possesses the ability to clean soiled surfaces. *Anionic Detergent* – A detergent that produces aggregates of negatively charged ions with colloidal properties. *Cationic Detergent* – A detergent that produces aggregates of positively charged ions with colloidal properties. *Nonionic Detergent* – A detergent that produces aggregates of electrically neutral molecules with colloidal properties. See ▶ Surfactants.

Deterioration \di-ˌtir-ē-ə-ˈrā-shən\ *n* (ca. 1658) A usually gradual process of permanent impairment in the appearance or application of physical properties. See ▶ Degradation.

Deuterium \dü-ˈtir-ē-əm *also* dyü-\ *n* [NL, fr. Gk *deuteros* second] (1933) Hydrogen 2; $_1^2$H; "heavy hydrogen." (Giambattista A, Rihcardson B, Richardson RC (2002) College Physics, 9th Rev, edn. McGraw-Hill Publishing Co., New York.)

Deuteron *n* Nucleus of the deuterium atom or the ion of deuterium. Its structure – one neutron and one proton.

Developed Dyes *n* Dyes that are formed by the use of a developer. The substrate is first dyed in a neutral solution with a dye base, usually colorless. The dye is then diazotized with sodium nitrate and an acid and afterwards treated with a solution of β-naphthol, or a similar substance, which is the developer. Direct dyes are developed to produce a different shade or to improve washfastness or lightfastness. See ▶ Dyes.

Developing *n* A stage in dyeing or printing in which leuco compounds, dyes, or dye intermediates are converted to the final, stable state or shade.

Developing Ink *n* A non-drying greasy ink composition specifically formulated for use in initial fixing or subsequent renewal of the image on a lithographic plate.

Devolatilization *n* The removal from a resin of a substantial percentage of some unwanted volatile matter such as water, solvent, or monomer. Vented extruders have been widely used for this task. See ▶ Extraction Extrusion.

Dew Point (Dew Temperature) *n* (ca. 1833) Temperature of the atmosphere at which the saturation vapor pressure equals the actual vapor pressure of the water vapor in the air and dew begins to form.

Dewaxing *n* Process of removing waxes from natural resins (dammars, shellacs, etc.).

Dexel *n* Cellulose acetate, manufactured by British Celanese, Great Britain.

Dexsil *n* Poly(carborane siloxane), manufactured by Olin, U.S.

Dextin Water-soluble adhesive or thickener prepared by the incomplete hydrolysis of starch by thermostatic (use of diastase), or acid methods. Syn: ▶ British Gum.

Dextrorotatory \❙dek-(❙)strō-❙rō-tə-❙tōr-ē\ *adj* (1878) To turn clockwise (sym. Is D) or toward the right; *esp*: rotating the plane of polarization of light toward the right, <-crystals> – compare ▶ Levorotatory (rotates light to the left). (Morrison RT, Boyd RN (1992) Organic chemistry, 6th edn. Prentice Hall, Englewood Cliffs)

DGEBA *n* Abbreviation for ▶ Diglycidyl Ether of Bisphenol.

D-Glass *n* Glass with a high boron content, used for fibers in laminates that require a precisely controlled dielectric constant.

DHA *n* Abbreviation for ▶ Dehydroacetic Acid.

DHP *n* Abbreviation for Dihexyl Phthalate.

DHXP *n* Same as DHP, preceding.

Di- A prefix meaning two or twice. The terms *bi-* and *bis-* are nearly equivalent, assigned with slight differences in meaning according to custom.

Diacetin *n* (glyceryl diacetate. (1,3-diacetin), and 1,2-diacetin, a mixture of isomers. A water-soluble plasticizer, and a solvent for cellulose nitrate and cellulose acetate.

Diacetone Alcohol *n* (4-hydroxy-4-methyl-2-pentanone) (A pleasant-smelling, colorless liquid, miscible with water and most organic liquids, used as a solvent for cellulosic, vinyl, and epoxy resins. Mol wt, 116.16; sp gr, 0.9406/20°C; bp, 169°C (760 mmHg); flp, 52°C (125°F).

Diacetyl Peroxide See ▶ Acetyl Peroxide.

Diafoam *n* A term sometimes used for ▶ Syntactic Foam that contains gas bubbles in addition to microspheres.

Diagonal (45°C) Flame Test \dī-❙a-gə-n°l\ *n* See ▶ Flammability Test.

Diagonal Grain *n* Grain in which the longitudinal elements form an angle with the axis of the piece as a result of sawing at an angle with the bark of the tree or log; a form of cross grain.

Diagonal-Grained Wood *n* Wood in which the fibers are at an angle with (that is, diagonal to) the axis of a piece as a result of sawing at an angle with the axis of the tree.

Dial \❙dī(-ə)l\ *n* [ME *dyal*, fr. ML *dialis* clock wheel revolving daily mfr. L *dies* day] (15c) In a circular-knitting machine, a circular steel plate with radically arranged slots for needles. A knitting machine equipped with both a dial and a cylinder can produce double-knit fabrics. (Tortora PG, Merkel RS (2000) Fairchild's dictionary of textiles, 7th edn. Fairchild, New York)

Diallyl Chlorendate *n* (DAC) A reactive monomer used as a flame-resisting agent in diallyl phthalate, epoxy, and alkyd resins. It can be used in the monomeric form (a high-viscosity liquid); or in the polymeric form, alone or in conjunction with other flame retardants.

Diallyl Esters *n* Series of unsaturated esters which polymerize rapidly to yield crosslinked resins in the presence of peroxides at relatively low temperatures (e.g., 40–100°C).

Diallyl Isophthalate *n*. A polymerizable monomer, used in laminating and molding.

Diallyl Maleate *n*. A monomer that polymerizes readily when exposed to light or temperatures above 50°C.

Diallyl Phthalate *n* (diallyl o-phthalate, DAP). In the monomeric form, DAP is a colorless liquid ester with a viscosity about equal to that of kerosene, widely used as a crosslinking monomer for unsaturated polyester resins, and as a polymerizable plasticizer for many resins. If polymerizes easily, either gradually or rapidly, increasing in viscosity until it finally becomes a clear, infusible solid. The name DAP is used for both the monomeric and polymeric forms. In the partly polymerized form, DAP is used in the production of thermosetting molding powders, casting resins, and laminates.

Diallyl Phthalate Resins *n* Laminating resin prepared by free radical polymerization of diallyl phthalate. The polymer is highly crosslinked, with good thermal stability and retention of electrical properties under conditions of wet and dry heat.

Di-Alphinyl Phthalate *n* A plasticizer derived by the esterification of primary aliphatic alcohols in the range C_7 to C_9. It is used in place of dioctyl phthalate in some applications.

Dialysis \dī-ˈa-lə-səs\ *n* [NL, fr. Gk, separation, fr. *dialyein* to dissolve, fr. *dia-* + *lyein* to loosen] (1861) The separation of different-sized molecules by transport in solution through a semipermeable membrane, driven by the difference in chemical potential between the liquids separated by the membrane. The membranes may be any of a wide range of materials, e.g., parchment or microporous films of cellulosic polymers.

Diamagnetic Materials \ˌdī-ə-mag-ˈne-tik\ Are those within which an externally applied magnetic field is slightly reduced because of an alteration of the atomic electron orbits produced by the field. Diamagnetism is an atomic-scale consequence of the Lenz law of induction. The permeability of diamagnetic material is slightly less than that of empty space.

Diamine \ˈdī-ə-ˌmēn\ *n* [ISV] (1866) A compound with two amino groups. Hexamethylenediamine, one of the intermediates in the manufacture of nylon 66 salt, is an example of this chemical type.

4,4′-Diaminodiphenylmethane *n* (DDM, methylenediamine, MD). A silvery, crystalline material obtained by heating formaldehyde aniline with aniline hydrochloride and aniline. It is used as a curing agent for epoxy resins and as an intermediate in making diisocyanates for urethane elastomers and foams by reaction with phosgene. Possible occupational hazards in the use of DDM are toxic hepatitis and liver damage.

4,4′-Diaminodiphenylsulfone *n* (DDS). An epoxy curing agent that gives the highest deflection temperatures among amine-cured epoxies.

Diamyl Phthalate *n* $C_6H_4(COOC_5H_{11})_2$. A plasticizer derived by esterification of phthalic anhydride with amyl alcohol, compatible with most vinyls, polymethyl methacrylate, and cellulosics.

Diaphragm Gate \dī-ə- fram\ *n* (web gate) In an injection or transfer mold, a gate that forms a complete thin web across the opening of the part, used in the molding of annular and tubular objects.

Diarylide Yellow A strong yellow toner used in many types of printing inks. See ▶ Benzidine Yellow.

Diatomaceous Calcite \ dī-ə-tə- mā-shəs kal- sīt\ *n* 2.4 $CaCO_3 \cdot SiO_2$ (Approximate). Liquid blend of $CaCO_3$ and diatomaceous silica that is mined from a single large deposit located in Kansas. Used as an extruder. Sp gr, 2.54; density, 21.2 lb/gal; O.A., 25/100 lb; particle size, 10 μm.

Diatomaceous Earth *n* (1883) See ▶ Diatomaceous Silica.

Diatomaceous Silica *n* $SiO_2 \cdot 7H_2O$. Pigment White 27 (77811). A form of hydrous silica, processed from natural diatomate, a sedimentary rock of varying degrees of consolidation that is composed essentially of the fossilized siliceous skeletal remains of single-cell aquatic plant organisms called diatoms. Refractive index. 1.42; density, 2.00 g/cm^3 (16.7 lb/gal); particle size, 6–10 μm. Syn: ▶ Diatomaceous Earth, ▶ Diatomite, and the obsolete names of ▶ Kieselguhr, ▶ Infusorial Earth, ▶ Tripoli.

Diatomite \di- a-tə- mīt\ *n* (1887) (diatomaceous earth, DE, kieselguhr, infusorial earth, siliceous earth, tripolite) The naturally occurring deposit of skeletons of small unicellular algae called *diatoms*, consisting of from 83% to 89% silica. Its many uses include fillers for plastics.

Diazo \dī- a-( )zō\ *adj* [ISV *diaz-, diazo-*, fr. *di-* + *az-*] (1878) -N = N- grouping. Diazo compounds are intermediates for manufacture of dyestuffs.

Diazo Solution Solution resulting from treatment of an aromatic primary amine with sodium nitrite and mineral acid. Such solutions are stable only below 5°C and are employed as one of the ingredients in the manufacture of azo colors.

DIBA *n* Abbreviation for ▶ Diisobutyl Adipate.

Dibasic \( )dī- bā-sik\ *adj* (1868) Pertaining to acids having two replaceable hydrogen atoms e.g., sulfuric acid, H_2SO_4, or to acid salts in which two of the three hydrogens have been replaced by metal(s), e.g., K_2HPO_4.

Dibasic Lead Phosphite *n* $2PbO \cdot PbHPO_3 \cdot 0.5H_2O$. A white, crystalline powder long used as a heat and light stabilizer for PVC and other chlorine-containing resins. It has good electrical properties, and acts as an antioxidant and screening agent for UV light. Less used today because of concern to keep lead out of the environment.

Dibasic Lead Stearate *n* $2PbO \cdot Pb(OOCC_{17}H_{35})_2$. A good heat stabilizer with lubricating properties. See also ▶ Lead Stearate.

Dibenzyl Ether $(C_6H_5CH_2)_2O$. A plasticizer for cellulose nitrate.

Dibenzyl Phthalate See ▶ Benzyl Phthalate.

Dibenzyl Sebacate *n* $(C_6H_5CH_2OOC)_2(CH_2)_8$. A nontoxic plasticizer, often used in vinyl compounds for lining container closures.

Diblock Polymers *n* Block copolymers consisting of two blocks, one of A repeating units and one of B repeating units. Thus its structure may be represented as
AAA......AAAABBBB......BBB

Diborane \-ˈbōr-ˌān\ *n* B_2H_6 (boron hydride, boroethane) A colorless gas that has been used as a catalyst in the polymerization of ethylene.

DIBP Abbreviation for ▶ Diisobutyl Phthalate.

2,3-Dibromo-2-Butene-1,4-Diol *n* (dibromobutenediol) $HOCH_2C(Br)=C(BR)CH_2OH$. A low-molecular-weight, chemically reactive, brominated primary glycol. It is used as a building block for condensation polymers that can be incorporated into a wide variety of polymers including esters, urethanes, and ethers. It is also used as a flame-retardant monomer for polyurethanes and thermoplastics, and as a substitute for methylene-bis-o-chloroniline (MOCA) in urethane foams.

Dibromoneopentyl Glycol *n* A high-melting solid, available in powder or flake form, used as a flame retardant for polyester resins. A more convenient liquid material is made by using dibromoneopentyl glycol to form a polyester alkyd that is dissolved in styrene, the resulting liquid being more easily used in polyester reactors. The material is also adaptable to urethane foams, polymeric plasticizers, and coating resins.

2,3-Dibromopropanol *n* A brominated alcohol used as a component in making fire-retardant urethane foams.

Dibutoxyethoxy Ethyl Adipate *n* $[C_4H_9OC_2H_4OC_2H_4COO(CH_2)_2]_2$. A plasticizer for cellulose nitrate, ethyl cellulose, polyvinyl butyral, and polyvinyl acetate.

Dibutoxyethyl Adipate *n* $(C_2H_4COOC_2H_4OC_4H_9)_2$. A primary plasticizer for PVC and many other resins, imparting low-temperature flexibility and UV resistance. It is widely used as a plasticizer for polyvinyl butyral in the interlayer of safety glass.

Dibutoxyethyl Phthalate *n* (DBEP) $C_6H_4(COOC_2H_4OC_4H_9)_2$. A primary plasticizer for vinyls, methacrylates, nitrocellulose, and ethyl cellulose, imparting low-temperature flexibility and UV resistance. Incorporation of up to 20% of DBEP into vinyl calendaring compounds eliminates defects such as streaks and blisters. In plastisols, DBEP imparts low initial viscosity and low fusion temperatures.

Dibutoxyethyl Sebacate *n* $(CH_2)_8(COOC_2H_4OC_4H_9)_2$. A primary plasticizer for PVC and PVAc, with low-temperature resistance.

Dibutyl Adipate \ ˈdī- ˈbyü-təl-\ n $(C_4H_9COO)_2(CH_2)_4$. A plasticizer for vinyl and cellulosic resins.

Dibutyl Butyl Phosphonate n $C_4H_9P(O)(OC_4H_9)_2$. A plasticizer and antistatic agent.

Dibutyl Fumarate \fyú- ˈmar- ˈnāt\ n $C_2H_2(COOC_4H_9)_2$. Derived from fumaric acid and used as a plasticizer for polyvinyl acetate, polyvinyl chloride, and PVC-AC copolymers.

Dibutyl Isosebacate n $C_8H_{16}(COOC_4H_9)_2$. A plasticizer for vinyls and other thermoplastics.

Dibutyl Phthalate n [*phthalic* acid + ^{1}-*ate*] (1925) (DBP) $C_6H_4(COOC_4H_9)_2$. One of the most widely used plasticizers for cellulose nitrate and other cellulose-ester and -ether lacquers and coatings. It is a primary plasticizer for many other resins, but its high volatility limits its use in vinyls. Sp gr. 1.0484 (20/20°C); flp, Cleveland Open Cup, 171°C (340°F); bp, 340°C.

Dibutyl Sebacate n (DBS) $(CH_2)_8(COOC_4H_9)_2$. A plasticizer, one of the most effective of the sebacate family. It has good low-temperature properties, low volatility, and is compatible with vinyl chloride polymers and copolymers, polyvinyl butyral and ethyl cellulose. It is nontoxic, suitable for uses in food wrappings.

Dibutyl Succinate \ ˈsək-sə- ˈnāt\ n $(CH_2)_2(COOC_4H_9)_2$. A plasticizer for cellulosic resins.

Dibutyl Tartrate \ˈtär-ˌtrāt\ n $(CHOH)_2(COOC_4H_9)_2$. A lubricant, plasticizer, and solvent for cellulosic resins.

Dibutyltin Bis(isooctylmercapto Acetate) n A stabilizer for rigid PVC, used primarily during the period from 1953 to 1970. Thereafter, improved butyltin and methyltin derivatives, and synergistic mixtures of them with other stabilizers, replaced this stabilizer for most uses.

Dibutyltin Di-2-Ethylhexoate n $(C_4H_9)_2Sn(OOCC_7H_{18})_2$. A white, waxy solid made by reacting dibutyltin oxide with 2-ethylhexoic acid. Used as a catalyst for silicone curing and in polyether foams.

Dibutyltin Dilaurate n (DBTDL) $(C_4H_9)_2Sn(OOCC_{11}H_{23})_2$. A lubricating stabilizer for vinyl resins, a catalyst for urethane foams and for condensation polymerizations. It is used in vinyl compounds when good clarity is needed, and also imparts excellent light stability. However, it degrades at high processing temperatures.

Dibutyltin Maleate n $[(C_4H_9)_2Sn(OOCCH)_2]_x$. A white amorphous powder used as a condensation-polymerization catalyst and a stabilizer for PVC. The molecular weight of the material varies, and grades of lower molecular weight tend to be volatile and produce gases. The higher-weight polymers are very effective in rigid PVC.

Dibutyltin Sulfide n $[(C_4H_9)_2SnS]_3$. An antioxidant and stabilizer for vinyl resins.

1,1-Dibutylurea n (N,N-dibutylurea) $NH_2CON(C_4H_9)_2$. A polymerizable compound. When copolymerized with simple urea and formaldehyde, permanently thermoplastic resins are obtained.

Dicapryl Adipate n $C_4H_8(COOC_8H_{17})_2$. A plasticizer for cellulosic and vinyl resins, yielding good low-temperature flexibility. Also compatible with polymethyl methacrylate and polystyrene.

Dicapryl Phthalate *n* [(DCP) di-(2-octyl) phthalate] $(C_8H_{17}COO)_2C_6-H_4$. A plasticizer for cellulosic and vinyl resins. It is similar to DOP and DIOP, but has low initial viscosity and is preferred in plastisols.

Dicapryl Sebacate *n* $(CH_2)_8(COOC_8H_{17})_2$. A plasticizer for vinyl resins and acrylonitrile rubbers, imparting good low-temperature flexibility.

Dicarboxylic Acid \ ▪dī- ▪kär- ▪bäk- ▪si-lik\ *n* Any of a large family of organic acids containing two carboxylic (–COOH) groups. Those of greatest importance in the plastics industry are the adipic, azelaic, glutaric, pimelic, sebacic, and succinic acids, esters of which are widely used as plasticizers that impart low-temperature flexibility. They are also used in the production of alkyd and polyester resins, polyurethanes, and nylons.

Dicers *n* One of the two basic types of cold cutting systems. It consists of a die, a cooling area, a drying area (if water was used), and a cutting chamber.

Dicetyl Ether \- ▪sē-təl ▪ē-thər\ (dihexadecyl ether) A mold lubricant.

2,4-Dichlorobenzoyl Peroxide *n* A crosslinking agent for silicone elastomers. It is sold as a 40% active paste dispersed in silicone fluid (Candox®).

Dichlorodifluoromethane \- ▪dī- ▪flúr-ə- ▪me- ▪thān\ *n* (1936) (Freon 12) CCl_2F_2. Long used as a refrigerant, aerosol propellant, and as a blowing agent for foamed plastics that was safer than inflammable hydrocarbons, this accused destroyer of stratospheric ozone is rapidly being phased out of all its former uses.

1,1-Dichloroethylene *n* Syn: ▶ Vinylidene Chloride.

α-Dichlorohydrin *n* CH$_2$ClCHOHCH$_2$Cl. A cellulosic-resin solvent.

Dichloromethane \(ˌ)dī-ˌklō-rō-ˈe-ˌthān\ *n* (1936) Syn: ▶ Methylene Chloride.

2,6-Dichlorostyrene *n* Cl$_2$C$_6$H$_3$CH=CH$_2$. A monomer and comonomer used mainly in plastics research.

Dichlorotetrafluoroethane *n* (Freon 114) ClF$_2$CCClF$_2$. A fluorocarbon blowing agent used when a low boiling point (3.6°C) is required.

Dichroism \ˈdī-(ˌ)krō-ˌi-zəm\ *n* (1819) (1) A property possessed by many doubly refracting crystals of exhibiting different colors when viewed from different directions. (2) Displaying different colors as a result of changes in concentration or thickness. Such changes do not result from changes in the absorption coefficients for the material, but rather on the change in perceived color resulting from more or less absorption across the visible spectrum as the concentration is changed. The term "dichroism" is frequently used (incorrectly) to describe this effect. Dichromism is predictable on the basis of the laws of subtractive colorant mixture.

Dicing *n* The process of cutting thermoplastic sheets (or square strands) into cubical pellets for further processing.

Di-compounds *n* See under other headings, such as ▶ Ethyl Oxalate, etc.

Dicumyl Peroxide *n* [C$_6$H$_5$C(CH$_3$)$_2$O]$_2$. A vulcanizing agent for elastomers, also used to crosslink polyethylene.

Dicyandiamide *n* (cyanoguanidine) H$_2$NC(=NH)NHCN. The widely used, but incorrect name for the dimmer of cyanamide. Cyanoguanidine is used mainly in the production of melamine, but also as a stabilizer for vinyl resins and curing agent for epoxy resins.

Dicyclohexyl Azelate *n* C$_6$H$_{11}$OOCC$_7$H$_{14}$COOC$_6$H$_{11}$. A plasticizer for PVC.

Dicyclohexyl Phthalate *n* (DCHP) C$_6$H$_4$(COOC$_6$H$_{11}$)$_2$. A plasticizer for PVC and many other resins. It imparts good electrical properties, low volatility, low water and oil absorption, and resistance to extraction by hexane and gasoline. In vinyls DCHP is usually combined with other plasticizers. In cellulose nitrate, polystyrene, polymethyl methacrylate, and ethyl cellulose it serves as a primary plasticizer.

DIDA *n* Abbreviation for ▶ Diisodecyl Adipate.

Didecyl Adipate *n* $C_4H_8(COOC_{10}H_{21})_2$. A plasticizer for PVC and cellulosics. Its most noteworthy properties are low-temperature flexibility, low volatility, and good electricals.

Didecyl Ether *n* $(C_{10}H_{21})_2O$. A processing and mold lubricant.

Didecyl Phthalate *n* (DDP) $C_6H_4(COOC_{10}H_{21})_2$. A primary plasticizer for vinyl resins, also compatible with polystyrene and cellulosics. It has the lowest specific gravity of the most common phthalate plasticizers, low volatility, and resistance to extraction by soapy water.

DIDG *n* Abbreviation for ▶ Diisodecyl Glutarate.

DIDP *n* Abbreviation for ▶ Diisodecyl Phthalate.

Die \ˈdī\ [ME *dī*, fr. MF *dé*] *n* (1) ▶ Extrusion Die. (2) The recessed block into which plastic material is injected or pressed, shaping the material to the desired form. The terms *mold cavity* or *cavity* are more often used. (3) Steel-rule die. See ▶ Die Cutting.

Die Adapter *n* (extrusion) See ▶ Adapter.

Die Blade (die lip) In extrusion, a deformable member attached to a die body that determines the slot opening and that may be adjusted to produce uniform thickness across the film or sheet being made.

Die Block *n* That part of an extrusion die that holds the forming bushing and core.

Die Body *n* In the U.S., same as ▶ Adapter. In Great Britain, the outer body or barrel of an extrusion die.

Die Cart *n* A sturdy, height-adjustable framework on casters designed to support a heavy extrusion die, such as a sheet die.

Die Characteristic Of an extrusion die, the relationship between rate of flow of melt through the die, the pressure drop through the die, and the viscosity of the melt. It is defined by the flow equation

$$Q = K \cdot \Delta P/\eta$$

in which Q = the volumetric flow rate, P = the pressure drop, η = the effective viscosity at the temperature and shear rate of the melt in the die, and K = the die characteristic. K is dependent on die dimensions and has the dimensions of volume. The die characteristic is linked to the ▶ Screw Characteristic by ΔP, which must be the same for both, whereas η usually will not be the same for both.

Dieckmann Reaction *n* Cyclization by way of a base catalyzed intramolecular casein type reaction of esters with active delta or epsilon ethylenes.

Dieckman reaction

Die Core *n* The tapered element in an extrusion die for pipe or tubing that guides the material to the webs of the spider. *Sometimes called the Torpedo or Spreader.*

Die Cutters *n* Cuts shapes from sheet stock by sharply striking it with a shaped knife-edge, known as steel rule die.

Die Cutting *n* (blanking, clicking, drinking) The process of cutting shapes from sheets of plastic by pressing a shaped knife edge through one or several layers of sheeting. The dies are often called *steel-rule dies*, and pressure is applied smartly by hydraulic or mechanical means.

Die-Entry Angle *n* At the entrance to the land of an extrusion die, the angle included between the adjacent, inside die surfaces. In many rheometer orifices and in some profile dies, the entry angle is 180°, the extreme of abruptness in actual use. Streamlined-entry dies stretch out the reduction and change of cross-sectional shape over a distance, so the included angle at the die lip may be 20–60°. Extreme gradualness is designed into high-speed wire-coating dies, where the approach to the lip is made in several states of decreasing taper, the final included angle being only 6–10°.

Die Gap *n* (die-lip opening) In film and sheeting dies, the perpendicular distance between the lips of the die land at any point along the width of the die, usually measured after the die has been brought up to temperature but before extrusion has begun. During extrusion, internal pressure inside the die causes the gap to enlarge slightly, more at the center than at the ends of a center-fed die, necessitating adjustments to attain uniform thickness across the film or sheet.

Die Land *n* In an extrusion die, the land is the portion of the die wherein the dimensions of the opening are constant from a point within the die to the discharge point, giving the extrudate its final shape.

Dielectric \ˌdī-ə-ˈlek-trik\ *n* [*dia-* + *electric*] (1837) A material with electrical conductivity less than 10^{-6} siemens/cm (1 μS/cm), thus so weakly conductive that different parts of a sheet can hold different electrical charges. In radio-frequency heating, the term dielectric is used for the material being heated. The term is also used for the nonconductive material separating the conductive elements of a capacitor. Polymeric materials are widely employed as dielectrics. The two most important properties of a dielectric are its ▶ Dielectric Constant and ▶ Dielectric Strength, defined below. Values for some polymeric and other dielectrics at 60 Hz and room temperature are listed here.

Material	Dielectric constant	Dielectric strength, kV/mm
air	1.00054	0.8
alumina*	3.5	1.6–6.3
Pyrex glass	4.5	13
bond paper	3.5	14
polyethylene	2.3	18–39
polypropylene	2.1–2.7	18–26
nylon 6/6	3.6–4.0	12–16
glass-reinforced nylon 6/6	4.0–4.4	19
crystal polystyrene	2.5–2.65	20–28
acrylic resin	3.3–3.9	16
phenoxy resin	4.1	16–20
thermoplastic polyurethane	6–8	33–43
GP phenolic	5–10	12–17
epoxy/glass	5.5	14

*At 1 MHz

(Harper CA (ed) (2002) Handbook of plastics, elastomers and composites, 4th edn. McGraw-Hill, New York; Ku CC, Liepins R (1993) Electrical properties of polymers. Hanser, New York; Carley JF (ed) (1993) Whittington's dictionary of plastics. Technomic, Lancaster)

Dielectric Absorption *n* An accumulation of electrical charges within the body of an imperfect dielectric material when it is placed in an electric field. (Ku CC,

Liepins R (1987) Electrical properties of polymers. Hanser, New York)

Dielectric Breakdown Voltage *n* (breakdown voltage, disruptive voltage) The voltage at which electrical breakdown of a specimen of electrical insulating material between two electrodes occurs under prescribed conditions.

Dielectric Constant *n* (1875) (permittivity constant) Between any two electrically charged bodies there is a force (attraction or repulsion) that varies according to the strength of the charges, q_1 and q_2, the distance between the bodies, r, and a characteristic of the medium separating the bodies (the dielectric) known as the dielectric constant, ε (Ku CC, Liepins R (1987) Electrical properties of polymers. Hanser, New York). The force is given by the

$$F = q_1 \cdot q_2 / (\varepsilon \cdot r^2)$$

For a vacuum, $\varepsilon = 1.0000$, values for some other materials are listed above. In practice the dielectric constant of a material is found by measuring the capacitance of a parallel-plate condenser using the material as the dielectric, then measuring the capacitance of the same condenser with a vacuum as the dielectric, and expressing the result as a ratio between the two capacitances (*Handbook of Chemistry and Physics, 52nd Ed.*, ed. R. C. Weast, The Chemical Rubber Co., Boca Raton, FL). When the dielectric is a polymeric material whose atoms or molecules may change their positions in an alternating electric field (a *polar* material), frictional energy is dissipated as heat and is characterized by the DISSIPATION FACTOR. (Carley JF (ed) (1993) Whittington's dictionary of plastics. Technomic, Lancaster)

Dielectric Heat Sealing *n* A sealing process sed for fusing vinyl films and other thermoplastics films with sufficient dielectric loss whereby two layers of film are heated by ▶ Dielectric Heating and pressed together between two electrodes, an applicator and a platen, the films serving as the dielectric of the so-formed condenser (Harper CA (ed) (2002) Handbook of plastics, elastomers and composites, 4th edn. McGraw-Hill, New York). The applicator may be a pinpoint electrode as in "electronic sewing machines," a wheel, a moving belt, or a contoured blade. Frequencies employed range up to 200 MHz, but are usually 30 MHz or less to avoid interference problems. See also ▶ Heat Sealing.

Dielectric Heating *n* (1944) (electronic heating, RF heating, radio-frequency heating, high-frequency heating, microwave heating) The process of heating poor (but polar) conductors of electricity (dielectrics) by means of high-frequency fields. At frequencies above 10 MHz sufficient heat for rapid sealing and welding of many plastics can be generated by low, safe voltages. The process of dielectric heating consists of placing the material to be heated between two shaped electrodes that are connected to a high-frequency current supply. These electrodes act as the plates of a capacitor and the material serves as the dielectric separating them. As the field changes polarity, charge-bearing atoms or groups of the dielectric undergo reorientation in an effort to keep their positive poles toward the electrode that is momentarily negative, thus generating molecular friction that is dissipated as heat. The theoretical rate of heating is given by the equation

$$P = 2\pi \cdot f \cdot C \cdot V^2 \tan \delta$$

where P = the power input (W), f = the electrical field frequency (Hz), C = the capacitance (F), V = the voltage difference between the electrodes (V), and ·tan δ = the loss factor (Ku CC, Liepins R (1987) Electrical properties of polymers. Hanser, New York). The actual rate of heating will be somewhat less because of nonuniformity of the field, air gaps between electrodes and material, and losses to the surroundings. Dielectric heating is most effective for materials such as PVC and phenolics that have high loss factors because of their numerous polar groups. Nonpolar plastics with low loss factors, such as polystyrene and polyethylene, are impractical to heat dielectrically (Harper CA (ed) Handbook of plastics, elastomers and composites, 4th edn. McGraw-Hill, New York). See also ▶ Dielectric Heat Sealing, ▶ Microwave Heating.

Dielectric Loss *n* Energy dissipated as heat within a polar material subjected to a rapidly alternating, strong electric field (Ku CC, Liepins R (1987) Electrical properties of polymers. Hanser, New York). See ▶ Dielectric Heating.

Dielectric Phase Angle *n* The angular difference in phase between the alternating voltage (usually sinusoidal) applied to a dielectric and the resulting current. The angle is often symbolized by θ, the cosine of which is the ▶ Power Factor. (Ku CC, Liepins R (1987) Electrical properties of polymers. Hanser, New York)

Dielectric Properties of Polymers *n* This term usually means the lack of electrical conductivity or insulating value. (Ku CC, Liepins R (1987) Electrical properties of polymers. Hanser, New York)

Dielectrics or Insulators or Non-conductors *n* A class of bodies supporting an electric strain. A charge on one part of a non-conductor is not communicated to any other part.

Dielectric Strength *n* (electric strength) A measure of the voltage required to puncture an insulating material, expressed in volts per mil of thickness (SI: V/mm). The voltage is the root-mean-square voltage difference between the two electrodes in contact with opposite surfaces of the specimen at which electrical breakdown occurs under prescribed test conditions. (Ku CC, Liepins R (1987) Electrical properties of polymers. Hanser, New York; Weast RC (ed) (1971) Handbook of chemistry and physics, 52nd edn.The Chemical Rubber Co., Boca Raton)

Die Lines *n* In extrusion and blow molding, longitudinal marks caused y damaged die surfaces or by extrudate buildup on dies.

Die-Lip Buildup *n* In extruding film and sheet, the gradual accumulation of a bead of resin on the face of the die, parallel to the slit. Similar buildup can occur on dies for shapes.

Diels-Alder Polymers *n* Polymers created as a result of a Diels-Alder reaction, by which a 1,3-diene reacts via addition to another unsaturated molecule (the dienophile) to form a cyclic adduct. (Morrison RT, Boyd RN (1992) Organic chemistry, 6th edn. Prentice Hall, Englewood Cliffs; Smith MB, March J (2001) Advanced organic chemistry, 5th edn. Wiley, New York)

Die Manifold See ▶ Manifold.

Diene Monomers \ˈdī-ˌēn ˈmä-nə-mər\ *n* Any of a family of monomers based on unsaturated hydrocarbons having two double bonds.

Diene Polymer *n* Any of a family of polymers based on unsaturated hydrocarbons having two double bonds (*diolefins*). When the double bonds are separated by only one single bond, as in 1,3-butadiene ($CH_2=CH-CH=CH_2$), the diene and its double bonds are said to be *conjugated*. In an *un*conjugated diene the double bonds are separated by at least two single bonds and act more independently. The family includes polymers of butadiene, isoprene, cyclopentadiene, and copolymers with ethylene and propylene. (Morrison RT, Boyd RN (1992) Organic chemistry, 6th edn. Prentice Hall, Englewood Cliffs; Smith MB, March J (2001) Advanced organic chemistry, 5th edn. Wiley, New York)

Diene Polymers *n* The family of polymers and copolymers based on unsaturated hydrocarbons or diolefins with two double bonds which includes ethylene, propylene, isoprene, butadiene, and cyclopentadiene.

Diene Value or Number *n* Amount of maleic anhydride (expressed as equivalents of iodine) which will react with 100 parts of oil under specific conditions. It is a measure of the conjugated double bonds in the oil. *Also known as the Maleic Anhydride Value or Maleic Value.*

Die Plate *n* (1) In injection molds, a member that is attached to the fixed or to the moving head of the press; *mold plate*, (2) In extrusion, especially of pellets or shapes, the die plate is that part of the die assembly that is bolted to the outlet of the die body and contains the orifices that form the melt into continuous strands or a particular cross-sectional shape.

Die Pressure *n* In extrusion, the pressure of the melt entering the die.

Dies, Coextrusion *n* Dies of nested configuration. Individual polymers flow through separate mandrel passages and are combined in the primary land area. Adjacent polymer layers must exhibit similar flows and adhesion characteristics to form the coextrusion.

Dieseling See ▶ Burn Mark.

Dies, Engraving See ▶ Dies, Hot Stamping.

Dies, Extrusion *n* Extrusion converts thermoplastics into a continuous stream of melt which is then shaped by an extrusion die into uniform cross-sectional shape. The main classes of dies are for film, sheet, coating, pipe, and profiles.

Dies, Hot Stamping *n* Method of depositing a decorative layer on the surface of a polymer, whereby a hot metal or rubber diem which bears the required design or lettering is placed in contact with the medium to be applied and the surface to be decorated. The application of heat melts the release coating and activates the adhesive. After a short contact, the die pressure is released and the film carrier separated from the plastic surface, leaving a decorative coating transferred in the area of the die.

Die Spider *n* In extrusion, the legs or webs supporting the die core within the head of an in-line pipe, tubing, or blown-film die, In many pipe dies, the spider legs are cored to permit application of air or water for cooling the mandrel.

Dies, Pultrusion *n* Polymer thermosetting process whereby fibers impregnated with a mixed solution of polymer and the necessary additives are pulled through a heated die. The fibers entering the die are generally saturated with the solution but are solid when exiting from the machine. The process can produce solid, open-sided, hollow shapes that can be cut to length and packaged for shipment.

Die Stamping See ▶ Intaglio.

Die Swell See ▶ Extrudate Swelling.

Die-Swell Ratio *n* (extrudate-swelling ratio) In extrusion, particularly in extruding parisons for blow molding, the ratio of the outer parison diameter or parison wall thickness to, respectively, the outer diameter of the parison die or the die gap. The ratio is affected by the polymer type, its temperature, the die geometry, and the extrusion rate. Some writers have defined the die-swell ratio as the ratio of the *cross section* of the extrudate shortly after emergence to that of the die opening.

2,2-Diethoxyacetophenone *n* (DEAP) A photoinitiator used for curing acrylate coatings, either in an inert atmosphere or in air with UV light. It absorbs impinging light.

Diethoxyethyl Adipate *n* $(CH_2)_4(COOC_2H_4OC_2H_5)_2$. A plasticizer for cellulose resins.

Diethoxyethyl Phthalate *n* $C_6H_4(COOC_2H_4OC_2H_5)_2$. A plasticizer for cellulosic and vinyl resins.

Diethyl Adipate *n* $(CH_2)_4(COOC_2H_5)_2$. A plasticizer for cellulosic resins.

Diethylaluminum Chloride *n* A colorless liquid that bursts into flame instantly upon contact with air (pyrophoric) and reacts violently with water. It is used as a catalyst in olefin polymerization.

3-Diethylaminopropylamine *n* An epoxy curing agent.

Di(2-ethylbutyl) Azelate *n* A plasticizer for PVC, its copolymers, and cellulose esters. It is very compatible and efficient in vinyls, finding its largest use in high-clarity film and sheeting.

Diethyl Carbonate *n* \ˈkär-bə-ˌnāt\) (ethyl carbonate) A solvent for cellulosic and many other resins.

Diethylene Glycol Bis-(Allyl Carbonate) *n* The monomer for some polycarbonate polymers.

Diethylene Glycol Dibenzoate *n* plasticizer for cellulosic resins, polymethyl methacrylate, polystyrene,

PVC, polyvinyl acetate, and other vinyls. It imparts good stain resistance.

Diethylene Glycol Dipelargonate *n* [CH₃(CH₂)₇COOC₂H₄]₂O. A simple diester of pelargonic acid used mainly as a secondary plasticizer for vinyl resins, but also as a plasticizer for cellulosics. Within the limits of its compatibility it provides economical low-temperature flexibility.

Diethylene Glycol Dipropionate *n* A plasticizer for cellulosic plastics.

Diethylene Glycol Distearate *n* (C₁₇H₃₅COOC₂H₄)₂O. A plasticizer for cellulose nitrate and ethyl cellulose.

Diethylene Glycol Monoacetate *n* HO(CH₂)₂O(CH₂)₂OOCCH₃. A solvent for cellulose nitrate and cellulose acetate.

Diethylene Glycol Monobutyl Ether *n* C₄H₉OCH₂CH₂OCH₂CH₂OH. A solvent with a high boiling point, used in coatings when very slow drying rates are desired. It is also useful as a dispersant in vinyl organosols, and as an intermediate fore the production of plasticizers. See ▶ Butyl Carbitol.

Diethylene Glycol Monobutyl Ether Acetate See ▶ Butyl Carbitol Acetate.

Diethylene Glycol Monolaurate See ▶ Diglycol Laurate.

Diethylene Glycol Monoricinoleate See ▶ Diglycol Ricinoleate.

Diethylenetriamine *n* (DETA, 2,2′-diaminodiethylamine) A pungent liquid providing fast cures with epoxy resins, even at room temperature.

Diethyl Ether \(▎)dī- ▎e-thəl-\ *n* (ca. 1930) (1) A light volatile flammable liquid used as a sovlent and anesthetic. (2) Any of various organic compounds characterized by an oxygen atom attached to two carbon atoms.

Di(2-ethylhexyl) Adipate *n* (dioctyl adipate, DOA) A primary plasticizer for vinyls, cellulose nitrate, polystyrene, and ethyl cellulose. In vinyls, DOA is often used in combination with phthalate and other plasticizers, imparting good resilience, low-temperature flexibility, and resistance to extraction by water. It is FDA-approved for use in vinyl food-packaging films.

Di(2-ethylhexyl) Azelate *n* (dioctyl azelate, DOZ) $(CH_2)_7[COOCH_2-CH(C_2H_5)C_4H_9]_2$. A plasticizer for vinyl chloride polymers and copolymers. It is one of the most compatible of the low-temperature, monomeric plasticizers, has low volatility, and imparts low extractability by water and soapy water. This plasticizer has been approved for food-contact use.

Di(2-ethylhexyl) 2-Ethylhexyl Phosphonate *n* C_8H_{17} $PO(OC_8H_{17})_2$. A plasticizer and stabilizer.

Di(2-ethylhexyl) Hexahydrophthalate *n* (dioctyl hexahydrophthalate) A light-colored liquid, a plasticizer for vinyls.

Di(2-ethylhexyl) Isophthalate *n* (dioctyl isophthalate, DIOP) A primary plasticizer for PVC, most notable for low volatility and its resistance to marring by nitrocellulose lacquers, in addition to good general-purpose properties. It is also compatible with polyvinyl butyral, vinyl chloride-acetate copolymers, cellulosic resins, polystyrene, and chlorinated rubber.

Di(2-ethylhexyl) Phthalate *n* (dioctyl phthalate, DOP) $C_6H_4[COO-CH_2CH(C_2H_5)C_4H_9]_2$. The most widely used plasticizer for PVC, also compatible with ethyl cellulose, cellulose nitrate, and polystyrene. It is generally recognized as imparting the best all-around good properties vinyls, and is often used as the standard against which other plasticizers are evaluated. DOP has been approved by the FDA for use in packaging films for greaseless foodstuffs of high water content.

Di(2-ethylhexyl) Sebacate *n* (dioctyl sebacate) A plasticizer for vinyl chloride polymers and copolymers, cellulosic plastics, polystyrene, and polyethylene. It imparts good low-temperature properties.

Di(2-ethylhexyl) Succinate *n* (dioctyl succinate) A plasticizer.

Di(2-ethylhexyl)-4-Thioazelate *n* C$_8$H$_{17}$OOC(CH$_2$)$_2$CS(CH$_2$)$_4$COOC$_8$–H$_{17}$. A plasticizer for cellulose nitrate, ethyl cellulose, polymethyl methacrylate, polystyrene, and vinyl resins.

Diethyl Oxalate \ˈäk-sə-ˌlāt\ See ▶ Ethyl Oxalate.

Diethyl Phthalate *n* (ethyl phthalate) A plasticizer and solvent for nearly all thermoplastics and coumarone resins. It has been approved by FDA for use in food packaging.

Diethyl Sebacate *n* A plasticizer with good low-temperature properties, compatible with PVC and many other thermoplastics.

Diethyl Succinate *n* plasticizer for cellulosic resins.

Diethyl Tartrate *n* A solvent and plasticize for cellulosic resins.

1,1-Diethylurea *n* A white solid polymerizable with simple urea and formaldehyde to form permanently thermoplastic resins.

Diethyl Zinc *n* (1952) C$_4$H$_{10}$Zn. A volatile pyrophoric liquid organometallic compound used to catalyze polymerization reactions and to deacidify paper.

Differential Refractometer *n* An instrument used in connection with a chromatographic column in Size-exclusion Chromatography of polymers (in dilute solution). Two streams, one pure solvent, the other the eluting polymer solution, pass through the instrument, whose signal, proportional to the difference ion refractive indices of solution and solvent, is interpreted as polymer concentration. The instrument is also used to determine the rate of change of refractive index with concentration (dn/dc), an important parameter in the computation of weight-average molecular weights from light-scattering measurements in dilute polymer solutions.

Differential Scanning Calorimeter *n* (DSC) An instrument that measures the rate of heat evolution of absorption of a specimen while it is undergoing a programmed temperature rise (Gooch, 1997). A recorder displays the data as a trace of increase in heat per increase in temperature (dq/dT) versus temperature. An example of a DSC thermogram of polypropylene is shown. The DSC has been used to study curing reactions and related properties of thermosetting resins, and heats of decomposition of resins. (An example of a DSC instrument is the PerkinElmer Diamond DSC.

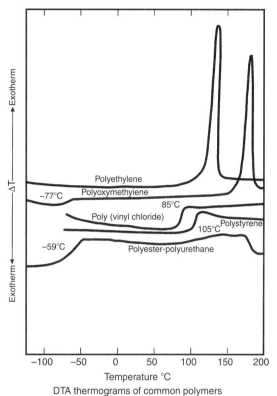

DTA thermograms of common polymers

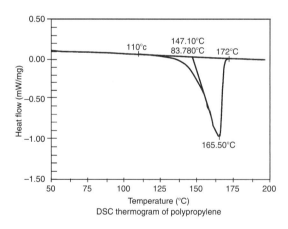

DSC thermogram of polypropylene

Differential Thermal Analysis *n* (DTA) An analytical method in which the specimen material and an inert substance are heated concurrently in separate minipans and the difference in temperature between the two is recorded, along with the temperature of the inert substance (Gooch, 1997). DTA has been useful; in the study of phase transitions and curing and degradation reactions in polymers. An example of a DTA thermogram is shown. An example of a DTA instrument is the PerkinElmer Diamond TG/DTA. (Courtesy of PerkinElmer, Inc.)

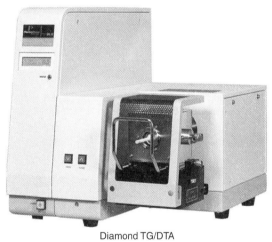

Diamond TG/DTA

Differentiating Solvent *n* A solvent which will discriminate among the strengths of acids or bases.

Diffraction \di-**ˈ**frak-shən\ *n* [NL *diffraction-*, *diffractio*, fr. L *diffringere* to break apart, fr. *dis-* + *frangere* to break] (1671) Dispersion of light into a spectrum when it strikes an aperture or a slit of width in the same order of magnitude as the wavelength of the light. Diffraction depends on the wave nature of light and occurs in all wave motions. If monochromatic light strikes the edge of an opaque obstacle, bands of light and dark are observed near the edges of the beam. Diffraction is thus a special case of interference. See ▶ Dispersion, ▶ Light and ▶ Diffraction Grating.

Diffraction Grating n (1867) Device used to disperse a beam of electromagnetic radiation into its constituent wavelengths, i.e., for producing a spectrum. Grating may be prepared by ruling equidistant parallel lines onto a glass surface (transmission grating) or a metal surface (reflection grating). If s is the distance between the rulings, d the angle of diffraction, then the wavelength where the angle of incidence is 90° is (for the nth order spectrum),

$$\lambda = \frac{s \sin d}{n}$$

If i is the angle of incidence, d the angle of diffraction, s the distance between the rulings, n the order of the spectrum, the wavelength is,

$$\lambda = \frac{8}{n}(\sin i + \sin d)$$

Diffused Light \di-**ˈ**fyüz-\ n [ME *diffused*, pp., fr. L *diffusus*, pp] Nondirectional light.

Diffuse Reflectance n The ratio of the flux reradiated by diffuse reflection. See ▶ Reflectance, Diffuse and ▶ Specular Reflectance Excluded.

Diffuse Transmittance See ▶ Transmittance.

Diffusion n The movement of a material in the body of a plastic. If the concentration (mass of solid per unit volume of solution) at on surface of a layer of liquid is d_1 and at the other surface d_2, the thickness of the layer h and the area under consideration A, then the mass of the substance which diffuses through the cross-section A in time t is,

$$m = \Delta A \frac{(d_2 - d_1)t}{h}$$

where Δ is the coefficient of diffusion.

Diffusion, Autoacceleration n In some vinyl polymerizations, as the reaction approaches completion and the viscosity of the reaction medium rises, there is a rising rate of increase of molecular weight of the chains that have not yet been terminated. This autoacceleration phenomena greatly affects the diffusivity of the polymer solution.

Diffusion Bonding n An experimental method of joining materials in which extremely flat and finely finished surfaces are clamped together under high vacuum for a period of hours or days, sometimes at a high temperature, allowing the atoms of each member to diffuse into the other. As of early 1992, the method had been used only with inorganic crystalline materials.

Diffusion Constant n Symbol D. The constant of proportionality in Fick's laws of diffusion between the rate of diffusion and the concentration gradient (Fick's first law) and the variation of concentration with time and concentration gradient (Fick's second law).

Diffusion Couple n An assembly of two materials in such intimate contact that each diffuses into the other.

Diffusivity n (1) (diffusion coefficient) The constant of proportionality in Fick's first law of diffusion, which states that the mass of molar rate of transport (flux) of one molecular species into another is equal to the diffusivity times the gradient of concentration. Several related units are in use, e.g., cm^2/s, ft^2/h. The SI unit is m^2/s, corresponding to flux in $mol/(s \cdot m^2)$ and gradient in $(mol/m^3)/m$. See also ▶ Fick's Law. (2) ▶ Thermal Diffusivity. Diffusivity or coefficient of diffusion is also given by Δ in the equation

$$\frac{dQ}{dt} = -\Delta \left(\frac{dc}{dx}\right) dydz$$

where dQ is the amount passing through an area $dy\,dz$ in the direction of x in a time dt where dc/dz is the rate of increase of volume concentration in the direction of x. Dimensions, $[l^2 t^{-1}]$.

Diffusivity of Heat n It is given by Δ in the equation

$$\frac{dH}{dt} = -\Delta sd \frac{dT}{dx} dydz$$

where dH is the quantity of heat passing through the area $dy\,dz$ in the direction of x in a time dt. The rate of variation of temperature along x is given by dT/dx, s is specific heat and d, density. Dimensions, $[l^2 t^{-1}]$.

Digester \-**ˈ**jes-tər\ n (1614) Jacketed reaction vessel used to decompose, soften, or cook substances at high pressure and temperature.

Diglycidyl Ether of Bisphenol A *n* (DGEBA) The main constituent of most commercial epoxy resins prior to curing, DGEBA is formed by the reaction of excess epichlorohydrin with bisphenol A in the presence of aqueous sodium hydroxide.

Diglycol Laurate \-ˈglī-ˌkól ˈlór-ət\ (diethylene glycol monolaurate) A plasticizer for ethyl cellulose, cellulose nitrate, polyvinyl butyral, and vinyl chloride-acetate copolymers.

Diglycol Ricinoleate *n* (diethylene glycol Monoricinoleate) $CH_3(CH_2)_5-CH(OH)CH_2CH=CH(CH_2)_7COOC_2H_4OC_2H_4OH$. A plasticizer for ethyl cellulose and cellulose nitrate.

Dihalide *n* A compound containing two halogen atoms (of valence = ~1) per molecule.

Dihexyl Phthalate (DHXP) See Di(2-ethyl butyl) phthalate.

Dihexyl Sebacate *n* A plasticizer for vinyls.

Dihydrodiethyl Phthalate *n* A plasticizer for PVC and cellulose nitrate, with partial compatibility with other thermoplastics.

Dihydroxy Anthraquinone Lake \-ˌan(t)-thrə-kwi-ˈnōn-\ *n* See ▶ Alizarin Red.

Diisobutyl Azelate *n* $(CH_3)_2CHCH_2OOC(CH_2)_7$. A plasticizer for cellulosics, polymethyl methacrylate, PVC, and polyvinyl acetate.

Diisobutyl Adipate *n* (DIBA) A plasticizer for most synthetic resins, including cellulosics, PVC, and other vinyls, and FDA-approved for use in food-packaging films. In vinyls, it is a very active solvent and lowers processing temperatures to levels that permit elimination or lower percentages of stabilizers.

Diisobutyl Aluminum Chloride *n* [$(CH_3)_2CHCH_2$] AlCl. A catalyst for polymerizing olefins.

Diisobutyl Ketone \-ˈkē-ˌtōn\ *n* (2,6-dimethyl-4-heptanone) A high-boiling ketone with moderate solvent power for cellulose nitrate and vinyl copolymers. Having limited solvency for PVC, it is used as a viscosity modifier in organosols.

Diisobutyl Phthalate *n* (DIBP) $C_6H_4[COOCH_2CH(CH_3)_2]$. A plasticizer for vinyls, cellulosics, and polystyrene.

Diisocyanate *n* Any compound containing two isocyanate (—NCO) groups, used in the production of polyurethanes. Many methods have been reported for synthesizing diisocyanates, but the one most widely used is reacting phosgene with an amine in a solvent. Toluene diisocyanate (TDI), the most commonly used, is an 80–20 mixture of 2,4- and 2,6-toluene diisocyanate isomers. Also used are diphenylmethane-4,4′-diisocyanate (MDI), a modified toluene diisocyanate, and polymethylene polyphenyl isocyanate (PAPI). The diisocyanates are key ingredients in the production of urethane foams, fibers, coatings, and solid elastomers. See also ▶ Isocyanate and ▶ Polyurethane.

Diisodecyl Adipate *n* (DIDA) ($-C_2H_4COOC_{10}H_{21})_2$. A plasticizer for PVC in lower concentrations, e.g., up to 30 phr, at which it imparts low-temperature flexibility and resistance to lacquer marring. It is often used in combination with phthalate and phosphate plasticizers. DIDA is completely compatible with vinyl chloride-acetate copolymers, cellulose acetate-butyrate with high butyral content, cellulose nitrate, ethyl cellulose, and chlorinated rubber. In polystyrene, it may be used up to 25 phr.

Diisodecyl-4,5-Epoxy-Tetrahydrophthalate *n* A plasticizer for PVC that also acts as a stabilizer and fungistat. It is compatible with cellulose nitrate, ethyl cellulose, polymethyl methacrylate, polystyrene, and other vinyls polymers and copolymers.

Diisodecyl Glutarate *n* $(CH_2)_3(COOC_{10}H_{21})_2$. A plasticizer for PVC having low-temperature properties equal to those of dioctyl adipate but with lower volatility and greater resistance to soapy water.

Diisodecyl Phthalate *n* (DIDP) $C_6H_4(COOC_{10}H_{21})_2$. A general-purpose plasticizer for vinyl resins, imparting good water resistance and suitable for processing at high temperatures due to its low volatility. It is also compatible with most synthetic resins, e.g., cellulose nitrate, ethyl cellulose, and polystyrene.

Diisononyl Phthalate *n* (DINP) $C_6H_4(COOC_9H_{19})_2$. A plasticizer for PVC, cellulosics, and polystyrene. It has lower volatility than dioctyl phthalate with equivalent low-temperature performance and poorer efficiency.

Diisooctyl Adipate *n* (DIOA) $(-C_2H_4COOC_8H_{17})_2$. A primary plasticizer for vinyls, cellulose nitrate, polystyrene, polymethyl methacrylate, and ethyl cellulose. Its performance is similar to that of dioctyl phthalate.

Diisooctyl Azelate *n* (DIOZ) $C_7H_{14}(COOC_8H_{17})_2$. A plasticizer for cellulosics resins and polymers and copolymers of vinyl chloride. In vinyls it imparts good low-temperature properties and other characteristics similar to those obtained with dioctyl azelate.

Diisooctyl Fumarate *n* $(=CHCOOC_8H_{17})_2$. A plasticizer for PVC.

Diisooctyl Isophthalate *n* $C_6H_4(COOC_8H_{17})_2$. A plasticizer for PVC, also compatible with cellulose nitrate, ethyl cellulose, polystyrene, and other vinyl resins.

Diisooctyl Monoisodecyl Trimellitate *n* A plasticizer for cellulosics and vinyl resins.

Diisooctyl Phthalate *n* (DIOP) $C_6H_4(COOC_8H_{17})_2$. A primary plasticizer for PVC, ethyl cellulose, cellulose nitrate, and polystyrene. In vinyls, its performance is similar to that of dioctyl phthalate except that it is slightly less volatile than DOP and produces better viscosity characteristics in plastisols. DIOP is FDA-approved for food-packaging materials and medical applications involving contact with water, but not with fats.

Diisooctyl Sebacate *n* (DIOS) A plasticizer for vinyl and other resins.

Diisopropylene Glycol Salicylate \sə-ˈli-sə-ˌlāt\ *n* $HOCH_2CH(CH_3)CH$ $(CH_3)CH_2O-OCC_6H_4OH$. An ultraviolet absorber in plastics.

Dilatancy \dī-ˈlā-tᵊn(t)-sē\ *n* (1885) Flow characterized by reversible, instantaneous increase in viscosity with increasing shear rate. The opposite of *pseudoplasticity*. Dilatancy in a pigment-vehicle system or pigment dispersions commonly results from high pigment loadings. The curve of the plot of shear stress versus shear rate is nonlinear with shear stress increasing faster than the shear rate. *Also known as Shear Thickening.*

Dilatometer \ˌdi-lə-ˈtä-mə-tər\ *n* [ISV] (ca. 1883) An instrument for measuring the volume changes of a liquid or solid as the sample's temperature changes. A simple liquid dilatometer consists of a small bulb holding a known mass of sample and sealed to a graduated capillary of known diameter. As the sample is heated, its expansion is indicated by its rise up the capillary. From these measurements the ▶ Volume Coefficient of Thermal Expansion can be estimated. (The inverse principle is used in liquid-in-glass thermometers.)

Dilauryl Ether *n* (didodecyl ether A lubricant for plastics processing.

Dilinoleic Acid *n* $C_{34}H_{62}(COOH)_2$. An unsaturated, dibasic acid used as a modifier in alkyd, nylon, and polyester resins.

Diluent \ˈdil-yə-wənt\ *n* [L *diluent-, diluens*, pp of *diluere*] (ca. 1721) A substance that dilutes another substance. In an organosol, a diluent is a volatile liquid such as naphtha that has little or no solvating effect on the resin, but serves to lower the viscosity of the mix, and is evaporated during processing. In the paint industry, a diluent is any substance capable of thinning paints, varnishes, etc. The term is sometimes also used for a liquid added to a thermosetting resin to reduce its viscosity; and for an inert powdered substance added to an elastomers or resin merely to increase the volume.

(Compare: ▶ Filler.) See also ▶ Extruder and ▶ Thinner.

Dilute-Solution Viscosity *n* (solution viscosity) (1) A catchall term that can mean any of the interrelated and quantitatively defined viscosity ratios of dilute polymer solutions or their absolute viscosities. (2) The kinematic viscosity of a solution as measured by timing the rate of efflux of a known volume of solution, by gravity flow, through a calibrated glass capillary that is immersed in a temperature-controlled bath. Two common types of viscometer are the Ostwald-Fenske and Ubbelohde. From the viscosities of the solution η and the solvent η_o, and the solution concentration c, five frequently mentioned "viscosities" (viscosity ratios, actually) can be derived, as follows:

Relative viscosity = $\eta_r = \eta/\eta_o$;
Specific viscosity = $\eta_{sp} = \eta_r - 1$;
Reduced viscosity = $\eta_{red} = \eta_{sp}/c$;
Inherent viscosity = $\eta_{inh} = (\ln \eta_r)/c$;
Intrinsic viscosity = $[\eta]$ = the limit as $c \to 0$ of η_{sp}/c, = the limit as $c \to 0$ of η_{inh}.

The intrinsic viscosity, because it is extrapolated to zero concentration from a series of measurements made at different concentrations, is independent of concentration. However, different solvents yield different intrinsic viscosities with the same polymer, so the solvent used must be identified. Some ASTM tests for viscosities of dilute polymer and plastics solutions are D 789, D 1243, D 1601, D 2857, and D 4603, all in Section 08. D 445 and D 446, describing the proper use of the viscometers mentioned above, are in Section 05.

Dilution Ratio *n* (1) As used in the surface-coatings industry, dilution ratio is the volume ratio of diluent to solvent in a blend of theses two constituents that just fails to completely dissolve 8.00 g of nitrocellulose in 100 mL of the blend. The procedure is described in ASTM D 1720, Section 06. It is used to determine the most economical, yet adequate amount of high-cost active solvent required in a nitrocellulose-lacquer system. (2) In most other contexts, dilution ration is the quotient of the concentration of the undiluted solute divided by that of the diluted solution, both concentrations in the same units.

Dimension \də-ˈmen(t)-shən *also* dī-\ *n* [ME, fr. MF, fr. L *dimension-*, dimensio, fr. *dimetiri* to measure out, fr. *dis-* + *metiri* to measure] (14c) A geometric element in a design, such as length or angle, or the magnitude of such a quantity.

Dimensional Formulae *n*\-ˈfór-myə-lə\ If mass, length, and time are considered fundamental quantities, the relation of other physical quantities and their units to these three may be expressed by a formula involving the symbols *l*, *m* and t respectively, with appropriate exponents. For example; the dimensional formula for volume would be expressed, $[l^3]$; velocity $[lt^{-1}]$; force $[m\ lt^{-2}]$. Other fundamental quantities used in dimensional formulae may be indicated as follows: θ, temperature, ε the dielectric constant of a vacuum; μ, the magnetic permeability of a vacuum.

Dimensional Restorability *n* The ability of a fabric to be returned to its original dimensions after laundering or dry cleaning, expressed in percent. For example, 2% dimensional restorability means that although a fabric may shrink more than this in washing, it can be restored to within 2% of its original dimensions by ordinary home pressing methods.

Dimensional Stability *n* The ability of a substance (wood, plastic, etc.) to retain its shape when subjected to varying degrees of temperature, moisture, pressure, or other stress.

Dimer \ˈdī-mər\ *n* [ISV] (ca. 1926) (1) A molecule formed by union of two identical simpler molecules. For example, C_4H_8 is a dimer of C_2H_4, as N_2O_4 is of NO_2. (2) A substance composed of dimmers.

Dimer Acid *n* A coined, generic term for high-molecular-weight, dibasic acids that combine and polymerize with alcohols and polyols to make plasticizers, etc. A trimer acid is analogous, having three acid groups.

Dimer Acids *n* Liquid polycarboxylic acids produced by polymerization of unsaturated fatty acids. Pure dimmer acid is a C_{36} dicarboxylic acid. Commercial dimmer acids are composed of predominantly C_{36} dicarboxylic acid with minor amounts of C_{18} monocarboxylic acid and varying proportions of C_{54} tricarboxylic acid.

Dimerization *n* State of polymerization when two similar molecules have united. The product is described as a dimmer.

Dimethoxyethyl Adipate *n* A plasticizer for cellulose-ester polymers.

Di(2-methoxyethyl) Phthalate *n* A plasticizer, especially for cellulose acetate, but compatible, too, with other cellulosics, polystyrene, and vinyls.

Dimethyl Acetamide *n* (DMAC) $CH_3CON(CH_3)_2$. A colorless liquid, a solvent for resins and plastics, a catalyst, and an intermediate.

2-Dimethylamino Ethanol *n* (DMAE) A colorless liquid derived from ethylene oxide and dimethylamine, a catalyst for urethane foams. It has little odor and toxicity, and resists staining.

3-Dimethylaminopropylamine *n* A colorless liquid used as a curing agent for epoxy resins.

Dimethylaniline *n* (DMA) The term usually means the tertiary amine, N,N-dimethylaniline, through ring-substituted isomers are known. This amine is useful as an accelerator in polyester molding and sprayup.

Di(methylcyclohexyl) Adipate *n* A plasticizer compatible with most thermoplastics.

Di(methylcyclohexyl) Phthalate *n* A plasticizer for cellulosics, polystyrene, PVC, and other vinyl resins.

Dimethylformamide *n* (DMF) A colorless and very active solvent for PVC, nylon, polyurethane, and many other resins and elastomers, with fairly low volatility. Its strong solvent power makes it useful as a solvent booster in coating, printing, and adhesive work, and in paint strippers. Because it is toxic and readily absorbed through the skin, DMF must be handled, and disposed of, with care.

Dimethyl Glutarate *n* (DMG) A liquid chemical intermediate, a source of dicarboxylic acid, used in making plasticizers, polyester resins, synthetic fibers, films, adhesives, and solvents.

Dimethyl Glycol Phthalate *n* $C_6H_4(COOCH_2CH_2OCH_3)_2$. A solvent and plasticizer for cellulosics resins.

Dimethyl Polysiloxane See ▶ Polydimethylsiloxane.

2,2-Dimethyl-1,3-Propanediol See ▶ Neopentyl Glycol.

Dimethyl Sebacate *n* $[-(CH_2)_4COOCH_3]_2$. A solvent and plasticizer for cellulosics and vinyl resins, also compatible with most other thermoplastics.

Dimethyl Sulfoxide *n* (DMSO) $(CH_3)_2SO$. An active polar solvent useful for dissolving such polar polymers as polyacrylonitrile and for certain polymerization reactions.

Dimethyl Terephthalate *n* (DMT) C_6H_4–1,4-$(COOCH_3)_2$. A white crystalline solid generally obtained by the oxidation of *p*-xylene. The carbomethoxy groups of DMT are typical of those attached to a benzene ring, and their ready participation in alcoholysis reactions is the basis for most of the uses of the material. DMT is used in making polyethylene terephthalate (PET) and polyester fibers therefrom.

Dimethyl Urea \yú-ˈrē-ə\ $(CH_3NH)_2CO$. Primary condensation product of urea and formaldehyde. When condensed in the presence of alcohols, it forms

Dimethylisobutylcarbinyl Phthalate *n* A plasticizer for most common thermoplastics.

Dimethyl Ketone Syn: ▶ Acetone.

Dimethylol Urea *n* A colorless crystalline material resulting from the combination of urea and formaldehyde in the presence of salts or alkaline catalysts, the first or A-stage of urea-formaldehyde resin.

Dimethyl Phthalate *n* A nontoxic plasticizer for most common thermoplastics, but with limited compatibility with PVC. Plasticizer for nitrocellulose and cellulose acetate, resins, lacquers, plastics. Colorless, odorless, lightfast, stable, nontoxic liquid. Sp gr, 1.189; bp, 282°C; flp, 149°C (300°F).

oil-soluble resins. With mildly acid salts, it may be used as an adhesive.

Diminution of Pressure at the Side of a Moving Stream *n* If a fluid of density *d* moves with a velocity *v*, the diminution of pressure due to the motion is (neglecting viscosity),

$$p = \frac{1}{2}dv^2$$

Dimity \ˈdi-mə-tē\ *n* [alter. of ME *demyt*, fr. ML *dimitum*, fr. MGk *dimitos* of double threat, fr. Gk *di-* + *mitos* warp thread] (1570) A sheer, thin, spun cloth that sometimes has cords or stripes woven in. It is used for aprons, pinafores, and many types of dress goods.

Dimorphous \(ˌ)dī-ˈmór-fəs\ *adj* [Gk *dimorphos* having two forms, fr. *di-* + *-morphos*-morphous] (1832) Material which can exist in two distinct crystalline forms, having different melting points.

Dimple \ˈdim-pəl\ *v* [ME *dympull*; akin to OHGr *tumphilo* whirlpoor, OE *dyppan* to dip] (1602) A depression or indentation on a surface. Syn: ▶ Sink Mark.

DIN *n* (1) Abbreviation for Deutsches Industrie Norm (German Industry Standard). (2) Also abbreviation for Deutsches Institut Für Normung (formerly DNA).

DIN Color Difference Equation *n* Color difference equation based on the DIN Color Order System. It is described in English by Richter in *J. Opt. Soc. Am.*, **45**, 223.

DIN Color System *n* Official German Standard Color System (DIN 6164). The color solid is described in terms of hue (DIN – Farbton), abbreviated F or T, saturation (DIN – Sättigung), abbreviated S, and relative lightness (DIN – Dunkelstufe), abbreviated D.

Dingler's Green See ▶ Chromium Oxide Green.

Di-*n*-hexyl Adipate *n* $(CH_2)_4(COOC_6H_{13})_2$. An important low-temperature plasticizer for synthetic rubbers and several plastics, including some cellulosics, PVC, and polyvinyl acetate.

Di-*n*-hexyl Azelate *n* (DNHZ) $(CH_2)_7(COOC_6H_{13})_2$. A plasticizer for cellulose and vinyl resins. It has been approved for food-contact use, has low volatility and good compatibility.

Dinitraniline *n* Base which, when diazotized and coupled with β-napthol, forms Permanent Red 2G.

Dinitraniline Orange *n* Pigment Orange 5 (12075). A monoazo pigment dyestuff prepared by coupling diazotized dinitro aniline to beta naphthol. It bleeds in organic solvents and has fair heat resistance. Its lightfastness in masstone is good. The good chemical resistance, freedom from toxic metals and high color intensity of dinitraniline orange encourage its use over the less expensive chrome and molybdate oranges in some specialized finishes not requiring good bake or bleed resistance.

Dinitrosopentamethylenetetramine *n* (DNPT) A blowing agent widely used for foam rubber, but of limited use in the plastics industry due to its high decomposition temperature and unpleasant residual odor.

Dinitrosoterephthalamide *n* (DNTA) A chemical blowing agent for vinyls, liquid polyamide resins, and silicone rubbers. It is especially noted for its low decomposition exotherm.

Dinking See ▶ Die Cutting.

Di-*n*-octyl, *n*-Decyl Phthalate *n* (DNODP) A mixed plasticizer for PVC and several other thermoplastics.

Di-*n*-octyltin Maleate Polymer *n* Like the preceding tin stabilizer, this one, too, has been FDA-approved for use in food-packaging compositions up to 3 phr.

Di-*n*-octyltin-*S*,*S*′-bis(Isooctyl Mercaptoacetate) *n* A stabilizer for PVC that has been approved for use in food-grade bottles up to 3 phr when made to certain purity specifications.

Dinonyl Adipate *n* (DNA) A non-alcohol ester used as a plasticizer for cellulosic, acrylic, styrene, and vinyl polymers.

Dinonyl Phthalate *n* (DNP) A general-purpose plasticizer for vinyl resins, with low volatility and good electrical properties.

DINP *n* Abbreviation for ▶ Diisononyl Phthalate.
DIOA *n* Abbreviation for ▶ Diisooctyl Adipate.
Dioctyl- See Di(2-ethylhexyl)- for several compounds for which this shorter prefix is more commonly used.
Dioctyl Ether *n* A mold and processing lubricant.

Dioctyl Fumarate *n* (DOF) An unsaturated plasticizer for vinyl resins.

Di(2-octyl) Phthalate *n* (DOP) See ▶ Dicapryl Phthalate from which, by addition polymerization, is prepared the transparent, water-white polymer CR-39, one of the allyl-resin family. Since it is difficult and tricky to handle, the resin has found little commercial use aside from optical applications. For these, its excellent optical properties and resistance to scratching, and its low density – half that of glass – have given it wide use in eyeglasses. Other uses include optical filters, instrument windows, welders' masks, and large windows in atomic-energy plants.

Dioctyl Terephthalate *n* (DOTP) Although the physical properties of DOTP are similar to those of DOP (the ortho isomer), DOTP is less volatile, imparts slightly better low-temperature flexibility, and is more resistant to lacquer marring.

Dioctyltin Stabilizer See ▶ Organotin Stabilizer.

Diofan *n* Dispersion of copolymers of vinylidene chloride. Manufactured by BASF, Germany.

Diol \ˈdī-ˌōl\ *n* [ISV] (1923) An acronym for dihydric alcohol, i.e., an alcohol containing two hydroxyl (–OH) groups. Ethylene glycol, $HOCH_2CH_2OH$, and 1,5-pentanediol. $HO(CH_2)_5OH$, are examples.

Diolefin Polymer \dī-ˈō-lə-fən\ *n* See ▶ Diene Polymer.

Diolefin Resins *n* More generally known as aromatic petroleum residues. By-products from the cracking of petroleum.

Diolen *n* Fiber from poly(ethylene terephthalate). Manufactured by Glanzstoff, Germany.

DIOP *n* Abbreviation for ▶ Diisooctyl Phthalate.

DIOS *n* Abbreviation for ▶ Diisooctyl Sebacate.

Dioxan (Dioxane) \dī-ˈäk-ˌsän\ *n* [ISV] (1912) Cyclic 1,4 diethylene dioxide. $CH_2CH_2OCH_2CH_2O$. Bp, 101°C; sp gr, 1.035/20°C; fp, 11°C (52°F); refractive index, 1.423; vp, 26 mmHg/20°C. A water-miscible solvent for cellulose esters.

1,4-Dioxane *n* (diethylene ether, diethylene dioxide, dioxyethylene ether) $OCH_2CH_2OCH_2CH_2$. A solvent for cellulose esters and other plastics. Dioxane is rarely used today because of its suspected carcinogenicity.

Di-o-xenyl Phenyl Phosphate *n* $(C_6H_5C_6H_4O)_2(C_6H_5O)$ PO. A plasticizer for cellulosics, polystyrene, and vinyls.

DIOZ *n* Abbreviation for ▶ Diisooctyl Azelate.

Dip \ˈdip\ *n* (1599) The angle measured in a vertical plane between the direction of the earth's magnetic field and the horizontal.

Dip *v* [ME *dippen*, fr. OE *dyppan*; akin to OHGr *tupfen* to ash, Lithuanian *dubus* deep] (before 12c) (1) Immersion of a textile material in some processing liquid. The term is usually used in connection with a padding or slashing process. (2) The rubber compound with which tire cords and other in-rubber textiles are treated to give improved adhesion to rubber.

Dipcoat *n* A paint or plastic coating which is applied by completely immersing an article in a tank of the coating.

Dip Coating *n* A coating process wherein the object to be coated, preheated or at room temperature, depending on the materials, is dipped into a tank of fluid resin, solution, or dispersion, withdrawn and subjected to further heat or drying to solidify the deposit. See ▶ Fluidized-Bed Coating for a similar process employing powdered resin. See ▶ Coating, ▶ Dip.

Dip Dyeing See ▶ Dyeing.

Dipentene *n* (Commercial Product) Optically inactive form of the monocyclic terpene hydrocarbon limonene. Commercial dipentenes contains a substantial portion of other monocyclic and bicyclic, as well as some oxygenated terpenes having closely related boiling ranges. They are generally obtained by fractional distillation from the crude oils recovered in the several commercial methods of utilizing pine wood, also by isomerization during the chemical processing of terpenes. The four kinds of commercial dipentene are: (1) *Steam-distilled*

Dipentene — From the crude oleoresinous extract used for the processing of related steam-distilled wood naval stores. (2) *Sulfate Dipentene* — From the crude condensate of the vapors generated in the digestion of wood in the sulfate paper pulp process. (3) *Destructively Distilled Dipentene* — From the lighter portions of the oil, recovered during the destructive distillation of pine wood. (4) *Chemically Processed Dipentene* — Recovered as a by-product in connection with the chemical treatment and conversion of other terpenes. Used as a solvent, but also has anti-skinning properties. Sp gr, 0.850; flash point, 52°C (Cleveland Open Cup); boiling range, 175°C (347°F) to 188°C (370°F). (5) Pure Compound, *di*-limonene, *p*-mentha-1,8-diene $C_{10}H_{16}$, acyclic diene, a colorless liquid, sp gr, 0.84; bp, 176°C; molecular weight, 136.12, with an orange/lemony odor, used as a solvent for coumarone and alkyd resins, rubber, and natural resins.

Dip Forming *n* (dip molding) A process similar to dip coating, except that the fused, cured or dried deposit is stripped from the dipping mandrel. As most frequently used for making vinyl-plastisol articles, the process comprises dipping into the plastisol a preheated form shaped to the desired inside dimensions of the finished article allowing the plastisol to gel in a layer of the desired thickness against the form surface; withdrawing the coated form; heating the deposit to fuse the layer; cooling and stripping off the deposit. Some articles may be inverted after stripping so that the textured inside surface becomes the external surface of the finished article.

Dip-Grained Wood *n* Wood which has single waves or undulations of the fibers, such as occur around knots and pitch pockets.

Diphenyl \(▮)dī- ▮fe-nºl\ n Liquid whose vapor is used as a heat transfer agent in chemical processes. At low pressures, the temperatures obtainable with diphenyl vapor are higher than those obtainable with steam. Syn: ▶ Biphenyl.

Diphenylamine *n* [ISV] (1872) (DPA) $(C_6H_5)_2NH$. A crystalline solid. Used as a stabilizer for several plastics. Also, a weak secondary base used as a dyestuff intermediate and as an indicator.

Diphenyl Carbonate *n* $(C_6H_5O)_2CO$. The monomer from which polycarbonates are produced.

Diphenyl Decyl Phosphate *n* A nearly colorless liquid used as a stabilizer for vinyl and polyolefin resins.

Diphenyl Ether *n* A plasticizer. Bp, 252°C; mp, 28°C.

Diphenyl Ketone See ▶ Benzophenone.
Diphenyl Mono-o-Xenyl Phosphate *n* $(C_6H_5O)_2$ $(C_6H_5C_6H_4O)PO$. A plasticizer for cellulosics plastics, polystyrene, and, with limited compatibility, vinyls.
Diphenyl Octyl Phosphate *n* (DPOP) $(C_6H_5O)_2$ $(C_8H_{17}O)PO$. A flame-retardant plasticizer for PVC and cellulosics resins. It has been approved by DDA for use in food packaging.

Diphenyl Phosphate *n* A permanent plasticizer. Also called DPP.

Diphenyl Phthalate *n* (DPP) A powder that melts at 75°C. It is used as a solid plasticizer for rigid PVC, cellulosics, and other resins.

Diphenyl Xylenyl Phosphate *n* A plasticizer for cellulose acetate-butyrate, cellulose nitrate, polystyrene, PVC, and vinyl chloride-acetate copolymers.

Diphosphoglyceric Acid \(ˌ)dī-ˈfäs-fō-gli-ˌser-ik-\ *n* (1959) A diphosphate of glyceric acid that is an important intermediate in photosynthesis and in glycolysis and fermentation.

Dip Molding See ▶ Dip Forming.

Dipole \ˈdī-ˌpōl\ *n* [ISV] (1912) (1) A combination of two electrically or magnetically charged particles of opposite sign that are separated by a small distance. (2) Any system of charges, such as a circulating electric current, having the property (a) that no net force acts upon it in a uniform field; or, (b) a torque proportional to $\sin \theta$, where θ is the angle between the dipole axis and a uniform field, *does* act on the charges. In polymers, atoms such as Cl bearing negative charges induce opposite charge in neighboring atoms, thus creating dipoles. Tiny movements of these dipoles in reaction to the changing directions of rapidly alternating fields are the basis for ▶ Dielectric Heating, and (c) it produces a potential which is proportional to the inverse square of the distance from it.

Dipole Bonds *n* Chemical bonds formed when the intramolecular interactions are polar covalent (uneven sharing of electron pair). The strength of the dipole–dipole interaction increases with the polarity of the participating molecules.

Dipole–Dipole Forces *n* Forces between polar molecules.

Dipole Moment *n* (1926) The product of the magnitude of the charge at one end of a dipole times the distance between the opposite charges. A mathematical entity; the product of one of the charges of a dipole unit by the distance separating the two dipolar charges. In terms of the definition of a dipole (2), the dipole moment **p** is related to the torque **T**, and the field strength **E** (or **B**) through the equation $\mathbf{T} = \mathbf{p} \times \mathbf{E}$.

Dipole Moment, Molecular *n* It is found from measurements of dielectric constant (i.e., by its temperature dependence, as in the *Debye equation for total polarization*) that certain molecules have permanent dipole moments. These moments are associated with transfer of charge within the molecule, and provide valuable information as to the molecular structure.

Dip Penetration *n* The degree of saturation through a tire cord after impregnation with an adhesive.

Dip Pickup *n* The amount of adhesive applied to a tire cord by dipping, expressed as a percentage of the weight of the cord before dipping.

Dipping *n* Method of application in which the complete immersion of an article in the coating is followed by draining. Paints, lacquers, and stains designed for such use are usually designated "dipping." This process may be carried out either at ordinary or elevated temperatures. See ▶ Coating, ▶ Dip.

Dipropylene Glycol *n* (2,2′-dihydroxydipropyl ether) $(CH_3CHOH-CH_2)_2O$. A high-boiling glycol ether with a low order of toxicity, widely used as a solvent and chemical intermediate. As a solvent it is used with cellulose acetate and nitrate, and is one of the few known solvents for polyethylene. Thus it is used in screening tests to identify polyethylene. As an intermediate, dipropylene glycol reacts with dibasic acids to form alkyd resins, polyester plasticizers, and urethane-foam intermediates.

Dipropylene Glycol Dibenzoate A plasticizer for PVC, also compatible with most common thermoplastics, imparting good stain resistance.

Dipropylene Glycol Monosalicylate *n* (salicylic acid, dipropylene glycol monoester) A light-colored oil used in ultraviolet-screening agents and plasticizers.

Dipropyl Ketone *n* A stable, colorless liquid, a solvent for many resins. Bp, 143°C; flp, 49°C (120°C); vp. 5 mmHg/20°C.

Dipropyl Oxalate \-ˈäk-sə-ˌlāt\ *n* An active, high-boiling (211°C) solvent.

Dipropyl Phthalate *n* A plasticizer for cellulose acetate and cellulose acetate-butyrate.

Dipropyl Succinate *n* Plasticizer. Bp, 246°C; sp gr, 1.016/4°C.

Diprotic Acid *n* An acid with two available H^+, or two denotable protons.

Dip Treating *n* The process of passing fiber, cord, or fabric through an adhesive bath, followed by drying and heat-treating of the adhesive-coated fiber to obtain better adhesion.

Di-*p*-xylylene *n* (DPX, DPXN) $(-CH_2C_6H_4CH_2-)_2$. The stable dimmer of *p*-xylylene, a white powder. When heated to 600°C in an evacuated chamber, the monomer is regenerated and instantly polymerizes to form a tough, impervious film on any cold surface between the heating pot and the vacuum source. See ▶ Parylene.

Diradicals \-ˈra-di-kəls\ *n* Molecular species having two unpaired electrons, in which at least two different electronic states with different multiplicities (electron-paired [singlet state] or electron-unpaired [triplet state]) can be identified. E.g., $H_2C\text{--}CH_2C\cdot H_2$ propane-1,3-diyl (trimethylene).

Direct Coating *n* The simplest method of coating, this procedure involves spreading the coating with a knife. The moving fabric substrate is usually supported by a roller or a sleeve. The gap between the knife and the fabric determines coating thickness.

Direct Dyes *n* A class of dyestuffs that are applied directly to the substrate in a neutral or alkaline bath. They produce full shades on cotton and linen without mordanting and can also be applied to rayon, silk, and wool. Direct dyes give bright shades but exhibit poor washfastness. Various aftertreatments are used to improve the washfastness of direct dyes, and such dyes are referred to as "aftertreated direct colors." See ▶ Dyes.

Direct Esterification *n* In the production of polyethylene terephthalate, the process in which ethylene glycol is reacted with terephthalic acid to form bis-*p*-hydroxyethyl terephthalate monomer with the generation of water as a by-product.

Direct Gate *n* A gate that has the same cross section as the runner, i.e., the absence of a distinguishable gate.

Directionally Oriented Fabrics *n* Rigid fabric constructions containing inlaid warp or fill yarns held in place by a warp-knit structure. Used in geotextiles, coated fabrics, composites, etc.

Directional Reflectance See ▶ Reflectance, Directional.

Direction of Twist See ▶ Twist, Direction of.

Direct Lighting *n* Lighting which is so controlled by a reflector that the major portion of the light directly reaches the spot to be illuminated without redirection, diffusion or reflection.

Direct Printing See ▶ Printing.

Dirt \ˈdərt\ *n* [ME *drit*, fr. ON; akin to OE *drītan*, to defeat] (13c) Disfiguring foreign material other than microorganisms on or embedded in a dried coating. *Also called Soil.*

Dirt Collection *n* The accumulation of dust, dirt, and other foreign matter on a paint surface.

Dirt Resistance *n* The ability of a coating to resist soiling by foreign material, other than microorganisms, deposited on or embedded in the dried coating.

Disbond *v* In an adhesive-bonded joint, to separate at the bond surface. *n* Such as a separation.

Discharge-Inception Voltage *n* In a dielectric-strength test, the voltage at which discharges begin in the voids within the specimen.

Discharge Printing See ▶ Printing.

DISCO *n* An epoxy-based composite prepreg containing discontinuous but oriented fibers, nearly as strong as unidirectional composites of the same fiber content, and formable into complex shapes and over corners.

Discoloration \(ˌ)dis-ˌkə-lə-ˈrā-shən\ *n* (1642) (1) Any change from an initial color possessed by a plastic. (2) A lack of uniformity in color where color should be uniform over the whole area of a plastic object. In the second sense, where they are applicable, one may use the more definite terms *mottle*, *segregation*, or *two-tone*. Discoloration can be caused by inadequate blending of ingredients, by overheating, exposure to light, irradiation, or chemical action.

Discolored Pick See ▶ Mixed End or Filling.

Discontinuities See ▶ Holidays.

Discontinuous Phase See ▶ Dispersed Phase.

Dished *n* Showing a symmetrical concave distortion of a flat or curved surface of a plastic object, so that, as normally viewed, it appears more concave than its design calls for. Opposite of domed. See also ▶ Warp and ▶ Sink Mark.

Disk-and-Cone Agitator *n* A mixing device comprised of disks or cones rotating at speeds between 20 and 60 rev/s or higher. The disks/cones displace fluid contacting their surfaces by centrifugal force. They are used in preparing pastes and dispersions.

Disk Extruder *n* Any of a variety of novel devices in which pellets or powders are melted by contact with a heated, rotating disk and scraped off the disk into a screw pump or other pressure-developing machine; or devices in which friction of pellets between rotating disks and nearby stationary surfaces causes the plastic to melt, thence to be discharged to a pressure-developing machine for shaping into a product. One such device, the "Diskpack" extruder, has been developed and marketed by the Farrel Group of Emhart Corporation. See also ▶ Extruder, ▶ Elastic-Melt.

Disk Feeder *n* A horizontal, flat or grooved, rotating disk at the bottom of a hopper feeding a continuous extruder that controls the feed rate by varying the speed of rotation of the disk, or by varying the clearance between the disk and the scraper that directs the feed from the disk into the feed throat of the extruder.

Disk Gate *n* See ▶ Diaphragm Gate.

Disk Test *n* An in-rubber test used to predict the fatigue resistance of tire cords and other industrial yarns.

Dislocation \dis-()lō-kā-shən\ *n* A structural defect in which the lattice planes in a crystal is incomplete or warped. *Also known as a Line Defect.*

Disorder, Crystalline *n* Defect in the lattice of a crystal, characterized by local faults in the atomic arrangement.

Dispersant \di-spər-sənt\ *n* (1941) A dispersing agent, often of a surface active chemical, that promotes formation of a dispersion or maintains a state of dispersion by preventing settling or aggregation.

Dispersant *n* In an organosol, a liquid component that has a solvating or peptizing action on the resin, thus aiding in dispersing and suspending it. Additive that increases the stability of a suspension of powders (pigments) in a liquid medium. *Also known as Dispersing Agent.*

Dispersed *n* Finely divided or colloidal in nature.

Dispersed Phase *n* That phase in an emulsion or suspension which is broken down into droplets or discrete particles and dispersed throughout the other or continuous phase. *Also called Discontinuous Phase.*

Disperse Dyes *n* A class of slightly water-soluble dyes originally introduced for dyeing acetate and usually applied from fine aqueous suspensions. Disperse dyes are widely used for dyeing most of the manufactured fibers. See ▶ Dyes.

Disperse Phase *n* In a ▶ Suspension or ▶ Emulsion, the "disperse phase" refers to the particles of one material individually dispersed, more or less stably, in the continuously connected domain of another material, usually a liquid, known as the *continuous phase*.

Disperse System *n* Any system in which particles (of any size and state) are dispersed in a homogeneous medium. A colloidal solution represents a special kind of disperse system.

Dispersing Agent *n* A material added, usually in relatively small percentage, to a suspending medium to promote and maintain the separation of discrete, fine particles of solids or liquids. Dispersants are used, for example, in the wet grinding of pigments and for suspending water-insoluble dyes. See ▶ Dispersant.

Dispersion Curve *n* A graph of the refractive index (ordinate) as a function of wavelength of radiation (abscissa).

Dispersion, Degree of *n* Quantity varying reciprocally with aggregate size. The larger the aggregates, the lower the degree of dispersion.

Dispersion Forces *n* (1) Weak forces between atoms or molecules due to momentary fluctuations in their electronic charge-cloud distributions. Also called *London Forces*. (2) The force of attraction between molecules possessing no permanent dipole. The interaction energy is given by

$$U_p = -\frac{3}{4} h \frac{v_o \alpha^2}{r^6}$$

where h is Planck's constant, V_o a characteristic frequency of the molecule, r the distance between the molecules, and α the polarizability.

Dispersion, Light *n* (1) The separation of light into separate wavelengths by means of a dispersion device such as a prism or diffraction grating. Fundamentally, dispersion is related to different wavelengths of light in a given medium. (2) The variation of refractive index with color (or wavelength) of light. The spreading of white light into its component colors when passing through a glass prism is due to dispersion which, in turn, is due to the fact that the refractive index of transparent substances is lower for long wavelengths than for short wavelengths. A measure of dispersion is v defined as:

$$v = \frac{n_D - 1}{n_F - n_C}$$

where n_D = refractive index at 589 nm (yellow); n_F = 486 nm (blue); n_C = 656 nm (red).

See ▶ Dispersion Curve.

Dispersion, Particles and Liquids *n* (14c) (1) A system of two or more phases comprising one or more finely divided materials distributed in another material. Types of dispersions are *emulsions* (liquids in liquids), *suspensions* (solids in liquids), *foams* (gases in liquids or solidified liquids), and *aerosols*, (liquids in gases). In the plastics industry, the term dispersion usually denotes a finely divided solid dispersed in a liquid or in another solid. Examples are fillers and pigments in molding compounds, plastisols and organosols. This term is used loosely in many ways with special meanings which are often implied but not defined. Since the term is decidedly generic and includes all types of colloidal systems, it should not be used on the assumption that it refers to a desirable feature of one type of dispersion. For example, in paint and printing ink technology a "good" dispersion usually means that a pigment is finely divided and deflocculated in the vehicle. Where aggregates are present in the dispersed phase, the material is considered without exception a "poor" dispersion regardless of the state of flocculation. Where the pigment particles are finely divided but flocculated, the material has been called variously a "good" and a "poor" dispersion. (2) A mixing process in which the particles or globules of the discontinuous phase are reduced in size. This may be accomplished by breaking solid particles, as in ball-milling, or by shearing viscous regions, as in kneading or extrusion. (3) Two-phase system in which one phase, called the dispersed phase, is permanently distributed as small particles through the second phase called the continuous phase. See also ▶ Segregation.

Dispersion, Pigment *n* Suspension of pigment particles uniformly in a medium such as a paint vehicle, plastic matrix, etc. The process of dispersing the pigment involves the separation of individual pigment particles, and coating them with the medium.

Dispersion Resin *n* A special type of PVC resin with very small spherical particles, usually 1 μm or less in diameter, permitting them to be mixed with plasticizers by simple stirring techniques. They are used in compounding plastisols and organosols. See ▶ Organosol and ▶ Plastisol.

Dispersion Resins *n* Used as the primary vinyl resin in the production of plastisols and organosols. These resins are fine, white powders, with an average particle size of 1 μm and a bulk density of approximately 15–20 lb per cubic foot. The fine particle size, combined with the dense nature of the particle allows these products to be easily dispersed in plasticizers and solvents. The specific selection of a dispersion resin is primarily based on the fusion, rheological and appearance requirements of the end product and the manufacturing process.

Dispersions *n* Systems of two or more phases comprising one or more finely divided materials distributed in another material. Types of dispersion are emulsions (liquids in liquids), suspensions (solids in liquids), foams (gases in liquids or solidified liquids), and aerosols (liquids in gases).

Dispersive Mixing *n* Any mixing process in which the principal action is reduction of the size of particles or interlayers. The progress of the process is judged by two criteria, ▶ Intensity of Segregation and ▶ Scale of Segregation.

Dispersive Power *n* If n_1 and n_2 are the indices of refraction for wavelengths λ_1 and λ_2 and n the mean index or that for sodium light, the dispersive power for the specified wavelength is

$$\bar{\omega} = \frac{n_2 - n_1}{n - 1}$$

Dispersoid \-ˈspər-ˌsóid\ *n* (1611) Particles of a dispersion.

Displacement \di-ˈsplā-smənt\ *n* (1611) A reaction in which an elementary substance displaces and sets free a constituent element from a compound.

Displacement Angle *n* In filament winding, the angle whose tangent equals the quotient of the advancement distance of the winding ribbon on the equator after one complete turn, divided by (π × the equatorial diameter).

Displacement or Elongation *n* (at an instant) The distance of a vibrating or oscillating particle from its position of equilibrium.

Disproportionation \-ˌpōr-shə-ˈnā-shən\ *n* (ca. 1929) A reaction in which one substance acts simultaneously as an oxidizing agent and a reducing agent; auto-oxidation. It can also be written as the termination by chain transfer between macroradicals, to produce a saturated and an unsaturated polymer molecule.

Dissipation \ˌdi-sə-ˈpā-shən\ *n* The loss modulus in a plastic part when imparted with rapid, cyclic changes (or even reversals) of stress. The product of mechanical dissipation is heat, which can raise the temperature of the part and cause it to weaken, creep rapidly, or even fail prematurely. Dissipation can also apply to electrical systems, whereby a material with small dissipation will tend to better insulate heat. This is a desirable property in electrical insulations for high-frequency applications because it minimizes the waste of electrical energy as heat.

Dissipation Factor *n* (electrical) The ratio of the conductance of a capacitor in which the test material is the dielectric to its susceptance; or the ratio of its parallel reactance to its parallel resistance. Most plastics have a low dissipation factor, a desirable property in electrical insulations for high-frequency applications because it minimizes the waste of electrical energy as heat. On the other hand, for those plastics with high dissipation factors, e.g., PVC and phenolic, the property provides a fast and economical method of even heating using microwaves and is the basis of electronic preheating of thermosetting molding powders and of electronic sealing of flexible-vinyl films. See ▶ Dielectric Heating.

Dissipation Factor *n* (mechanical) The ratio of the loss modulus to the modulus of elasticity in a plastic part undergoing rapid, cyclic changes (or even reversals) of stress. As in electrical dissipation, the product of mechanical dissipation is heat, heat that can raise the temperature of the part and cause it to weaken, creep rapidly, or even fail prematurely.

Dissociation \(ˌ)di-ˌsō-sē-ˈā-shən\ *n* (1611) The splitting apart of a molecule to form two fragments; the reaction of an electrolyte with a solvent to form ions. *Sometimes called Ionization.*

Dissociation Constant n, K_{diss} The equilibrium constant for a dissociation equilibrium.

Dissymmetry \(ˌ)di(s)-ˈsi-mə-trē\ n (1845) Present in a molecule that does not have mirror image symmetry but does have a translational or rotational axis of symmetry.

Distearyl Ether (dioctadecyl ether) A mold lubricant.

Distemper \dis-ˈtem-pər\ n [obs. *distemper*, v., dilute, mix to produce distemper, fr. ME, fr. MF *destemprer*, fr. L *dis-* + *temperare*] (1632) Heavily pigmented, matt drying composition, capable of being thinned with water, in which the binding medium consists essentially of either glue or casein or similar sizing material.

Distemper Blush n A flat brush, five to ten inches wide, well packed, with long bristle.

Distemper Colors n Colorants suitable for tinting distempers. The two chief requirements are lightfastness in reduced form and fastness of alkali.

Distemper, Oil-Bound See ▶ Oil-Bound Distemper.

Distensibility \-ˈsten(t)-sə-bəl\ *adj* [*disten-* (fr. L *distensus*, pp of *distendere*) + *-ible*] (ca. 1828) Ability to be stretched. See ▶ Elongation.

Distillation \ˌdis-tə-ˈlā-shən\ n (14c) Process of separation consisting of vaporizing a liquid and collecting the vapor, which is then usually condensed to a liquid.

Distillation Range n Temperature range over which a mixture of liquids is distilled.

Distinctness-of-Image Gloss n The sharpness with which image outlines are reflected by the surface of an object.

Distortion \di-ˈstȯr-shən\ n (1581) (1) A change in the shape of a solid body, often associated with temperature differences or stress gradients within the body. (2) An apparent change of shape as perceived through an optically imperfect, transparent membrane or reflected from an imperfect mirror.

Distressing n In antiquing, the furniture process designed to produce a more authentic look of age or craftsmanship. It attempts to simulate the small visible marks of abuse picked up by a piece of furniture during a long period of use. In simulating natural marring or aging of some kind, it may involve the use of gouges, hammers, chains, blunt or pointed instruments such as common car keys, a belt buckle or sharp pencil point. Other effects may be achieved with a fine brush or a pointed wood instrument dipped into ink or glaze and placed on a surface. If no natural counterpart to an effect exists, it is not authentically aged and should be considered a novelty finish or accent piece.

Distributing Roller n A rubber covered roller which conveys ink from the fountain to the ink drum of a rotating press.

Distribution Law n A substance distributes itself between two immiscible solvents so that the ratio of its concentrations in the two solvents is approximately a constant (and equal to the ratio of the solubilities of the substance in each solvent). The above statement requires modification is more than one molecular species is formed.

Distribution Length n In fibers, a graphic or tabular presentation of the proportion or percentage (by number or by weight) of fibers having different lengths.

Distributive Mixing n Any mixing operation in which the principal action is convection rather than dispersion, resulting in elements of the different phases being more intimately blended, but without much size reduction. Typical is blending of masterbatch pigmented pellets with virgin pellets prior to extrusion or molding, by tumbling.

Di-*tert*-Butyl Peroxide n A member of the alkyl peroxide family, used as an initiator in vinyl chloride polymerization, polyester reactions, and as a crosslinking agent. A stable liquid used as a catalyst for polymerizations at high temperatures of a variety of olefin and vinyl monomers, e.g., ethylene, styrene, and styrenated alkyds.

Ditetrahydrofurfuryl Adipate n A heterocyclic plasticizer for cellulose acetate-butyrate.

Ditridecyl Phthalate n (DTDP) $C_6H_4(COOC_{13}H_{27})_2$. A primary plasticizer for PVC, also compatible with cellulosics and polystyrene. In vinyls, it imparts resistance to high temperatures and to extraction by hot soapy water, excellent flexibility, and anti-fogging properties.

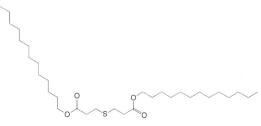

Ditridecyl Thiodipropionate *n* $(C_{13}H_{27}OOCCH_2CH_2)_2S$. A stabilizer, plasticizer, and softening agent.

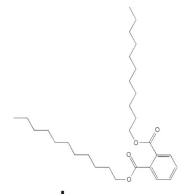

Diundecyl Phthalate *n* (DUP) A plasticizer characterized by low volatility and good low-temperature properties compared to other phthalates.

Divergent Die \də-ˈvər-jənt-\ *n* A die for hollow articles in which the internal channels leading to the orifice increase in cross section toward the lip.

Divided Threadline Extrusion *n* Spinning of two separate threadlines from one spinneret.

Divinyl B Syn: ▶ Butadiene.

Divinylbenzene *n* (DVB, vinylstyrene) A monomer derived from styrene, used in making ion-exchange resins, synthetic rubbers, and casting resins. The commercial product contains 55% mixed *m*- and *p*-isomers, the rest being ethylvinylbenzenes. It is often used along with styrene as a reactive monomer in the production of polyester resins, to which it imparts a higher degree of crosslinking and superior chemical resistance.

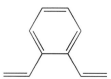

Divinyl Polymerization, Polymers *n* Polymerization of a monomer which contains two vinyl groups. When the two double bonds are conjugated (as in 1,3-dienes) the polymerization is called diene polymerization.

DIY *n* Abbreviation for "DO IT YOURSELF."

Dl-Limonene See ▶ Dipentine.

DLVO Theory *n* A dispersion stabilizing theory in which zeta potential is used to explain that as two particles approach one another their ionic atmospheres begin to overlap and a repulsion force is developed. The theory was prepared by Dergaquin and Landau in the Soviet Union and Verwey and Overbeck in the Netherlands, thus the DLVO.

DMA *n* Abbreviation for ▶ Dimethylaniline.

DMEP *n* Abbreviation for ▶ Di(2-methoxyethyl) phthalate.

DMF *n* Abbreviation for ▶ Dimethyl Formamide.

DMG *n* Abbreviation for ▶ Dimethyl Glutarate.

DMSO *n* Abbreviation for ▶ Dimethyl Sulfoxide.

DMT *n* Abbreviation for Dimethyl Terephthalate.

DNA *n* Abbreviation for ▶ Dinonyl Adipate.

DNHZ *n* Abbreviation for ▶ Di-*n*-hexyL Azelate.

DNODA *n* Abbreviation for Di-*n*-octyl-*n*-decyl Adipate. See ▶ *n*-Octyl-*n*-decyl Adipate.

DNODP *n* Abbreviation for Di(*n*-octyl-*n*-decyl) Phthalate.

DNP *n* Abbreviation for ▶ Dimethyl Phthalate.

DNP *n* Abbreviation for ▶ Dinonyl Phthalate.

DNPT *n* Abbreviation for ▶ Dinitrosopentamethylenetetramine.

DNTA *n* Abbreviation for ▶ Dinitrosoterephthalamide.

DOA *n* Abbreviation for ▶ Di(2-ethylhexyl) Adipate.

Dobby \ˈdä-bē\ *n* [perhaps fr. *Dobby*, nickname for *Robert*] (1878) (1) A mechanical attachment on a loom. A dobby controls the harnesses to permit the weaving of geometric figures. (2) A loom equipped with a dobby. (3) A fabric woven on a dobby loom.

Doctor \ ▎däk-tər\ *n* [ME *doctour* teacher, doctor, fr. MF & ML; MF, fr. ML *doctor*, fr. L, teacher, fr. *docēre* to teach] (1) Device for spreading a thin film of even thickness on base material. In the paint industry, it is used to prepare paint and varnish films of even and predetermined thickness. (2) In gravure printing and coating, a blade that scrapes off the excess ink or lacquer from the surface of the etched cylindrical roll just prior to printing, leaving the "cells" filled with ink or lacquer. {*doctor knife* G Abstreifmesser n; Rakelmesser n, F racle f, S cuchilla de recubrimiento, cuchilla f, I raschiatore m; racla f}.

Doctor-Bar Blade *n* A scraper mechanism that regulates the amount of adhesive on the spreader rolls or on the surface being coated.

Doctor Mark or Streak *n* Streak or ridge in coated fabrics caused by a damaged doctor blade. Also called *Knife Mark*.

Doctor Roll *n* Roller mechanism which is revolving at a different surface speed, or in an opposite direction, resulting in a wiping action for regulating the adhesive supplied to the spreader roll.

Doctor Streak *n* A defect in printed fabrics consisting of a wavy white or colored streak in the warp direction. It is caused by a damaged or improperly set doctor blade on the printing machine.

Doctor Test *n* Method for detecting the presence of mercaptan sulfur in gasolines, naphthas, and kerosenes.

Dodecanoic Acid See ▶ Lauric Acid.

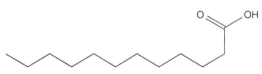

4-Dodecyloxy-2-Hydroxbenzophenone *n* An ultraviolet inhibitor for polyethylene and polypropylene, also suggested as suitable for PVC, polystyrene, polyesters, and surface coatings such as those based on cellulosic and acrylic resins.

Doeskin Finish \ dō- ▎skin\ *n* A soft low nap that is brushed in one direction. Cloth with this type of finish is used on billiard tables and in men's wear.

DOF *n* Abbreviation for ▶ Dioctyl Fumarate.

DOFF A set of full bobbins produced by one machine (a roving frame, a spinning frame, or a manufactured filament-yarn extrusion machine).

Doffer *n* (1) The last or delivery cylinder of the card from which the sheet of fibers is removed by the doffer comb. (2) An operator who removes full bobbins, spools, containers, or other packages from a machine and replaces them with empty ones.

Doffer Comb *n* A reciprocating comb, the teeth of which oscillate close to the card clothing of the doffer to strip the web of fibers from the card.

Doffer Loading *n* Fibers imbedded so deeply into the doffer wire clothing that the doffer comb cannot dislodge them to form a traveling web.

Doffing *n* The operation of removing full packages, bobbins, spools, roving cans, caps, etc., from a machine and replacing them with empty ones.

Dogbone *n* A slang term for the dumbbell shape of tensile-test specimens having relatively wide ends for gripping and a narrower, gage-length center within which all plastic deformation and breakage occur. See ▶ Tensile Bar.

Dogskin See ▶ Orange Peel (coatings defect).

Doily \ ▎dói-lē\ *n* [*Doily* or *Doyley fl* 1711 London draper] (1711) In filament winding, the planar reinforcement that is applied to a local area between windings to provide extra strength in an area where a cut-out is to be made; for example, a port opening.

DOIP *n* Abbreviation for Dioctyl Isophthalate. See ▶ Di (2-ethylhexyl) Isophthalate.

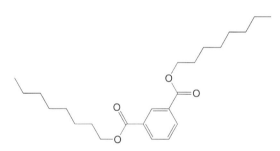

Dolomit \ ▎dō-lə- ▎mīt\ *n* [F, fr. *Déodat* de *Dolomieu* † 1801 F geoloist] (1794) $CaCO_3 \cdot MgCO_3$. A mineral having the empirical composition, 1 mole of calcium carbonate and 1 mole of magnesium carbonate. Any calcium carbonate rock containing 20% or more of magnesium carbonate. See ▶ Calcium Carbonate.

Domain \dō- ▎mān\ *n* [ME *domayne*, fr. MF *domaine*, *demaine*, fr. L *dominium*, fr. *dominus*] (15c) A region in a ferromagnetic substance within which the magnetic moments of all the atoms are aligned.

Domains *n* A term used in non-crystalline systems in which the chemically different sections of the chain separate, generating two or more amorphous phases.

Domed *n* Having a symmetrical convex protrusion in the surface. Opposite of dished.

Dominant Wavelength *n*, CIE Wavelength of spectrally pure light which, when additively mixed with the illuminant and adjusted for luminance, would match the color. It is expressed in wavelength units, nm (or μm), and is determined graphically from the chromaticity diagram or by suitable computation. The dominant wavelength of the nonspectral colors, the purples, is described in terms of the dominant wavelength of the green, to which they are the additive complements, followed by the letter c. Thus, a dominant wavelength designation of 520c describes a purple which is additively complementary to a green having a dominant wavelength of 520 nm. Dominant wavelength describes the hue of a color, although the visual evaluation of constant hue does not necessarily coincide with constant dominant wavelength.

Domino \ˈdä-mə-ˌnō\ *n* Early French wallpaper made in small sheets, originally imitating marble, later patterned.

Donegal \ˌdä-ni-ˈgól\ *n* A tweed fabric with colorful slubs woven in, donegal is used for suits and coats.

Doorframe *n* An assembly built into a wall consisting of two upright members (jambs) and a head (lintel) over the doorway; encloses the doorway and provides support on which to hang the door.

DOP *n* Abbreviation for Dioctyl Phthalate. See ▶ Di(2-ethylhexyl) Phthalate.

Dope dōp\ *n* [D *doop* sauce, fr. *dopen* to dip; akin to OE *dyppan* to dip] (1807) (1) A solution of a cellulosic plastic, historically cellulose nitrate, used for treating fabrics. (2) Cellulose ester lacquer for adhesive or coating purposes. See ▶ Spinning Solution.

Dope Cotton *n* Designation of high viscosity nitrocellulose, i.e., 20–30, 40, and 70 s nitrocellulose. See ▶ Nitrocellulose.

Dope Dyed See ▶ Dyeing, Mass-Colored.

Doping *n* The addition of controlled small amounts of a foreign substance to an otherwise pure substance.

Doppler Effect \ˈdä-plər\ *n* [Christian J. *Doppler*] (1905) (Light) The apparent change in the wavelength of light, produced by the motion in the line of sight of either the observer or the source of light.

Doppler Effects *n* Effects on the apparent frequency of a wave train produced (1) by motion of the source toward or away from the stationary observer, and (2) by motion of the observer toward or from the stationary source; the motion in each case being with reference to the (supposedly stationary) medium. For sound waves, the observed frequency f_o, in cycles/s, is given by

$$f_o = \frac{v + w - v_o}{v + w - v_s} f_s$$

where v is the velocity of sound in the medium, v_o is the velocity of the observer, v_s is the velocity of the source, w is the velocity of the wind in the direction of sound propagation, and f_s is the frequency of source. For optical waves

$$f_o = f_s \sqrt{\frac{c + v_r}{c - v_r}}$$

where v_r is the velocity of the source relative to the observer and c is the speed of light.

Dormer \ˈdór-mər\ *n* [MF *dormeor*, fr. L *dormitorium*] (1592) Gable-like projection or a window set upright in a sloping roof, or the roofed projection in which this window is set.

DOS *n* Abbreviation for Dioctyl Sebacate. See ▶ Di(2-ethylhexyl) Sebacate.

Dosimeter \dō-ˈsi-mə-tər\ *n* [LL *dosis* + ISV *-meter*] (1938) detector worn by workers to measure the amount of an environmental agent, usually radioactivity, but sometimes noise or noxious gases, to which the worker has been exposed during a working shift or longer period of time.

Dot *n* The individual element of a halftone printing plate.

DOTP *n* Abbreviation for ▶ Dioctyl Terephthalate.

Dotted Swiss *n* (ca. 1924) A sheer cotton or cotton blend fabric with small dot motif, dotted swiss is used for dress goods, curtains, baby clothes, etc.

Double Back *n* A secondary backing glued to the back of carpet, usually to increase dimensional stability.

Double Bond *n* A type of covalent bond (pi or π bond in C=C), common in organic chemistry, in which two pairs of electrons are shared between two elements. The double bond may be symbolized either by ":" or "=," as in ethylene, $CH_2:CH_2$, or dimethyl ketone, $(CH_3)_2C=O$. Elements possess a definite number of valency linkages or bonds, and these are employed in uniting with other elements. When all these bonds are completely employed in such unions, saturated compounds result. If, however, a residual bond or linkage remains, this may be involved with another in forming a reactive or unsaturated double bond. If a hydrogen atom is removed from each of the carbon atoms, two spare linkages occur, and an unsaturated double bond results. Double bonds are centers of great reactivity, and are able to accept addition of suitable elements such as hydrogen, chlorine, etc., to yield fully saturated compounds. The reactivity of double bonds in drying oils is responsible for their drying (autoxidation) and polymerizing properties.

Double-Bond Addition *n* Chemical reaction whereby a double-bonded molecule (typically an alkene) is reacted with a reagent to form a single product. When an alkene undergoes addition, two groups add to the carbon atoms of the double bond, and the carbons become saturated.

$$H_2C\!=\!\!CH_2 + X\!-\!Y \longrightarrow \overset{\overset{X}{|}}{H_2C}\!-\!\overset{\overset{Y}{|}}{CH_2}$$

Double Bond, Carbon–Carbon *n* The sharing of two electron pairs between adjacent carbon atoms. The carbon–carbon double bond (pi or π bond) is the most reactive part of an alkene, but cannot rotate around its primary axis.

Double Cloth Construction *n* Two fabrics are woven in the loom at the same time, one fabric on top of the other, with binder threads holding the two fabrics together. The weave on the two fabrics can be different.

Double Decomposition *n* Double decomposition consists of a simple exchange of the parts of two substances to form two new substances.

Double End *n* Two ends woven as one in a fabric. A double end may be intentional for fabric styling, or accidental, in which case a fabric defect results.

Double Helix *n* (1954) Physical structure of DNA, whereby two complementary strands are joined by hydrogen bonds between the base pairs. This double strand coils into a helical arrangement. See ▶ DNA.

Double-Hung Sash *n* Window divided into two sections, one lowering from the top and the other rising from the bottom.

Double Knit Fabric *n* (1895) A fabric produced on a circular-knitting machine equipped with two sets of latch needles situated at right angles to each other (dial and cylinder).

Double Pick See ▶ Mispick.

Doubler *v* (13c) In filament winding, a local area with extra wound reinforcement, either wound integrally with the part or wound separately and bonded to the part.

Double-Ram Press *n* A press for injection or transfer molding in which two distinct systems of the same kind (hydraulic or mechanical), or of different kinds, create respectively the injection or transfer force and the clamping force.

Double Refraction *n* (1831) The refraction of light in two slightly different directions to form two rays or vector components. Each ray is polarized, and their vibration directions are perpendicular to each other. Furthermore, each ray has a different velocity, and therefore a different refractive index. See ▶ Birefringence.

Double-Screw Extruder See ▶ Extruder, Twin-Screw.

Double Seam Can *n* Can formed by interlocking the edges of both the end and the body, double folding and rolling firmly together.

Double Selvage See ▶ Rolled Selvage.

Double-Shot Molding *n* (two-shot molding, insert molding, two-color molding, over-molding) A process for making two-color or two-material parts by means of successive molding operations. The basic process includes the steps of injection molding one part, transferring this part to a second mold as an insert, and molding the second material against the first. Examples of parts made by double-shot molding are computer keys, pushbuttons, telephone-keypad buttons, and other such products in which indicia must resist heavy wear and remain permanently legible. In a modification of the process, cups and the like with differently colored insides and outsides are made automatically by means of a machine equipped with two injection molders and a swinging platen carrying two cup cores indexed around a central tie-bar, bringing the molds into position for each successive shot. This arrangement permits simultaneous molding of both shots. Some of these products are made today by ▶ Coinjection.

Double-Skinned Sheet *n* A plastic sheet consisting of a relatively thick center section bonded on each face to thinner layers that differ in color or composition from the center section. Today such sheets are usually made by coextrusion of the three plastic layers and, in many cases, two additional adhesive layers between the core and the skin.

Double Spread See ▶ Spread.

Doubletone Ink *n* A type of printing ink which produces the effect of two-color printing with a single impression. These inks contain a soluble toner which bleeds out to produce a secondary color.

Double Weave *n* A fabric woven with two systems of warp or filling threads so combined that only one is visible on either side. Cutting the yarns that hold the two cloths together yields two separate cut pile fabrics.

Doubling *n* (1) A process for combining several strands of sliver, roving, or yarn in yarn manufacturing. (2) The process of twisting together two or more singles or plied yarns, i.e., plying. (3) A British term for twisting. (4) The term doubling is sometimes used in a sense opposite to singling. This is unintentional plying. (5) A yarn, considerably heavier that normal, produced by a broken end becoming attached to and twisting into another end.

Dough \dō\ *n* [ME *dogh*, fr. OE *dāg*; akin to OHGr *teic* dough, *fingere* to shape, Gk *teichos* wall] (dough-molding compound) A term sometimes used for a reinforced-plastic mixture of dough-like consistency in an uncured or partly cured state. A typical dough consists of polyester resin, glass fiber, calcium carbonate, lubricants and catalysts. The compounds are formed into products either by hand layup or, more usually, by compression molding.

Douppioni *n* A rough or irregular yarn made of silk reeled from double or triple cocoons. Fabrics of douppioni have an irregular appearance with long, thin slubs. Douppioni-like yarns are now being spun from polyester and/or rayon staple.

Dowel \daú(-ə)l\ *n* [ME *dowle*; akin to OHGr *tubili* plug, LGk *typhos* wedge] (14c) (dowel pin) A hardened steel pin, usually having a slight taper, used to maintain alignment between two or more parts of a mold or machine.

Dowel Bushing *n* A hardened steel insert in the portion of a mold that receives the dowel pin.

Downdraft Metier *n* A dry-spinning machine in which the airflow within the drying cabinet is in the same direction as the yarn path (downward).

Downgrade *n* In quality control, the lowering of the grade and/or value of a product due to the presence of defects.

Downtwister *n* A cap, ring, or flyer twisting frame.

Downtwisting *n* A process for inserting twist into yarn in which the yarn passes downward from the supply package (a bobbin, cheese, or cone) to the revolving spindle. The package or packages of yarn to be twisted are positioned on the creel, and the ends of yarn are led downward through individual guides and stop motions to the positively driven feed roll and from there to the revolving take-up package or bobbin, which inserts twist.

Dowtherm® *n* The trade name (Dow Chemical Co.) of a liquid heat-transfer medium consisting of biphenyl and phenyl ether in eutectic ratio. High-boiling and stable, it is less used today than formerly because of its suspected carcinogenicity. Bp, 258°C. Used for heat transfer.

DOZ *n* Abbreviation for Dioctyl Azelate. See ▶ Di(2ethylhexyl) Azelate.

DP *n* Abbreviation for ▶ Degree of Polymerization.

DPA *n* Abbreviation for ▶ Diphenylamine.

DPCF *n* (DPCP) Abbreviation for ▶ Cresyl Diphenyl Phosphate, a plasticizer.

DPP *n* Abbreviation for ▶ Diphenyl Phthalate.

Draft *n* A slight taper in a mold wall, proceeding inward from the parting surface to the bottom of the cavity, designed to facilitate removal of the molded object from the mold. When the taper is reversed, tending to impede removal of the article, the term *back draft*, or *reverse draft* is employed. In weaving, a pattern or plan for drawing-in.

Draft Angle *n* In a mold, the angle in the profile plane with its vertex at the bottom of the mold or cavity, between the side of the mold and the vertical plane. In most cases, 1° is adequate for smooth ejection of parts from the mold.

Drafting See ▶ Drawing (1).

Draft Ratio *n* The ratio between the weight or length of fiber fed into various machines and that delivered from the machines in spun yarn manufacture. It represents the reduction in bulk and weight of stock, one of the most important principles in the production of yarn from staple fibers.

Drag *n* (Brush Drag) Resistance encountered when applying a coating by brush. The British term for this condition is "gummy."

Drag Flow *n* (1) In general, the laminar flow of a viscous liquid that is contained between two surfaces, one of which is moving relative to the other. (2) In the metering section of an extruder screw, the rate of drag flow is the component of total material flow in the down-channel direction caused by the relative motion between the screw and cylinder. If the screw were discharging freely, the output rate would be equal to the drag-flow rate. (3) In a wire-coating die, the flow generated by the relative motion of the wire through the stationary die. In both extruders and wire-coating dies, pure drag flows almost never occur, but are altered in complex ways by opposing or augmenting pressure fields and the non-Newtonia nature of the melts.

Dragged In Filling See ▶ Pulled-In Filing.

Dragon's Blood *n* Dark red resinous exudation, obtained from the surface of the fruit of the *Calamus draco* and similar species. Used to some extent as a colorant in coatings.

Drainage Fabrics See ▶ Geotextiles.

Dralon *n* Poly(acrylonitrile). Manufactured by Bayer, Germany.

Drape \drāp\ *v* [prob. back-form. fr. *drapery*] (1847) (1) With reference to plastics films and coated fabrics, their ability to hang without creases and to form graceful folds when used as draperies, shower curtains, and the like. (2) In sheet thermoforming, the ability of the preheated sheet to conform to the mold under the influence of gravity.

Drape-Assist Frame *n* In sheet thermoforming, a frame made from thin wires or thick bars shaped to the periphery of the depressed areas of the mold and suspended above the sheet to be formed. During forming, the assist frame drops down, drawing the softened sheet tightly into the mold and thereby preventing webbing between high areas of the mold and permitting closer spacing in arrays of multiple molds.

Drape Forming *n* (drape vacuum forming, drape thermoforming) Forming a thermoplastic sheet into three-dimensional articles by clamping the sheet in a movable frame, heating the sheet, then lowering it to drape over the high points of a male mold. Vacuum is then applied to complete the forming. See also ▶ Sheet Thermoforming.

Draw Back *n* A crossed end; an end broken during warping that when repaired was not free or was tied in with an adjacent end or ends overlapping the broken end. The end draws or pulls back when unwound on the slasher. Also see ▶ Sticker, (1).

Draw Crimping See ▶ Draw-Texturing.

Drawdown *n* (1) A film of ink deposited on paper by a smooth edged blade to evaluate the undertone and masstone of the ink. Syn: ▶ Pulldown. (2) A thin film of even thickness on base material cast by means of a drawdown bar. Syn: ▶ Casting. (3) In extrusion, the process of pulling the extrudate away from the die at a lineal speed greater than the average velocity of the melt in the die, thus reducing extrudate's cross-sectional dimensions. The term is also used by blow molders to denote the decrease in parison diameter and wall thickness due to gravity.

Draw Down *n* The amount by which manufactured filaments are stretched following extrusion. Also see ▶ Drawing, (2).

Drawdown Bar *n* Rectangular metal bar designed to deposit a specified thickness of wet coating film on test panels or other substrates.

Drawdown Ratio *n* In extrusion of fiber spinning, the ratio of the cross-sectional area of the die opening to that of the finished product. In making sheet or cast film, where the sheet width is nearly equal to the width of the die opening, the ratio of the thickness of the die opening to that of the final sheet is sometimes spoken of as the drawdown ratio.

Draw Frame Blends *n* Blends of fibers made at the draw frame by feeding in ends of appropriate card sliver. This method is used when blend uniformity is not a critical factor.

Drawinella *n* Cellulose triacetate., manufactured by Wacker, Germany.

Drawing *n* The process of stretching a thermoplastic filament, sheet, or rod to reduce its cross-sectional area and/or to improve its physical properties by ▶ Orientation.

Drawing *n* (1) The process of attenuating or increasing the length per unit weight of laps, slivers, slubbings, or rovings. (2) The hot or cold stretching of continuous filament yarn or tow to align and arrange the crystalline structure of the molecules to achieve improved tensile properties.

Drawing In *n* In weaving, the process of threading warp ends through the eyes of the heddles and the dents of the reed.

Drawn Tow *n* A zero-twist bundle of continuous filaments that has been stretched to achieve molecular orientation. (Tows for staple and spun yarn application are usually crimped.)

Drawout *n* A method of application in which the coating is applied, usually on paper, by spreading the pigment-vehicle mass evenly with a broad knife. Useful for examination of masstone, undertone, opacity, printing strength, and in some cases, fadeometer ratings of pigments.

Draw Process *n* Process for shaping a tubular or solid cylindrical specimen. The process involves passing the material through a die where it is then gripped by a suitable device which can pull it forward on a mechanical or hydraulic bench. A reduction in the diameter (solid specimens) and wall thickness (tubular specimens) results. Multi-die drawing is necessary in wire production.

Draw Ratio *n* (1) A measure of the degree of stretching during the orientation of a fiber or filament, expressed as the ratio of the cross-sectional area of the undrawn material to that of the drawn material. (2) In monofilament manufacture, the filament is wrapped several times around a vertical, driven roll, passed through a warming oven, then wrapped again around a second roll running faster than the first one. In this way the filament is stretched and oriented and its cross section is reduced. The ratio of the surface speed of the second, faster roll to that of the first equals the draw ratio.

Draw Resonance *n* A phenomenon occurring in film and filament extrusion in which the extrudate is drawn into a quenching bath at a certain critical speed that creates a cyclic pulsation in the cross-sectional area of the extrudate. The pulsation increases with rising drawing speed until the filament or film eventually breaks at the bath surface. Draw resonance has been observed while extruding polypropylene, polyethylene, and polystyrene.

Draw Sizing *n* A system linking drawwarping and sizing in a continuous process. A typical system includes the following elements: (1) creel, (2) eyelet board, (3) warp-draw machine, (4) intermingler, (5) tension compensator and break monitor, (6) sizing bath, (7) dryers, (8) waxing and winding units.

Draw-Texturing *n* In the manufacture of thermoplastic fibers, the simultaneous process of drawing to increase molecular orientation and imparting crimp to increase bulk.

Draw-Twisting *n* The operation of stretching continuous filament yarn to align and order the molecular and crystalline structure in which the yarn is taken up by means of a ring-and-traveler device that inserts a small amount of twist (usually ¼ to ½ turn per inch) into the drawn yarn.

Draw-Warping *n* A process in which a number of threadlines, usually 800 to 2,000 ends of POY feedstock, are oriented under essentially equal mechanical and thermal conditions by a stretching stage using variable speed rolls, then directly wound onto the beam. This process gives uniform end-to-end properties.

Draw-Winding *n* The operation of stretching continuous filament yarn to align or order molecular and crystalline structure. The drawn yarn is taken up on a parallel tub or cheese, resulting in a zero-twist yarn.

Dressed Lumber *n* Lumber having one or more of its faces planed smooth. *Also called Dressed Stuff and Surfaced Lumber.*

Dressing Compound *n* Bituminous liquid used hot or cold for dressing the exposed surface of roofing felt.

Dribbling or Drooling *n* A condition sometimes occurring in injection molding between shots in which melt drips from the withdrawn nozzle. Drooling was a common problem with 6/6 nylon in the 1950s because of depolymerization caused by moisture pickup in feed hoppers and overheating in the old, torpedo-type heating cylinders.

Drier *n* (1528) A composition which accelerates the drying of oil, paint, printing ink, or varnish. They consist mainly of metallic salts which exert a catalytic effect on the oxidation and polymerization of the oil vehicles employed and are available in both solid and liquid forms. *Also called Siccative.*

Drier Absorption See ▶ Drier Dissipation.

Drier Dissipation *n* A loss in catalytic power of a drier due to a physical absorption or a chemical reaction with certain pigments.

Driers Drop-Weight Test *n* (falling-weight test) Any test of impact resistance in which known weights are dropped once or repeatedly on the test specimen. Examples are ASTM D 4272 (plastic films), D 3029, D 4226, and D 4495 (rigid PVC sheet and parts), and F 736, Section 15, (polycarbonate sheet). Also see ▶ Free-Falling-Dart Test.

Drift See ▶ Overspray and ▶ Creep.

Drill *n* (1743) A strong denim-like material with a diagonal 2 × 1 weave running toward the left selvage. Drill is often called khaki when it is dyed that color.

Drop *n* One vertical descent of a scaffold.

Drop Black *n* Another name for bone black. See ▶ Bone Black.

Drop Ceiling *n* A form of decoration in which the ceiling paper is brought down to a suitable depth on the walls of a room and divided from the sidewall by a border or molding. Gives the illusion of a lower ceiling.

Drop Cloth *n* Cover used to protect floors and furniture, etc., from paint spillage and droppings.

Dropped Stitches *n* A defect in knit cloth characterized by recurrent cuts in one or more wales of a length of cloth.

Drop Siding *n* An exterior wall cladding of wooden boards (or strips of other material, such as aluminum or vinyl), which are tongued and grooved or rabbeted and overlapped so that the lower edge of each board interlocks with a groove in the board immediately below it. *Also known as Novelty Siding and Rustic Siding.*

Drop Stitch *n* (1) An open design made in knitting by removing some of the needles at set intervals. (2) A defect in knit fabric.

Drop Wires *n* A stop-motion device utilizing metal wires suspended from warp or creeled yarns. When a yarn breaks, the wire drops, activation the switch that stops the machine.

Drum Coloring See ▶ Dry Coloring.

Drum Extruder *n* A plasticating machine having a rotating cylindrical element inside a concentric or eccentric housing. Pellets or powder are fed into the gap between drum and housing at the top. Shear action melts the solids and the melt exists through a slot about 270° around from the feed point, usually passing into a pressure-developing device that can form an extruded product. A wiper bar above the die prevents melted material from recirculating.

Drum Tumbler *n* A device used to mix plastic pellets with color concentrates and/or regrind. The materials are charged into cylindrical drums that are tumbled end-over-end or rotated about an inclined axis for a time sufficient to thoroughly blend the ingredients.

Dry *vt* To change the physical state of an adhesive on an adherend by the loss of solvent constituents by evaporation or absorption, or both. A film is considered dry when it feels firm to the finger, using moderate pressure. See also ▶ Cure and ▶ Set.

Dry Back *n* The change in color or finish of an ink film as it dries.

Dry Blend *n* A dry, free-flowing mixture of resin powder, typically PVC, with plasticizers and other additives, prepared by blending the components in a large, closed mixing bowl with a high-speed rotor at its bottom, stopping the action at temperatures comfortably below the fluxing point. Dry blends are generally more economical feedstocks for extrusion than molding powders and pellets made by plasticating and extrusion, but in some cases have been difficult to process.

Dry-Blend *v* To combine ingredients, typically in a high-speed mixer, to produce a dry blend.

Dry Brush *v* Technique in which paint or ink is applied sparingly with a semidry brush.

Dry Bulk See ▶ Apparent Density.

Dry Cleaning *v* (1817) Removing dirt and stains from fabrics or garments by processing in organic solvents (chlorinated hydrocarbons or mineral spirits).

Dry Color *v* Term used loosely to describe dry colored pigments. A concentrated pigment or coloring matter suitable for grinding or dispersing and use in coatings.

Dry Coloring *v* The process of combining colorants to molding compounds and resin pellets by tumble-blending them with dyes, pigments, or color concentrates. This process enables custom molders and extruders to carry a large inventory of uncolored compound, preparing smaller batches of colored compounds to customers' specifications.

Dryer *n* (1528) A mechanical device designed to accelerate the drying of inks. Equipment used to remove moisture or water from plastic material or substrates during processing.

Dry Filling *n* The application of finishing chemicals to dry fabric, usually by padding. Dry forming: The production of fiber webs by methods that do not use water or other liquids, i.e., air-laying or carding.

Dry-Hard *n* Film is considered dry-hard when any mark left by the thumb, exerting maximum downward pressure on a film, is completely removed by a light polishing with a soft cloth.

Dry-Hiding *n* Increase in the hiding power of a paint which occurs in the drying process. It is most significant in nonglossy paints, which, after drying, have pigments or extender pigments protruding from the surface of the paint vehicle which scatter the incident light and add to the hiding.

Dry Hydrate *n* Calcium hydroxide; hydrated lime powder.

Dry Ice *n* (1925) Solidified carbon dioxide. See ▶ Carbon Dioxide.

Drying *n* (1) Process of change of coatings from the liquid to the solid state, due to evaporation of the solvent, physico-chemical reactions of the blinding medium, or a combination of these causes. (2) Process of removing moisture from pigments. It is carried out either in a current of hot air or in a vacuum.

Drying Cylinders *n* Any of a number of heated revolving cylinders for drying fabric or yarn. They are arranged either vertically or horizontally in sets, with the number varying according to the material to be dried. They are often internally heated with steam and Teflon coated to prevent sticking.

Drying of Ink *n* The conversion of an ink film to a solid state. This can be accomplished by any of the following means, either singly or in combination: oxidation, evaporation, polymerization, penetration, gelation, and precipitation.

Drying Oil *n* Any of several plant-derived polyunsaturated oils, such as linseed, oiticica, and tung, that, when exposed to air, form dry, tough, durable films. Linseed oil to be used in paints and varnishes today is usually boiled with cobalt or manganese salts of linoleic or naphthenic acids to shorten its "drying" time.

Drying Temperature See ▶ Temperature Drying.

Drying Time *n* (1) Time required for an applied film of a coating to reach the desired stage of cure, hardness, or nontackiness. (See ▶ Dry, ▶ Dry-Hard, ▶ Dust-Free Time, ▶ Set-to-Touch Time, ▶ Sand-Dry, ▶ Through-Dry, ▶ Dry-to-Handle Time, ▶ Touch-Dry, ▶ Tack-Free, ▶ Surface Drying, ▶ Dry-to-Recoat Time, and ▶ Dry-to-Sand.) (2) The time required for an ink to form a tack-free surface after being applied to the paper, or other printed surface.

Dry-Laid *n* Nonwovens Nonwoven web made from dry fiber. Usually refers to fabrics from carded webs versus air-laid nonwovens which are formed from random webs.

Dry Laminate *n* A laminate containing insufficient resin for complete bonding of the reinforcement.

Dry Layup *n* The construction of a laminate by layering preimpregnated, partly cured reinforcements in or on a mold, usually followed by bag molding or autoclave molding.

Dry Offset *n* An indirect letterpress process in which the ink is transferred from a relief plate to a blanket and then to the stock. *Also known as Letterset.*

Dry Purge *n* In extrusion, preparatory to shutting down operation after the feed hopper has been emptied, running the extruder until no more plastic emerges from the head end.

Dry Rot *n* (1795) Special type of brown rot, causing underspread damage in buildings. In the U.S., the causal organism is *Portia incrassate*.

Dry Spinning See ▶ Spinning.

Dry Spot *n* An imperfection in reinforced plastics, an area of incomplete surface film where the reinforcement has not been wetted with resin (ASTM D 883).

Dry Spray *n* Overspray or bounceback; sand finish due to spray particles being partially dried before reaching the surface.

Dry Strength *n* The strength of an adhesive joint determined immediately after drying or curing under specified conditions or after a period of conditioning in a standard laboratory atmosphere. See also ▶ Wet Strength and ▶ Strength, ▶ Dry.

Dry Strippable Paper See ▶ Strippable Coating.

Dry Tack See ▶ Tack, Dry.

Dry-Through *n* Film is considered dry-through when no loosening, detachment, wrinkling or other distortion of the film occurs when the thumb is borne downward while simultaneously turning the thumb through an angle of 90° in the plane of the film. The arm of the operator is kept in a vertical straight line from the wrist to the shoulder and maximum pressure is exerted by the arm.

Dry-to-Handle Time *n* Time interval between application and ability to pick up without damage. See ▶ Drying Time.

Dry-to-Recoat Time *n* Time interval between the application of the coating and its ability to receive the next coat satisfactorily.

Dry-to-Sand *n* That stage of drying when a coating can be sanded without the sandpaper sticking or clogging.

Dry-to-Touch Time *n* Interval between application and tack-free time. See ▶ Set-to-Touch Time.

Dry Wall *n* (1) An interior wall, constructed with a drywall finish material such as gypsum board or plywood. (2) In masonry construction, a self-supporting rubble or ashlar (squared building stone) wall built without mortar.

Dry Winding *n* Filament winding with preimpregnated roving, as distinguished from *wet winding* in which unimpregnated roving is pulled through a resin just prior to winding on a mandrel.

DSC Differential scanning calorimetry, measurement melting temperatures, glass transition temperature and heat of melting; lower temperature range than DTA. See ▶ Differential Scanning Calorimetry.

DSTDP *n* Abbreviation for Distearylthiodipropionate.

DTA *n* Differential thermal analysis measurement of melting temperature; generates a plot of "heat versus temperature" and semiquantiative, but higher temperature range than DSC. DTA is capable of measuring

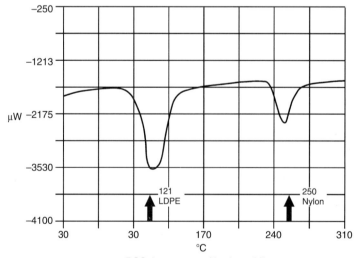

DSC thermogram of laminated film

events (e.g., melting, cyrstallization and decomposition) but not heat of melting or crystallization whereas DSC is quantitative and appropriate for thermally characterizing polymers. Abbreviation for ▶ Differential Thermal Analysis.

DTDP *n* Abbreviation for ▶ Ditridecyl Phthalate.

Dual-Sensor Control *n* An improved system for controlling cylinder and plastic temperatures in extruders. For each heating zone, there are two temperature sensors, one in a shallow well slightly beneath the heater, the other deep, just outside the lining layer and near the plastic. An average of the two signals is used to control the electrical heat input (Eurotherm/Welex).

Dubbing Out *n* (1) Filling in hollow and irregular surfaces and leveling walls with plaster before regular plasterwork. (2) Forming, very roughly, a plaster cornice, before the final plaster coat is applied.

Duck \ˈdək\ *n* [Du *doek* cloth; akin to OHGr *tuoh* cloth] (1640) A compact, firm, heavy, plain weave fabric with a weigh of 6–50 ounces per square yard. Plied yarn duck has plied yarn in both warp and filling. Flat duck has a warp of two single yarns woven as one and a filling of either single or plied yarn.

Duck Eye See ▶ Spinning.

Ductile Fracture \ˈdək-tᵊl\ *adj* [MF & L, MF, fr. *ductilis*, from *ducere*] (14c) (ductile rupture) The breaking or tearing, most commonly in tension, of a test specimen or part after considerable unrecoverable stretching (plastic strain) has occurred. Since the mode of fracture depends on conditions as well as material, the distinction between ductile and brittle fracture, which latter occurs after relatively little, recoverable strain, is not always clear. Low temperatures, especially below the glass transition (T_g), and high rates of strain favor brittle behavior, while the opposites favor ductile behavior.

Ductility *n* The amount of plastic strain that a material can undergo before rupture. A ductile material generally shows a ▶ Yield Point.

Ductor Roller *n* The roller which is in intermittent contact with the fountain roller and transfers ink to the distribution system of the press. On a lithographic press it is also the roller which transfers the fountain solution to the dampening rollers.

Dull \ˈdəl\ *adj* [ME *dul*; akin to OE *dol* follish, OIr *dall* blind] (13c) A term applied to manufactured fibers that have been chemically or physically modified to reduce their normal luster. Matte; opposite of bright; low in luster.

Dullness *n* (1) Lack of luster or gloss. (2) Colors of low chroma or saturation.

Dull Rubbing *n* Rubbing a dried film of a coating to dull finish, with an abrasive paper, pumice, steel wool and oil or water.

Dulmadge Mixing Section *n* Invented by F. E. Dulmadge, this mixing section, usually located near the end of an extruder or injection screw, is really three short sections separated by gaps with no flights. In each section, the regular screw flight is replaced by about 20 narrow, closely spaced flight segments with lead angles of about 70°. The melt approaching the first section is subdivided into 20 substreams that, on exiting, are swirled circumferentially before entering the second section, and again, on leaving that one, before entering the third. The main goal was to improve melt-temperature uniformity.

Dulong and Petit, Law of *n* The specific heats of the several elements are inversely proportional to their atomic weights. The atomic heats of solid elements are constant and approximately equal to 6.3. Certain elements of low atomic weight and high melting point have, however, much lower atomic heats at ordinary temperatures.

Dumas Nitrogen *n* A long used method for determining the percentage of nitrogen in organic materials, invented by A. Dumas in 1830. The sample was first oxidized with red-hot copper oxide, followed by reduction of $NO\chi$ to N_2 over hot copper. Following absorption of CO_2 in KOH solution and condensation and absorption of most of the water, the volume of N_2 was measured. The method has largely been supplanted by automated analyzers that directly determine C, H, and N spectroscopically and find O by difference.

Dumbbell *n* A piece of rubber cut in the shape of a dumbbell used for physical testing. See ▶ Dogbone.

Dumbells *n* A defect frequently seen in wet-formed non-woven fabrics; an unusually long fiber will become entangled with groups of regular-length fibers at each end, thus producing a dumbbell-shaped clump.

Dumont's Blue *np* See ▶ Cobalt Blue.

Dungaree \ˌdeŋ-gə-ˈrē\ *n* [Hindi *dūgrī*] (1673) A term describing a coarse denim-type fabric, usually dyed blue, that is used for work overalls.

Duplex Paper *n* Wallpaper which consists of two separate papers pasted together used to create a highly embossed effect.

Duplex Printing See ▶ Printing.

Duplicate Cavity Plate *n* A removable plate that retains cavities, used where two-plate operation is necessary for loading inserts, etc.

Durability *n* Degree to which paints and paint materials withstand the destructive effect of the conditions to which they are subjected.

Durable Press *n* A term describing a garment that has been treated so that it retains its smooth appearance, shape, and creases or pleats in laundering. In such garments no ironing is required, particularly if the garment is tumble-dried. Durable press finishing is accomplished by several methods; two of the most common are the following: (1) A fabric that contains a thermoplastic fiber and cotton or rayon may be treated with a special resin that, when cured, imparts the permanent shape to the cotton or rayon component of the fabric. The resin-treated fabric may be precured (cured in finishing and subsequently pressed in garment form at a higher temperature to achieve the permanent shape) or postcured (not cured until the finished garment has been sewn and pressed into shape). In both cases, the thermoplastic fiber in the garment is set in the final heat treatment. This fiber, when heat-set, also contributes to the permanence of the garment shape, but the thermoplastic component of the blend is needed for strength since the cotton or rayon component is somewhat degraded by the durable-press treatment. (2) Garments of a fabric containing a sufficient amount of a thermoplastic fiber, such as polyester, nylon, or acrylic, may be pressed with sufficient pressure and time to achieve a permanent garment shape. Also see ▶ Ease-of-Care, ▶ Permanent Finish, and ▶ Wash-and-Wear.

Duranit *n* Butadiene/styrene copolymer. Manufactured by Hüls, Germany.

Durene *n* (durol, 1,2,4,5-tetramethylbenzene) $C_6H_2(CH_3)_4$. A substance occurring in coal tar, but usually prepared from xylene and methyl chloride in the presence of $AlCl_3$. It has been patented (U.S. 4,000,120) as an additive to make packaging films of polyolefins and polystyrene photodegradable under direct action of sunlight.

Durethan *n* Polyamides or polyurethanes. Manufactured by Bayer, Germany.

Durette *n* Fiber of iosphthalic acid and *m*-phenylene diamine. Manufactured by Monsanto, U.S.

Durometer \dú-ˈrä-mə-tər *also* dyu\ *n* [L *durus* hard] (ca. 1879) An instrument used for measuring the hardness of rubber, plastic, or protective coatings. See ▶ ASTM D 2240, also ▶ Indentation Hardness.

Durometer Hardness See ▶ Indentation Hardness.

Dust \ˈdəst\ *n* [ME, fr. OE *dūst*; akin to OHGr *tunst* storm, and prob. to L *fumus* smoke] (before 12c) One of the size gradings of various gums, being composed of very small pieces, almost as fine as dust.

Dust Free *n* (1) Descriptive of the stage in the drying of a paint or varnish film at which dust will no longer stick to the surface. (2) That stage of drying when cotton fibers, that have been dropped onto the film from a height of one inch, can be removed by blowing lightly across the surface of the film. Also referred to as ▶ Cotton-free or ▶ Cotton-free Dry.

Dust-Free Time *n* The time required for a freshly applied paint or compound to form a skin on its surface so that dust will not adhere to it.

Dusting Bronzes See ▶ Bronzing.

Dust-Resistant *n* A term applied to a fabric that has been tightly woven so that it resists dust penetration.

Dutch Metal *n* Leaves of bright brass which are used for overlaying in the same manner in which gold leaf is applied.

Dutch Pink *n* Tint of red. The name is also used for certain yellow lakes, prepared from quercitron, Persian berries, or similar natural yellow coloring matters.

Dutch Process *n* Old name for stack process for manufacture of white lead.

Dutral *n* Ethylene/propylene copolymer, manufactured by Montecatini, Italy.

DVB *n* Abbreviation for ▶ Divinylbenzene.

Dwell \ˈdwel\ *vt* [ME, fr. OE *dwellan* to go astray, hinder, akin to OHGr *twellen* to tarry] (13c) (1) A pause in the application of pressure to a mold, made just before the mold has completely closed, to allow the escape of gas from the molding material. (2) In filament winding, the time that the traverse mechanism is stationary while the mandrel continues to rotate to the appropriate point for a new traverse to begin. (3) In heat sealing, dwell time is the time during which pressure and heat (or microwave energy) are applied to the area to be sealed.

Dwell Time *n* The time during a process in which a particular substance remains in one location (e.g., the time during which molten polymer remains in a spinning pack.)

Dye \ˈdī\ *n* [ME *dehe*, fr. OE *dēah, dēag*] (before 12c) An intensely colored substance that imparts color to a substrate to which it is applied (Merriam-Webster's Collegiate Dictionary (2004), 11th edn. Merriam-Webster, Springfield). Retention of the dye in the substrate may be by means of adsorption, solution, mechanical bonding, or by ionic or covalent chemical bonding. The dye substance is devoid of crystal structure. Dyes used for coloring plastics usually dissolve in the plastic melt, unlike ▶ Pigments which remain dispersed as undissolved particles. (Complete Textile Glossary (2000) Celanese Acetate LLC. Three Park Avenue,

New York; Vincenti R (ed) (1994) Elsevier's textile dictionary. Elsevier Science and Technology Books, New York)

Dye, Acid See ▶ Acid Dyes.

Dye, Basic See ▶ Basic Dye.

Dye Fleck *n* (1) An imperfection in fabric caused by residual undissolved dye. (2) A defect caused by small sections of undrawn thermoplastic yarn that dye deeper that the drawn yarn.

Dyeing *n* To add soluble colorants that either form a chemical bond with the substrate or become closely associated with it by a physical process in order to change the color. Below is a listing of different types of fabric dyeing and a definition of each. (Vincenti R (ed) (1994) Elsevier's textile dictionary. Elsevier Science and Technology Books, New York)

Dyeing, Chain *n* A method of dyeing yarns and fabrics of low tensile strength of tying them end-to-end and running them through the dyebath in a continuous process.

Dyeing, Cross *n* A method of dyeing blend or combination fabrics to two or more shades by the use of dyes with different affinities for the different fibers.

Dyeing, High-Temperature *n* A dyeing operation in which the aqueous dyebaths are maintained at temperatures greater than 100°C by use of pressurized equipment. Used for many manufactured fibers.

Dyeing, Ingrain *adj* (1766) – Term used to describe yarn or stock that is dyed in two or more shades prior to knitting or weaving to create blended color effects in fabrics.

Dyeing, Jet *n* High temperature piece dyeing in which the dye liquor is circulated via a Venturi jet thus providing the driving force to move the loop of fabric.

Dyeing, Mass-Colored *n* A term to describe a manufactured fiber (yarn, staple, or tow) that has been colored by the introduction of pigments or insoluble dyes into the polymer melt or spinning solution prior to extrusion. Usually, the colors are fast to most destructive agents.

Dyeing, Muff A *n* form of yarn dyeing in which the cone has been removed.

Package Dyeing See ▶ Dyeing, ▶ Yarn Dyeing.

Dyeing, Pad *n* A form of dyeing whereby a dye solution is applied by means of a padder or mangle.

Dyeing, Piece *n* The dyeing of fabrics "in the piece," i.e., in fabric form after weaving or knitting as opposed to dyeing in the form of yarn or stock.

Dyeing, Pressure *n* Dyeing by means of forced circulation of dye through packages of fiber, yarn, or fabric under superatmospheric pressure.

Dyeing, Reserve *n* (1) A method of dyeing in which one component of a blend or combination fabric is left undyed. The objective is accomplished by the use of dyes that have affinity for the fiber to be colored but not for the fiber to be reserved. (2) A method of treating yarn or fabric so that in the subsequent dyeing operation the treated portion will not be dyed.

Dyeing, Skein *n* The dyeing of yarn in the form of skeins, or hanks.

Dyeing, Solution *n* See ▶ Dyeing, Mass-Colored.

Dyeing, Solvent *n* A dyeing method based on solubility of a dye in some liquid other than water, although water may be present in the dyebath.

Dyeing, Space *n* A yarn-dyeing process in which each strand is dyed with more that one color at irregular intervals. Space dyeing produces an effect of unorganized design in subsequent fabric form. The two primary methods are knit-de-knit and warp printing.

Dyed, Space *n* See ▶ Dyeing, ▶ Mass-Colored.

Dyeing, Stock *n* The dyeing of fibers in staple form.

Dyeing, Thermal Fixation *n* A *process* for dyeing polyester whereby the color is diffused into the fiber by means of dry heat.

Dyeing, Union *n* A method of dyeing a fabric containing two or more fibers or yarns to the same shade so as to achieve the appearance of a solid colored fabric.

Dyeing, Yarn *n* The dyeing of yarn before the fabric is woven or knit. Yarn can be dyed in the form of skeins, muff, packages, cheeses, cakes, chain-wraps, and beams.

Dyes, Basic *n* A class of positive-ion-carrying dyes known for their brilliant hues. Basic dyes are composed of large-molecule, water-soluble salts that have a direct affinity for wool and silk and can be applied to cotton with a mordant. The fastness of basic dyes on these fibers is very poor. Basic dyes are also used on basic-dyeable acrylics, modacrylics, nylons, and polyesters, on which they exhibit reasonably good fastness.

Dyes, Cationic See ▶ Dyes, Basic.

Dyes, Fiber-Reactive Dyes *n* A type of water-soluble anionic dye having affinity for cellulose fibers. In the presence of alkali, they react with hydroxyl groups in the cellulose and thus are liked with the fiber. Fiber-reactive dyes are relatively new dyes and are used extensively on cellulosics when bright shades are desired.

Dyeing, Gel *n* Passing a wet-spun fiber that is in the gel state (not yet at full crystallinity or orientation) through a dyebath containing dye with affinity for the fiber. This process provides good accessibility of the dye sites.

Dyes, Macromolecular *n* A group of inherently colored polymers. They are useful both as polymers and as dyes with high color yield. The chromophores fit the recognized CI classes, i.e., azo, anthraquinone, etc., although not all CI classes are represented. Used for mass dyeing, hair dyes, writing inks, etc. (Tortora PG (ed) (1997) Fairchild's dictionary of textiles. Fairchild Books, New York)

Dyes, Metallized *n* A class of dyes that have metals in their molecular structure. They are applied from an acid bath.

Dyes, Naphthol *n* A type of azo compound formed on the fiber by first treating the fiber with a phenolic compound. The fiber is then immersed in a second solution containing a diazonium salt that reacts with the phenolic compound to produce a colored azo compound. Since the phenolic compound is dissolved in caustic solution, these dyes are mainly used for cellulose fiber, although other fibers can be dyed by modifying the process. (Also see ▶ Dyes, ▶ Developed Dyes.)

Dyes, Premetallized *n* Acid dyes that are treated with coordinating metals such as chromium. This type of dye has much better wetfastness than regular acid dye. Premetallized dyes are used on nylon, silk, and wool.

Dyes, Sulfur *n* A class of water-insoluble dyes that are applied in a soluble, reduced form from a sodium sulfide solution and are then reoxidized to the insoluble form on the fiber. Sulfur dyes are mainly used on cotton for economical dark shades of moderate to good fastness to washing and light. They generally give very poor fastness to chlorine.

Dyes, Vat *n* A class of water-insoluble dyes which are applied to the fiber in a reduced, soluble form (leuco compound) and then reoxidized to the original insoluble form. Vat dyes are among the most resistant dyes to both washing and sunlight. They are widely used on cotton, linen rayon, and other cellulosic fibers. (Tortora PG (ed) (1997) Fairchild's dictionary of textiles. Fairchild Books, New York)

Dyeing Auxiliaries *n* Various substances that can be added to the dyebath to aid dyeing. They may necessary to transfer the dye from the bath to the fiber or they may provide improvements in leveling, penetration, etc. Also call dyeing assistants.

Dyeing Plastics *n* Process of adding color to plastics to make them attractive in appearance. Color is produced by introduction of either dyes or pigments, which produce color by selectively absorbing, transmitting, reflecting, and scattering specific areas of light energy from wave bands that constitute white light.

Dye Migration See ▶ Migration, (1).

Dye, Mordant *n* Term given to a range of dyestuffs, the colors of which are developed when precipitated onto bases to form lakes. Often a salt-forming reaction takes place between the dyestuff and the base.

Dye Pigments *n* Dyes that by nature are insoluble in water and can be used directly as pigments without any chemical transformation.

Dye Range *n* A broad term referring to the collection of dye and chemical baths, drying equipment, etc., in a continuous-dyeing line.

Dyes *n* Substances that add color to textiles. They are incorporated into the fiber by chemical reaction, absorption, or dispersion. Dyes differ in their resistance to sunlight, perspiration, washing, gas, alkalies, and other agents; their affinity for different fibers; their reaction to cleaning agents and methods; and their solubility and method of application. Various classes and types are listed below. Also see ▶ Colour Index (CI).

Dyes and Dyestuffs *n* Organic coloring substances. Pigment dyestuffs are virtually insoluble in water and in the usual paint and varnish vehicles. Some examples of pigment dyestuffs are toluidine red, phthalocyanine blue, and arylamide yellow. Some dyes are soluble in water and organic solvents and are classed according to their solubilities, for example, water-soluble dyes, spirit-soluble dyes, and oil-soluble dyes. (Harper CA (2000) Modern plastics encyclopedia. McGraw Hill Professional, New York)

Dye Sites *n* Functional groups within a fiber that provide sites for chemical bonding with the dye molecule. Dye sites may be either in the polymer chain or in chemical additives included in the fiber. (Kadolph SJJ, Langford AL (2001) Textiles. Pearson Education, New York)

Dye, Spirit-Soluble See ▶ Spirit-Soluble Dye.

Dyestuff *n* Colorant which can be fixed firmly to a material to be dyed so as to be more or less fast to light, water, soap, etc.

Dynamic Adhesion *n* The ability of a cord-to-rubber bond to resist degradation resulting from flexure.

Dynamic Fatigue *n* Usually the same as *fatigue*. The "dynamic" modifier is sometimes used by persons who think of creep and creep failure as "static fatigue." See ▶ Alternating Stress Amplitude. See entries at ▶ Fatigue.

Dynamic Mechanical Analyzer *n* An instrument that can test in an oscillating-flexural mode over a range of temperature and frequency to provide estimates of the "real," i.e., in-phase, and "imaginary," i.e., out-of-phase parts of the complex modulus. The real part is the elastic

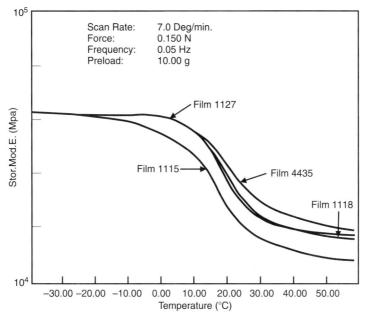

DMA thermograms of poly (styrene-co-butadient) copolymer films of different composition

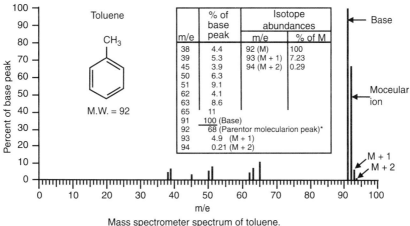

Mass spectrometer spectrum of toluene.
(Courtesy of John Wiley & Sons)

component, the imaginary part is the loss component. The square root of the sum of their squares is the complex modulus. With polymers, the components and the modulus are usually dependent on both temperature and frequency. ASTM D 4065 spells out the standard practice for reporting dynamic mechanical properties of plastics. An example of a DMA thermogram of different PerkinElmer, Inc., manufactures the Diamond DMA instrument. polymer films is shown (Sepe MP (1998) Dynamic mechanical analysis. Plastics Design Library, Norwich, New York). See also
▶ Mechanical Spectrometer.

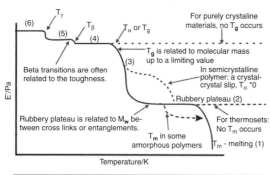

| (1) chain slippage | (3) gradual main chain | (5) bend and stretch |
| (2) large scale chain | (4) side groups | (6) local motions |

Dynamic Mechanical Properties *n* (1) The stress-strain properties of a material when subjected to an applied sinusoidally varying stress or strain. For a perfectly elastic material the strain response is immediate and the stress and strain are in phase. For a viscous fluid, stress and strain are 90° out of phase. (2) The mechanical properties of composites as deformed under periodic forces such as dynamic modulus, loss modulus and mechanical damping or internal friction. (Sepe MP (1998) Dynamic mechanical analysis. Plastics Design Library, Norwich, New York)

Dynamic Mechanical Spectrum *n* The information obtained from testing with a ▶ Mechanical Spectrometer. A plot or tabulation of complex modulus or its components vs frequency of oscillation or temperature or both. The mode of stress may be tensile/compressive, flexural, or torsional (shear). Because both abscissa (frequency) and ordinates can range widely, bilogarithmic plots are usual. (Sepe MP (1998) Dynamic mechanical analysis. Plastics Design Library, Norwich, New York)

Dynamic Stress *n* A stress whose magnitude and/or direction vary with time, typically cyclically and sinusoidally.

Dynamic Stress Relaxometer *n* An instrument that measures the relaxation response of an elastomeric material to a prescribed shear deformation over a range of temperature. Basic elements of the instrument are a cone-shaped stator cavity and a conical rotor, both electrically heated. The sample is placed in the stator, which rises to a position to form a constant specimen thickness, forcing out the excess material. After heating to the desired temperature, the rotor is rotated quickly through a small angle, to a known shear deformation. The subsequent drop-off torque, which results from the relaxation of stress within the sample, is recorded over the time it takes for it to decay.

Dynamic Valve *n* A device sometimes incorporated in an extruder head to control flow by adjusting the clearance between the conical elements, one stationary, the other rotating with the screw. Dynamic valves of various designs have also been used between the stages of two-stage, vented extruders as an aid to balancing the flow rates in the stages and preventing extrusion out the vent.

Dynamic Viscosity *n* (1) ▶ Absolute Viscosity as distinguished from ▶ Kinematic Viscosity. See also ▶ Viscosity. (2) In sinusoidally varying shear, the part of the stress in phase with the rate of strain, divided by the strain rate. (Shah V (1998) Handbook of plastics testing technology. Wiley, New York)

Dynapoint Process *n* A continuous computer-controlled process for manufacturing tufted carpets with intricate patterns from undyed yarn. The carpet is dyed as it is tufted and the colors and pattern are clearly visible through the primary backing of the carpet.

Dyne *n* In the (now deprecated) cgs system of units, the force required to accelerate a mass of one gram by one centimeter per second-squared. The dyne = 1×10^{-5} N. (Lide DR (ed) (2004) CRC handbook of chemistry and physics. CRC Press, Boca Raton)

Dynel *n* Vinyl chloride/acrylonitrile copolymer. Manufactured by Union Carbide, U.S.

Dypnone *n* (phenyl α-methyl styryl ketone, 1,3-diphenyl-2-butene- 1- one) $C_6H_5COCHC(CH_3)C_6H_5$. A plasticizer and ultraviolet absorber.

E

e \ē\ *n* {often capitalized, often attributive} (before 12c) (1) The base of natural logarithms, 2.71828.... (2) The charge on an electron, 1.60199×10^{-18} coulomb.

ΔE, Δe *n* The total color difference computed with a color difference equation. It is generally calculated as the square root of the sum of the squares of the chromaticity difference, ΔC, and the lightness difference, ΔL: $\Delta E = [(\Delta C)^2 + (\Delta L)^2]^{1/2}$.

E *n* (1) SI abbreviation for prefix EXA. (2) Symbol commonly used for modulus of elasticity in tension (see ▶ Modulus of Elasticity) (Sepe MP (1998) Dynamic mechanical analysis. Plastics Design Library, Norwich, New York); for activation energy in the ▶ Arrhenius Equation and for electric potential (Connors KA (1990) Chemical kinetics. Wiley, New York).

EA *n* Segmented polyurethane fiber.

EAA *n* Abbreviation for ▶ Ethylene–Acrylic Acid Copolymer.

Earth Pigment *n* Class of pigments which are usually mined directly from the earth, dried, generally Calcined and ground. Typical examples are red and yellow oxides of iron, yellow ochre, raw and burnt siennas, and raw and burnt umbers.

Ease-of-Care *n* A term used to characterize fabrics that, after laundering, can be restored to their original appearance with a minimum of ironing or other treatment. An ease-of-care fabric generally wrinkles only slightly upon laundering. Also see Durable Press and Wash and Wear (Vincenti R (ed) (1994) Elsevier's textile dictionary. Elsevier Science and Technology Books, New York).

Eave \ēvz\ *n* [ME *eves* (singular), fr. OE *efes*; akin to OHGr *obasa* portico, OE *ūp* up] (before 12c) Lower part of a roof projecting beyond the wall underneath. See ▶ Soffit.

Ebonite \e-be-nit\ *n* (1861) A hard material made by sulfur vulcanization of rubber.

Ebonize \-nīz\ *vt* (ca. 1828) To blacken with paint or stain to look like ebony.

Ebulliometry *n* Method of measuring molecular weight of polymers under 20,000; based on vapor pressure lowering and boiling point elevation (Pethrick RA and Dawkins JV (1999) Modern techniques for polymer characterization. Wiley, New York).

EC *n* Abbreviation for ▶ Ethyl Cellulose.
EC *n* Ethyl cellulose.
Ecology \i-kä-lə-jē-, e-\ *n* (Gr *Ökologie,* fr. *öko-* eco- + -logie –logy (1873) The interrelationships of living things to one another and to their environment.
Ecosystem \sis-təm\ *n* (1935) The interaction system of a biological community and its nonliving environment.
ECTFE See ▶ Poly(Ethylene–Chlorotrifluoroethylene).
EDC *n* Abbreviation for ▶ Ethylene Dichloride.
Eddy Current \e-dē kər-ənt\ *n* (1887) (Foucault current) The current induced in a mass of conducting material by a varying magnetic field. *Also called Focault Current.*
Edge Bead *n* In some cast-film and sheet-extrusion operations, the narrow border at the edge of the sheet, usually somewhat thicker, that must be trimmed off prior to winding or stacking the product.
Edge Crimping See ▶ Texturing.
Edge Dislocation *n* A dislocation in which a layer of particles in the crystal is incomplete.
Edge Grain *n* Wood or veneer so sawed that the annual rings form an angle of 45–90° with the surface of the piece. *Also called Vertical Grain.*
Edge Roll *n* The curl that develops on the edge of a single-knit fabric preventing it from lying flat.
Edge Runner Mill *n* Mill consisting of a horizontal, or substantially horizontal, circular pan which may be of cast iron or stone-lined. One or more circular rollers, which may also be of cast iron or suitable stone, are made to rotate edgewise around the pan. The material to be ground, which may be dry pigment or other powder or pigment paste, is disintegrated between the surfaces of the rollers and the pan. Sometimes referred to as ▶ Putty Chaser.
Edgewise *adv* (1677) Plane of the laminate perpendicular to the laminations.
Edistir *n* Poly(styrene). Manufactured by Montedison, Italy.
EDM *n* Abbreviation for ▶ Electrical-Discharge Machining.
EEA *n* Copolymer from ethylene and ethyl acrylate. Abbreviation for ▶ Ethylene–Ethyl Acrylate Copolymer.
Effective Modulus *n* Syn: ▶ Creep Modulus.
Effervescence \e-fər-ves\ *vt* [L *effervescere,* fr. *ex-* + *fervescere* to begin to boil, inchoative of *fervēre* to boil] (1784) The vigorous evolution of gas which accompanies some chemical reactions, e.g., treatment of a carbonate with certain mineral acids.
Efficiency \i-fi-shən-sē\ *n* (1633) (1) The efficiency of a machine or process, the ratio of the energy delivered to

that supplied. (2) The effectiveness of operation or production, particularly as related to some standard for same. (3) The reciprocal of the unit cost of items for sale (*cost efficiency, economy*). (Merriam-Webster's Collegiate Dictionary (2004), 11th edn. Merriam-Webster, Springfield, MA.).

Efficiency of Reinforcement *n* (fiber-efficiency factor) The percentage of fiber in a reinforced-plastic structure or part contributing to the property of concern. For example, in a unidirectionally reinforced bar, the theoretical efficiency for Young's modulus and fiber-direction tensile strength is 100%. For a sheet molded from chopped-strand mat with all fibers randomly oriented in the sheet plane, the efficiency for in-plane properties is 37%. With chopped strands randomly oriented in three dimensions, efficiency falls to 20%.

Efflorescence \-ˈre-sᵊn(t)s\ *n* (1626) An encrustation of soluble salts, commonly white, deposited on the surface of coatings, stone, brick, plaster, or mortar; usually caused by salts or free alkalies leached from mortar or adjacent concrete as moisture moves through it, also known as *laitance* (Weismantal GF (1981) Paint handbook. McGraw-Hill, New York) {G Ausblühen n, F efflorescence f, S eflorescencia f, I efflorescenza f}.

Effluent Limitations *n* Any restrictions, established by the government or by management, on quantities, rates and concentrations of chemical, physical, biological, and other constituents which are discharged from paint sources.

Efflux \ˈe-ˌfləks\ *n* [ML *effluxus*, fr. *effluere*] (1647) The process of flowing out, in any given system.

Efflux Viscometer *n* A cup type viscometer containing an orifice. Whatever precision and accuracy an efflux instrument may have is essentially dependent upon the dimension of the orifice. The closer an orifice resembles a capillary, the more accurate is the instrument. See ▶ Viscometer, ▶ Ford Cup.

Effusion \i-ˈfyü-zhən\ *n* (15c) The passage of a substance through a small orifice.

Eggshell \ˈeg-ˌshel\ *adj* (1835) (1) Gloss lying between semigloss and flat. Generally thought to be between 20 and 35 as determined with a 60° glossmeter. (2) An off white color.

Eggshell Finish *n* (1) Low sheen surface which exhibits surface reflectance (gloss) similar to that from an eggshell. Thus it is a gloss lying between flat and semi-gloss and is generally thought to be between 20 and 35, as determined with a 60° gloss meter. (2) Paint which exhibits eggshell gloss.

E Glass *n* A low-alkali ($Na_2O + K_2O = 0.6\%$) borosilicate glass, like Pyrex®, with good electrical properties, the most widely used glass-fiber reinforcement for plastics. Major constituents are: SiO_2, 54%; CaO, 17%; Al_2O_3, 15%; B_2O_3, 8%; MgO, 4.7%. The fibers are vulnerable to surface abrasion so are always sized before stranding. Average fiber properties are: density = 2.54 g/cm³; tensile modulus (axial) = 72 GPa; tensile strength = 3.5 GPa.

Egyptian Asphaltum \i-ˈjip-shən ˈas-ˈfól-təm\ *n* See ▶ Syrian Asphaltum.

Egyptian Blue *n* Blue pigment, of ancient origin, which is substantially a copper silicate, obtained by calcinations.

EHMWPE *n* Abbreviation for Extra-High-Molecular-Weight Polyethylene, any of a subfamily of linear PE resins having molecular weights in the range 250,000–1,500,000. See also ▶ Polyethylene.

Eicosanic Acid See ▶ Cosanic Acids. IUPAC Name: nonadecanoic acid.

Eilers Equation *n* A modification of the Einstein Equation relating the viscosity η_f of a Newtonian liquid filled with spherical particles to the viscosity of the pure liquid η_o and extending to concentrations above 10% of filler. It is:

$$\frac{\eta_n}{\eta_o} = \left[\frac{1 + 1.25 \cdot f}{1 - S \cdot f}\right]^2$$

where f is the volume fraction of spheres and S is an empirical coefficient usually between 1.2 and 1.3 (Sudduth RD (2003) J Appl Sci 48(1):25–36). See also ▶ Mooney Equation.

Einstein Equation *n* An equation relating the viscosity η_f of a sphere-filled, Newtonian liquid to that of the unfilled liquid η_o, for volume fractions f of spheres up to about 10%. It is:

$$\eta_{rel} = \eta_f/\eta_o = 1 + k_{E \cdot f}$$

where k_E is the Einstein coefficient = 2.5 for spheres. The Einstein equation has been extended to higher concentrations by adding terms in f^2, f^3, etc., and by correcting for the limiting volume fraction f_m that can be filled by uniform spheres. A versatile model of this type is

$$\eta_{rel} = \frac{1 + af + bf^2 + cf^3}{1 - (f/f_m)}$$

where a, b, and c are empirical constants and $f_m \approx 0.7$. The Mooney Equation is a form of this model for which b and $c = 0$ (Kamide K, Dobashi T (2000) Physical chemistry of polymer solutions. Elsevier, New York).

Einstein Theory for Mass–Energy Equivalence *n* The equivalence of a quantity of mass m and a quantity of energy E by the formula $E = mc^2$. The conversion factor c^2 is the square of the velocity of light (Serway RA, Faugh JS, Bennett CV (2005) College physics. Thomas, New York).

Ejection Ram *n* A small supplementary hydraulic ram fitted to a molding press to operate piece-ejection device.

Ejector Pin *n* (ejector sleeve, knockout pin, KO pin) A rod, pin, or sleeve that pushes a molding off a force or out of a cavity of a mold. Attached to an ejector bar or plate, it is actuated by the ejector rod(s) of the press or by auxiliary hydraulic or compressed-air cylinders.

Ejector Plate *n* A plate that backs up the ejector pins and holds the ejector assembly together.

Ejector-Return Pin *n* (return pin, surface pin, safety pin, position pushback) A projection, usually one of several that push back the ejector assembly as the mold closes.

Ejector Rod *n* A bar that actuates the ejector assembly when a mold is opened.

Elaidamide *n* $CH_3(CH_2)_7CH=CH(CH_2)_7CONH_2$. The amide of *trans*-9-octadecenoic acid, a steroisomer of ▶ Oleamide used in fractional percentages as a slip agent for polyethylene to be made into film.

Elaidic Acid *n* $CH_3(CH_2)_7CH=CH(CH_2)_7COOH$ (Octadec-9-enoic acid, *trans*-9-octadeconic acid). Properties: mp, 51°C; bp, 225°C/10 mm Hg; iodine value, 90.1.

Elastic \i-ˈlas-tik\ *adj* [NL *elasticus*, fr. LGk *elastos* ductile, beaten, fr. Gk *elaunein* to drive beat out; prob. akin to Gk *ēlythe* he went, OI *luid*] (1674) Ability of a substance to return to its approximate original shape or volume after a distorting force on the substance has been removed (Billmeyer FW, Jr (1984) Textbook of polymer science, 3rd edn. Wiley-Interscience, New York). See also ▶ Elasticity.

Elastic Compliance *n* The inverse of the Young's modulus.

Elastic Constant *n* Any of the several constants of a constitutive relationship between stress (of any mode) and strain in a material. For an isotropic material stressed in its elastic range, there are (at any temperature) four interrelated constants: tensile modulus, E, shear modulus, G, bulk modulus, B, and Poisson's ratio, μ. Two expressions of the relations are:

$$G = E/2(1+\mu) \text{ and } B = E/3(1-2\mu).$$

More constants are needed to define the behavior of nonisotropic materials (Shah V (1998) Handbook of plastics testing technology. Wiley, New York).

Elastic Deformation *n* A change in dimensions of an object under load that is fully recovered when the load is released. That part of the total strain in a stressed body that disappears upon removal of the stress. See also ▶ Plastic Deformation.

Elastic Design *n* Engineering design for load-bearing members based on the assumption that stress and strain are proportional and will be kept well within the elastic range, with working stresses set at half or less of the yield stress. Elastic design based on short-time test measurements may be applicable to the design of plastic products that will be loaded intermittently and for short periods. However, the universal phenomenon of creep in plastics means that, under sustained loads, *visco*elastic rather an elastic behavior is the norm. Even so, elastic-design methods can often be used by employing ▶ Creep Modulus and ▶ Creep Strength in place of short-time parameters (Shah V (1998) Handbook of plastics testing technology. Wiley, New York).

Elasticity *n* (1664) A property that defines the extent to which a material resist small deformations from which a material recovers completely when deforming force is removed. When the deformation is proportional to the applied load, the material is said to exhibit ▶ Hookean Elasticity. Elasticity equals stress divided by strain (Shah V (1998) Handbook of plastics testing technology. Wiley, New York; Elias HG (1977) Macromolecules, vol 1–2. Plenum Press, New York; Weast RC (ed) (1971) Handbook of chemistry and physics, 52nd edn. CRC Press, Boca Raton, FL).

Elasticity, Hookean *n* The simplest form of stretching elasticity is the Hookean law (e.g., a coiled steel spring). By taking the derivative of the free energy with respect to the relative area change one obtains from the Hookean law that the surface tension is proportional to the surface area (at fixed number of molecules) (Shah V (1998) Handbook of plastics testing technology. Wiley,

New York; Weast RC (ed) Handbook of chemistry and physics, 52nd edn. CRC Press, Boca Raton, FL).

Elasticized Fabric *n* A fabric that contains elastic threads. Such fabrics are used for girdles, garters, and similar items (Vincenti R (ed) (1994) Elsevier's textile dictionary. Elsevier Science and Technology Books, New York).

Elasticizer *n* A compounding additive that contributes elasticity to a resin such as chlorinated polyethylenes and chlorinated copolymers of ethylene and propylene are blended with PVC compositions to provide elasticity (Wickson EJ (ed) (1993) Handbook of polyvinyl chloride formulating. Wiley, New York; and Nass (ed) (1976) Encyclopedia of PVC. Marcel Dekker, New York).

Elastic Limit *n* The greatest stress that a material can experience which, when released, will result in no permanent deformation. With some polymers this limit can be well above the ▶ Proportional Limit (Billmeyer FS, Jr (1962) Textbook of polymer science. Wiley, New York; Elias HG (1977) Macromolecules, vol 1–2. Plenum Press, New York; and Miller ML (1966) Structure of polymers. Reinhold Publishing, New York; Brown R (1999) Handbook of physical polymer testing, vol 50. Marcel Dekker, New York). See also ▶ Yield Point.

Elastic-Melt Extruder See ▶ Extruder, ▶ Elastic-Melt.

Elastic Memory See Memory.

Elastic Moduli *n* (1904) (*Young's Modulus by Bending*), bar supported at both ends. If a flexure *s* is produced by the weight of mass *m*, added midway between the supports separated by a distance *l*, for a rectangular bar with vertical dimensions of cross-section *a* and horizontal dimension *b*, the modulus is,

$$M = \frac{mgl^3}{4sa^3b}$$

For a cylindrical bar of radius *r*,

$$M = \frac{mgl^3}{12\pi r^4 s}$$

For a bar supported at one end. In the case of a rectangular bar as described above,

$$M = \frac{4mgl^3}{sa^3b}$$

For a round bar supported at one end,

$$M = \frac{4mgl^3}{3\pi r^4 s}$$

(Brown R (1999) Handbook of physical polymer testing, vol 50. Marcel Dekker, New York).

Elastic Moduli *n* (*Modulus of Rigidity*) If a couple $C (= mgx)$ produces a twist of θ radians in a bar of length *l* and radius *r*, the modulus is

$$M = \frac{2Cl}{\pi r^4 \theta}$$

The substitution in the above formulae for the elastic coefficients of *m* in grams, g in centimeter per square second, *l*, *a*, *b*, and *r* in centimeter, *s* in centimeter, and *C* in dyne-centimeter will give moduli in dyns per square centimeter. The dimensions of elastic moduli are the same as of stress, $[m\, l^{-1} t^{-2}]$ (Weast RC (ed) Handbook of chemistry and physics, 52nd edn. CRC Press, Boca Raton, FL).

Elastic Moduli *n* (*Coefficient of Restitution*) Two bodies moving in the same straight line, with velocities v_1 and v_2 respectively, collide and after impact move with velocities v_3 and v_4. The coefficient of restitution is

$$C = \frac{v_4 - v_3}{v_2 - v_1}$$

(Weast RC (ed) Handbook of chemistry and physics, 52nd edn. CRC Press, Boca Raton, FL).

Elastic Modulus See ▶ Modulus of Elasticity.

Elastic Nylon See ▶ Nylon 6/10.

Elastic Polyamides *n* Elements that are combined by chemical processes into long-chain polymers that constitute the fiber-forming substance known as polyamides (Harper CA (ed) (2002) Handbook of plastics, elastomers and composites, 4th edn. McGraw-Hill, New York).

Elastic Recovery *n* That fraction of a given deformation that behaves elastically. A perfectly elastic material has a recovery of 100% while a perfectly plastic material has no elastic recovery. Elastic recovery is an important property in films used for stretch packaging because it relates directly to the ability of a film to hold a load together. Retention of the elastic-recovery stress over a period of time is also important (Shah V (1998) Handbook of plastics testing technology. Wiley, New York).

Elastic Solid *n* Solid in which, for all values of the shearing stress below the rupture (shear strength), the strain is fully determined by the stress regardless of whether the stress is increasing or decreasing.

Elastodynamic Extruder See ▶ Extruder, ▶ Elastic-Melt.

Elastomer \-tər-mər\ *n* [*elastic* + -o- + -*mer*] (ca. 1939) Generally, a material that at room temperature can be stretched repeatedly to at least twice its original length and, immediately upon release of the stress, returns with force to its approximate original length (Harper CA (ed) (2002) Handbook of plastics, elastomers and composites, 4th edn. McGraw-Hill, New York). Stretchability and deformation with returnability is one criteria by which materials called plastics in commerce are distinguished from elastomers and rubbers. More explicitly, polymers that stretch and recover reproducibly without permanent change in dimensions (or deformation) are rubber or elastic materials. Elastomers are useful and diverse substances that easily form various rubbery shapes. Many industries rely on parts made from elastomers, especially automobiles, sports equipment, electronics, electrical equipment, and assembly line factories (www.dupontelastomers.com). Another criterion is that, unlike thermoplastics that can be repeatedly softened and hardened by heating and cooling without substantial change in properties, most elastomers are given their final properties by mastication with fillers, processing aids, antioxidants, curing agents, and others followed by vulcanization (curing) at elevated temperatures that crosslinks the molecular chains and causes them to be thermosetting systems. However, a few elastomers are thermoplastic systems (Harper CA (ed) Handbook of plastics, elastomers and composites, 4th edn. McGraw-Hill, New York). Polymers usually considered to be elastomers, at least in some of their forms, are listed in the follow table (Elias HG et al. (1983) Abbreviations for thermoplastics, thermosets, fibers, elastomers, and additives. Polymer News 9:101–110; Carley JF (ed) (1993) Whittington's dictionary of plastics. Technomic Publishing, Lancaster, Pennsylvania) {G Elastomer n, F élastomère m, S elastómero m, I elastomero m}.

A beneficial property of elastomers is that they can be compounded or joined with other materials to strengthen certain characteristics. Other kinds of polymers may be installed next to various other materials, such as metal, hard plastic, or different kinds of rubber, with excellent adhesion (Harper CA (ed) (2002) Handbook of plastics, elastomers and composites, 4th edn. McGraw-Hill, New York; Skeist I (ed) (1990) Handbook of adhesives. Van Nostrand Reinhold, New York).

Elastomeric *adj* Rubberlike, and relating to or having the properties of elastomers.

Elaterite \i-ˈla-tər-ˌrīt\ *n* [Gr *Elaterit*, fr. Gk *elatēr*] (1826) Infusible asphaltic pyrobitumen of petroleum origin, sometime described as mineral rubber or mineral Caoutchouc.

Electret \i-ˈlek-trət\ *n* [*electr*icity + magn*et*] (1885) A disk of polymeric material that has been electrically polarized so that one side has a positive charge and the other a negative charge, analogous to a permanent magnet. Electrets may be formed of poor conductors such as polymethyl methacrylate, polystyrene, nylon, and polypropylene, by heating and cooling them in the presence of a strong electric field (Bar-Cohen Y (ed) (2001) Electroactive polymer (EAP) actuators as artificial muscles. SPIE Press, Bellingham, Washington; Ku CC, Liepins R (1987) Electrical properties of polymers. Hanser Publishers, New York).

Chemical name	Abbreviations
Fluorosilicone	FVSI
Hexafluoropropylene–vinylidene fluoride copolymer	FPM
Isobutene–isoprene copolymer	Butyl, GR-I
Organopolysiloxane	SI
Acrylic ester–butadiene copolymer	ABR, AR
Polybutadiene	BP, BR, CBR
Polychloroprene	CR
Polyepichlorohydrin	CO, CHR
Polyisobutene	PIB
Polyisoprene, natural	NR
Polyisoprene, synthetic	CI, IR, PIP
Polyurethane (polyester)	AU, PUR
Polyurethane (polyether)	EU
Polyurethane (polyether and polyester)	TPU
Styrene–butadiene copolymer	SBR, GR-S
Styrene–chloroprene copolymer	SCR
Polyethylene–butyl graft copolymer	TPO

Chemical name	Abbreviations
Acrylonitrile–chloroprene copolymer	NCR
Acrylonitrile–isoprene copolymer	NIR
Butadiene–acrylonitrile copolymer	GR-N, NBR, PBAN
Chlorinated polyethylene	CPE
Chlorosulfonated polyethylene	CSM, CSR, CSPR
Ethylene ether polysulfide	EOT
Ethylene–ethyl acrylate copolymer	EEA, E/EA
Ethylene polysulfide	ET
Ethylene–propylene copolymer	EPM, EPR
Ethylene–propylene–diene terpolymer	EPD, EPDM, EPT, EPTR
Fluoroelastomer (any)	FPM

Electrical Conductivity *n* (1) (Symbol is λ expressed in units of ohm^{-1}·m^{-1}) A measure of the ease of transporting electric charge from one point to another in an electric field (Ku and Liepins, 1987). (2) The reciprocal of resistivity (resistivity symbol is ρ expressed in units of ohms·m). (3) R = ρl/A, where R = resistance, ohms; ρ = resistivity, ohms·m; l = length of material, m; and A is area of cross-section of material, m^2 (Serway RA, Faugh JS, Bennett CV (2005) College physics. Thomas, New York; Emerson JA, Torkelson JM (eds) (1991) Optical and electrical properties of polymers: materials research society symposium proceedings, vol 24. Materials Research Society, Boston, Massachusetts, Ku CC, Liepins R (1987) Electrical properties of polymers. Hanser Publishers, New York; Seanor DA (1982) Electrical conduction in polymers. Academic Press, New York) {G elektrische Leitfähigkeit f, F conductivité électrique, conductivité f, S conductividad eléctrica, conductividad f, I conduttività elettrica, conduttività f}.

Electrical-Discharge Machining *n* (EDM, spark erosion) A method of machining molds and extrusion dies in which a conductive tool (often brass) has the inverse shape of the cavity or hole to be machined. A high-voltage DC difference is applied between the tool and the work piece. Capacitive discharge erodes steel from the work piece at about eight times the rate that the tool itself is eroded. Roughing and finishing tools are used. The process is accurate, produces good detail, can cut thin, deep slots, and can be used with hardened steels, thus averting the distortion sometimes caused by hardening after conventional machining. Metal removal is relatively slow (Carley JF (ed) (1993) Whittington's dictionary of plastics. Technomic Publishing).

Electrical Finish *n* A finish designed to increase or maintain electrical resistivity of a textile material.

Electrical Insulation *n* Material with very low conductivity which surrounds active electrical devices. Common electrical insulation chemicals are fluorine-containing polymers (Dissado LA, Fothergill CJ (eds) (1992) Electrical degradation and breakdown of polymers. Institution of Electrical Engineering (IEE), London; Ku CC, Liepins R (1987) Electrical properties of polymers. Hanser Publishers, New York).

Electrically Conductive Polymers *n* Electrical properties of polymers are their responses when an electric field is applied to them. Business-machine housings, structural components, and static-control accessories often require plastics that have some degree of electrical conductivity. Additives and fillers imparting such conductivity are metal powders, carbon black, carbon fibers, and metallized-glass fibers and spheres. Inherently conducting polymers are naturally conducting while filled polymers owe their conductivity to the electrical properties of the filler materials (Ku CC, Liepins R (1987) Electrical properties of polymers. Hanser Publishers, New York). See also ▶ Conducting Polymer.

Electrical conductivity of conductive polymers and other materials

Material	Dopant	Conductivity (ohm^{-1} cm^{-1})
Copper	–	6×10^5
Mercury	–	10^4
Polyacetylene	Arsenic pentafluoride	1,200
Poly (*p*-phenylene)	Arsenic pentafluoride	500
Polypyrrole	Iodine	100
Polypyrrole	Perchlorate[a]	40
Polyaluminophythalocyanine fluoride	Iodine	4.5
Polyphthalocyaninesilorane	Iodine	1.4
Polyphenylene sulfide	Arsenic pentafluoride	1
Germanium	–	2×10^{-2}
Polyurethane	–	10^{-9}
Epoxy	–	10^{-14}
Polystrene	–	10^{-16}

[a]Polypyrrole perchlorate film from electropolymerization of pyrrole in perchlorate solutions.
Source: Bhattacharya SK (1986) Metal-filled polymers. Marcel Dekker, p.58.

Electrical Measurements *n* Analog electronics deal with electrical signals that move through a continuous range of voltages. Digital electronics, on the other hand, deal with electrical signals that only assume fixed voltages (Giambattista A, Richardson R, Richardson RC, Richardson B (2003) College physics. McGraw-Hill Science/Engineering/Math, New York).

Electrical Resistance *n* (1) The electrical resistance is the ohmic resistance to the flow of electrical current (Giambattista A, Richardson R, Richardson RC, Richardson B (2003) College physics. McGraw-Hill Science/Engineering/Math, New York). Related properties of plastics are ▶ Insulation Resistance, ▶ Surface Resistivity, and ▶ Volume Resistivity. (2) The ability of plastics to withstand various electrical stresses. See ▶ Arc Resistance, ▶ Break-Down Voltage, ▶ Corona Resistance, ▶ Dielectric Strength.

Electric Field Intensity *n* The electric field intensity is measured by the force exerted on unit charge. Unit field intensity is the field which exerts the force of 1 dyne on unit positive charge. Dimensions, $\left[\varepsilon^{-\frac{1}{2}} m^{\frac{1}{2}} l^{-\frac{1}{2}} t^{-1}\right]$; $\left[\mu^{-\frac{1}{2}} m^{\frac{1}{2}} l^{-\frac{1}{2}} t^{-1}\right]$. The field intensity or force exerted on unit charge at a point distance r from a charge q in a vacuum

$$H = \frac{q}{r}$$

If the dielectric in the above cases is not a vacuum the dielectric constant ε must be introduced. The formula becomes

$$H = \frac{q}{\varepsilon r^2}$$

The value of ε is frequently considered unity for air. If the dielectric constant of a vacuum is considered unity the value for air at 0°C and 760 mm pressure is 1.000567 (Weast RC (ed) Handbook of chemistry and physics, 52nd edn. CRC Press, Boca Raton, FL).

Electric Strength *n* Another term for dielectric strength.

Electrochemical Cell *n* Any device which converts electrical into chemical energy, or vice versa.

Electroactive Polymers *n* (**EAP**) Polymers that respond to electrical and magnetic stimulus with a significant change in shape and size. An example of an electroactive polymer is poly(vinylidene fluoride) (PVDF) that has been used for pressure sensitive (piezoelectricity) devices such as pressure sensors. Also, some polymers respond to photonic and thermal stimulus (Bar-Cohen Y (ed) (2001) Electroactive polymer (EAP) actuators as artificial muscles. SPIE Press, Bellingham, Washington).

Electrochemical Equivalent *n* In an electrolytic cell, the mass of a metal deposited with the passage of 1 coulomb of electricity. In the SI system, the coulomb is defined as 1 A·s.) (Goldberg DE (2003) Fundamentals of chemistry. McGraw-Hill Science/Engineering/Math, New York).

Electrocoating *n* An organic finish which is often applied as a prime coat on steel. See ▶ Electrodeposition.

Electrocuring *n* Process which uses an electron beam to cure organic coatings applied to commercial products on a continuous production line.

Electrode \i-▪lek-▪trōd\ *n* (1834) A terminal member in an electrical circuit designed to promote an electrical field between it and another electrode. In the plastics industry electrodes are used in microwave heat sealing and surface treating of films. One of the electrodes may be a press platen or a roll (Goldberg DE (2003) Fundamentals of chemistry. McGraw-Hill Science/Engineering/Math, New York; Weast RC (ed) Handbook of chemistry and physics, 52nd edn. CRC Press, Boca Raton, FL).

Electrodeposition *vt* (1882) Method of paint application in which an article to be coated which is an electrical conductor is made one of the electrodes in a tank of water-thinned paint. The other electrode is generally a metal such as copper. The two electrodes are connected to a source of electric power, the polarity of the article to be coated being of the opposite sign to that on the particles in the liquid paint in the tank. The charged particles move towards the articles under the influence of the electric field, and when they give up their charge at the electrode (article) they are deposited and ultimately form a continuous film of paint (Weismantal GF (1981) Paint handbook. McGraw-Hill, New York).

Electrodeposition Coating The process of using a water-borne electrodeposition (E-Coat) paint process instead of a conventional organic solvent-based spray {G Elektrotauchlackierung f, F revêtement par électrodéposition, revêtement m, S revestimiento por electrodeposición, revestimiento m, I rivestimento per elettrodeposizione, rivestimento m}.

Electrode Potential *n* (1) The voltage associated with a half-reaction written, by convention, as a reduction; a reduction potential. (2) The difference in potential between an electrode and the immediately adjacent electrolyte referred to some standard electrode as zero potential.

Electroformed Mold *n* A mold made by electroplating a model which is subsequently removed from the metal deposit. The deposit is sometimes reinforced with cast or sprayed metal backings to increase its strength and rigidity. Such molds are used in slush casting of vinyl plastisols and other forming processes done at low pressures.

Electroforming *n* A method of making molds for plastics processes, usually those employing low or moderate pressures, in which a pattern made of preplated wax or flexible material is electroplated.

Electrokinetic Potential See ▶ Zeta Potential.

Electroless Plating *n* The deposition of metals on a catalytic surface from solution without an external source of current. The process is used as a preliminary step in preparing plastic articles for conventional electroplating. After cleaning or etching, the plastic surface is immersed in solutions that react to precipitate a catalytic metal *in situ*, for example first in an acidic stannous chloride solution, then into a solution of palladium chloride. Palladium is reduced to its catalytic metallic state by the tin. Another way of producing a catalytic surface is to immerse the plastic article in a colloidal solution of palladium followed by immersion in an accelerator solution. The electroless plating bath is a solution of several components, including nickel or copper salts, chelating agents, stabilizers, and reducers. Metal reduced from the salt plates on the active sites on the palladium surface, the plated plastic being removed from the bath when the thickness of the electroless deposit is from 3 to 7 μm. The plastic article thus treated can now be plated with nickel or copper by the electroless method, forming a conductive surface that then can be plated with other metals by conventional electroplating.

Electroless Plating Equipment *n* Equipment used in the deposition and formation of a continuous metallic film on a nonconductive plastic surface without the use of an electric current.

Electroluminescence \i-▪lek-trō-▪lü-mə-▪ne-s³n(t)s\ *n* (ca. 1909) Generation of light by high-frequency electrical discharge through a gas or by applying an alternating current to a phosphor {G Elektrolumineszenz f, F électroluminescence f, S electroluminiscencia f, I elettroluminescenza f}.

Electrolysis \-▪trä-lə-səs\ *n* (1834) The passage of an electric current through a medium to produce a chemical change. If a current i flows for a time t and deposits a metal whose electrochemical equivalent is e, the mass deposited is $m = eit$. The value of e is usually given for mass in grams, i in amperes and t in seconds.

Electrolyte *n* (1834) A substance which produces ions when dissolved solution or when fused, thereby becoming electrically conducting {G Elektrolyt m, F électrolyte m, S electrólito m, I elettrolito m}.

Electrolytic Cell *n* An electrochemical cell in which electrical energy is used to produce chemical change; a cell in which electrolysis takes place. See ▶ Cell, ▶ Electrolytic.

Electrolytic Dissociation or Ionization Theory *n* When an acid, base or salt is dissolved in water or any other dissociating solvent, a part or all of the molecules of the dissolved substance are broken up into parts called ions, some of which are charged with positive electricity and are called cations, and an equivalent number of which are charged with negative electricity and are called anions.

Electrolytic Solution Tension Theory *n* (Heimholtz Double Layer Theory) When a metal, or any other substance capable of existing in solution as ions, is placed in water or any other dissociating solvent, a part of the metal or other substances passes into solution in the form of ions, thus leaving the remainder of the metal or substances charged with an equivalent amount of electricity of opposite sign from that carried by the ions. This establishes a difference in potential between the metal and the solvent in which it is immersed.

Electrolytic White Lead *n* Form of white lead, chemically similar to the stack and chamber types, which is made by the electrolysis of a solution of a lead salt followed by treatment with carbon dioxide gas.

Electromagnetic Adhesive *n* An intimate blend of a material that absorbs electromagnetic energy with a thermoplastic of the same composition as the sections to be bonded. The adhesive is applied in the form of a liquid, a ribbon, a wire, or a molded gasket to one of the surfaces to be joined. The two surfaces are brought into contact, then the adhesive is rapidly heated by eddy currents induced by a high-frequency induction coil placed close to the joint. This melts the adhesive which, after cooling, bonds the surfaces together.

Electromagnetic Spectrum *n* (ca. 1934) The entire range of wavelengths or frequencies of electromagnetic radiation from the shortest gamma rays to the longest radio waves and including visible light.

Electromagnetic Welding See ▶ Induction Welding.

Electromagnetic Wave *n* (1908) One of the waves that are propagated by simultaneous periodic variations of electric and magnetic field intensity and that include radio waves, infrared, visible light, ultraviolet, X-rays, and gamma rays.

Electromotive Force *n* The difference in electric potential that causes current to flow in a circuit. The SI unit is the volt (V), defined as the difference of potential between two points of a conductor carrying a constant current of 1 ampere when the power dissipated between the two points equals 1 W. Thus, in SI, 1 V = 1 W/A.

Electromotive Force *n* This force is defined as that which causes a flow of current. The electromotive force of a cell is measured by the maximum difference of potential between its plates. The electromagnetic unit of potential difference is that against which 1 erg of work is done in the transfer of electromagnetic unit quantity. The **volt** is that potential difference against which 1 J of work is done in the transfer of 1 coulomb. One volt is equivalent to 10^8 electromagnetic units of potential. The *International* volt is the electrical potential which when steadily applied to a conductor whose resistance is one international ohm will cause a current of one international ampere to flow. The international volt = 1.00033 absolute volts. The electromotive force of a Weston standard cell is 1.0183 int. volts at 20°C. Dimensions, $\left[\varepsilon^{-\frac{1}{2}}m^{\frac{1}{2}}l^{-\frac{1}{2}}t^{-1}\right]$; $\left[\mu^{-\frac{1}{2}}m^{\frac{1}{2}}l^{\frac{3}{2}}t^{-1}\right]$.

Electromotive Series *n* A list of the metals arranged in the decreasing order of their tendencies to pass into ionic form by losing electrons.

Electron \i- ˈlek- ˌträn\ *n* [*electr-* + *²-on*] (1891) A (perhaps) fundamental subatomic particle with a very low mass and a unit negative electrical charge; found in the extranuclear region of an atom. The electron is a small particle having a unit negative electrical charge, a small mass, and a small diameter. Its charge is (4.80294 + .00008) × 10^{-10} absolute electrostatic units, it mass $\frac{1}{1837}$ that of the hydrogen nucleus, and its diameter about 10–12 cm. Every atom consists of one nucleus and one or more electrons. Cathode rays and Beta rays are electrons.

Electron An elementary subnuclear particle having a unit negative electrical charge, $1.60219 \cdot 10^{-19}$ coulomb, its mass (at rest) is 1/1837 that of a hydrogen nucleus, or $9.10953 \cdot 10^{-28}$ g, and its diameter is $5.6359 \cdot 10^{-13}$ cm. Every atom consists of a nucleus of protons and neutrons, and as many orbiting electrons as there are protons in the nucleus. Cathode rays and beta rays are electrons.

Electron Affinity *n* The quantity of energy released when a gaseous, isolated, ground-state atom (or, sometimes, ion) gains an electron.

Electron Beam *n* A stream of electrons in an electron optical system.

Electron Beam Crosslinking *n* E-beam crosslinking for plastics provides all of crosslinking's advantages – including tensile and impact strength, creep resistance, durability, solvent and chemical resistance, abrasion resistance, environmental stress crack resistance, and barrier properties – but none of the disadvantages of chemical crosslinking. E-beam crosslinking does not require any additives nor does it generate hazardous chemical by-products. This crosslinking method is energy efficient and the minimal amount of exposure time to the ebeam helps ensure high throughputs. Electron beam crosslinking is most frequently used for polyethylene and polyvinyl chloride products, but there is a growing amount of e-beam crosslinking of fluoropolymers including ETFE in molded parts and specialty wire and cable.

Electron-Beam Machines *n* The key to the technology is electrobeam (or E-beam) lithography. E-beam lithography is a technique for creating extremely fine patterns (much smaller than can be seen by the naked eye) required by the modern electronics industry for integrated circuits. Derived from the early scanning-electron microscopes, the technique consists of scanning a beam of electrons across a surface covered with a thin film, called a resist. The electrons produce a chemical change in this resist, which allows the surface to be patterned.

Electron-Beam Radiation *n* Magnetically accelerated electrons focused by electric fields have been used for crosslinking polyethylene in special applications such as wire coating, to improve modulus and temperature resistance. Treatment levels must be carefully controlled since overexposure will cause degradation. Electron beams are also used to cure epoxy-resin coatings, eliminating the need for photo-initiators.

Electron Capture *n* A mode of radioactive decay in which an electron from the extranuclear region, usually the K shell, is captured by a nucleus.

Electron Curing *n* A method for curing polymer matrix composites.

Electron-Deficient Compound *n* A compound in which insufficient electrons are available to bond all the atoms with conventional (two-center) covalent bonds.

Electron Diffraction *n* Used to identify crystalline substances based on the spacing of atomic planes within their structures.

Electronegativity *n* The relative tendency of a bonded atom to attract electrons to itself.

Electron Gas *n* (ca. 1929) The delocalized electrons in a metal.

Electron Lens *n* (1931) A device for focusing a beam of electrons by means of an electric or a magnetic field.

Electronic Heating See ▶ Dielectric Heating.

Electronic Treating See ▶ Corona-Discharge Treatment.

Electron Micrograph *n* (1934) An image formed on a photographic medium with an electron microscope, an instrument in which the subject is examined with an electron beam focused by electric fields. The beam may be transmitted through the specimen or, after precoating the surface to be examined with gold or platinum, reflected from it. Magnifications to 100,000 times are achievable with good resolution. Electron micrography has been a powerful tool in determining the structures of plastics crystals and the topography of fractured surfaces.

Electron Microprobe *n* An instrument that utilizes the bombardment of a small sample with a beam of high-energy electrons to determine the composition, elemental identification and quantification of a material (e.g., Al, O, etc.).

Electron Microscopy *n* (1932) Electron microscopy is applied to observe phase domain of a size of 50–1,000 ånström. This comes true with Transmission Electron Microscopy by applying dyeing techniques such as oxidizing the unsaturated domain with OsO_4 and RuO_4.

Electron Multiplier *n* (1936) A device that utilizes secondary emission of electrons for amplifying a current of electrons.

Electron Paramagnetic Resonance *n* A method to investigate the behavior of samples containing unpaired electrons (free radicals or compounds comprising an ion whose outer electronic shell is incomplete) in an applied magnetic field.

Electron Probe *n* (1962) A microprobe that uses an electron beam to induce X-ray emissions in a sample.

Electron Spectroscopy *n* A surface specific technique utilizing the emission of low energy electrons in the Auger process.

Electron Tube *n* (1922) An electronic device in which conduction by electrons takes place through a vacuum or a gaseous medium within a sealed glass or metal container and which has various uses based on the controlled flow of electrons.

Electron Volt *n* (1930) (eV) The kinetic energy acquired by any charged particle carrying unit electronic charge when it falls through a potential difference of 1 V. One electrovolt is equal to $1.60219 \cdot 10^{-19}$ J. Multiples of this unit in common use are the keV (10^3), MeV (10^6), and GeV (10^9). The GeV is also written BeV.

Electrophile \i-ˈlek-trə-ˌfīl\ *n* (1943) An atom or group of atoms which appears to seek electrons in its reactions.

Electrophoresis \-trə-fə-ˈrē-səs\ *n* [NL] (1911) The movement of suspended particles through a fluid or gel under the action of an electromotive force applied to electrodes in contact with the suspension.

Electrophoretic Deposition *n* A direct-current process analogous to electroplating, used to coat electrically-conductive articles with plastics, deposited from aqueous lattices or dispersions. The cathode may be a noncorrodible metal such as stainless steel, generally serving as the container in which the process is performed. The DC potential is usually under 100 V. The deposited coatings are baked to remove residual water. Among available polymer lattices suitable for the process are PVC, polyvinylidene chloride, acrylics, nylons, polyesters, polytetrafluoroethylene, and polyethylene.

Electroplating Chemicals *n* Copper, gold, silver, chromium, and nickel are generally used as the conductive metal for plating. Acrylonitrile–butadiene–styrene resins have been most widely used for electroplated articles. Others in commercial use for the process include cellulose acetate, some grades of polypropylene, polysulfones, polycarbonate, polyphenylene oxide, nylons, and rigid PVC.

Electroplating on Plastics *n* Articles of almost any of the common plastics can be plated by conventional processes used on metals after their surfaces have been rendered conductive by precipitation of silver or other conductor (see ▶ Electroless Plating). A layer of copper is usually applied first, followed by a final plating of gold, silver, chromium, or nickel. Acrylonitrile–butadiene–styrene resins have been most widely used for electroplated articles. Others in commercial use for the process include cellulose acetate, some grades of polypropylene, polysulfones, polycarbonate, polyphenylene oxide, nylons, and rigid PVC. See also ▶ Metallizing.

Electroplating Plastics *n* Plastics can be plated by conventional processes used for metals, after their surfaces have been rendered conductive by precipitation of silver or other conductive substance.

Electrostatic Coating *n* A coating that creates, electrical charges, and disperses particles or droplets toward the target by a variety of methods {G elektrostatisches Pulversprühverfahren n, F revêtement de poudre électrostatique, revêtement m, S recubrimiento por polvo electrostático, recubrimiento m, I rivestimento con polvere elettrostatica, rivestimento m}.

Electrostatic Detearing *n* Process of removing blobs and the thick edges of paint from an article which has been coated by dipping. The process consists of passing the dipped article, after a limited period of draining, over a grid at a high-electrical potential. The blobs and thick edges of paint are removed from the article by attraction to the grid.

Electrostatic Fluidized-Bed Coating *n* A process combining elements of the fluidized-bed method of coating and electrostatic spraying. Pointed electrodes are inserted through the porous bottom of a fluidized-bed container. When the bed is aerated in the usual manner, a potential of about 100 kV is applied between the electrodes and ground. The associated charge repels the fluidized plastic particles into the space above the bed, from which they are attracted to a grounded article to be coated. The article may be at room temperature when inserted in the powder bed, the coating temporarily adhering by electrostatic charge. Subsequent heating fuses the coating.

Electrostatic Forces *n* The forces that exist between particles which are electrically charged.

Electrostatic Printing *n* (**electrostatography**) Contactless printing by any of several methods based on electrostatic principles. *Electrophotography* depends on light or other electromagnetic energy and photo-semiconductors which are nonconductors of electricity in the dark and conductors when exposed to electromagnetic radiation. *Electrography* involves the use of a dielectric image, stencil, or facsimile scanning for source to form the image.

Electrostatic Printing *n* A printing process employing electrostatic charge to transfer powdered ink from an electrically charged stencil to a plastic film or sheet. The film to be printed is interposed between a grounded metal plate and the stencil. Areas corresponding to those *not* to be printed are masked on the stencil as in conventional screen printing. The powdered ink is brushed on the back side of the screen, where it receives a charge propelling it toward the grounded plate as an image cloud until intercepted by the film. Post-heating is usually required to fuse the ink to the substrate.

Electrostatic Spray Coating *n* A spraying process that employs electrical charges to direct the paths of atomized particles to the work surface. Dry plastic powders are charged with the static electricity as they emerge from a spray gun, the nozzle of which is attached to the negative terminal of a high-voltage DC power supply. The charged particles are attracted to the grounded object, which must be at least slightly electrically conductive. The powder coating is subsequently heated to obtain a smooth, homogeneous layer.

Electrostatic Spraying *n* (1) A process where paint spray is blown through an electrostatic field. (2) A system of applying paint in which the sprayed paint droplets are given an electrical charge that results in their attraction to the grounded work piece.

Electrostatic Spraying *n* Methods of application spraying in which an electrostatic potential is created between the article to be coated and the atomized paint particles. The charged particles of paint are attracted to the article being painted and are then deposited and discharged. The electrostatic potential is used in some processes to aid the atomization of the paint.

Element \ˈe-lə-mənt\ *n* [ME, fr. OF & L; OF, fr. L *elementum*] (13c) A pure substance composed of atoms each having the same atomic number (number of protons). An element cannot be chemically decomposed. Substances which cannot be decomposed by the ordinary types of chemical change, or made by chemical union.

Elemental Analyses *n* Method of substrate characterization, by which molecules are broken down to their individual elements.

Elementary Process A single step of a reaction mechanism.

Elemi Gum \ˈe-lə-mē ˈgəm\ *n* [NL *elimi*, prob. fr. Arabic al *lāmi* the elemi] (1543) Any of various natural oleoresins derived from certain tropical trees, especially *Canarium Iuzonicum* of the Philippines, and used in making varnishes and inks. It is soluble in most organic solvents, and used chiefly to impart elasticity and adhesion to lacquers and varnishes.

Eleomargaric Acid *n* Another name for ▶ Eleostearic Acid.

Eleostearic Acid *n* CH$_3$(CH$_2$)$_3$(CH=CH)$_3$(CH$_2$)$_7$COOH. Ocatadeca-9,11,13-trienoic acid. *cis,trans,trans*,9,11,13-Octadecatrienoic acid. Principal constituent acid of tung oil, characterized by the presence of three conjugated double bonds. Two forms are known, namely α and β types, with mps of 49°C and 71°C, respectively.

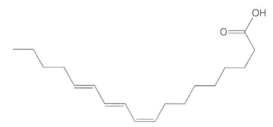

Elimination Reaction *n* A reaction in which atoms or groups on adjacent atoms in a molecule are removed to leave a double or triple bond between the atoms.

Ellis Model *n* A three-constant model of pseudoplastic flow that merges Newton's law of flow, applicable at very low shear rates, with the ▶ Power Law at high rates and provides a smooth transition between the two. For one-dimensional flow the equation is:

$$-\frac{dv_z}{dx} = \frac{\tau_{xz}}{\eta_o}\left[1 + \left[\frac{\tau_{xz}}{\tau_{1/2}}\right]^{\alpha-1}\right]$$

where v_z is the z-directed velocity perpendicular to coordinate χ, $\tau_{\chi z}$ is the opposing shear stress, η_o is the zero-shear (Newtonian) viscosity, a is an exponent larger than 1, approximately equal to the reciprocal of the flow-behavior index and $\tau_{1/2}$ is the shear stress at which the viscosity (= shear stress/shear rate) is half η_o.

Elmendorf Tear Strength *n* The Elmendorf tear tester, originally developed to test papers and fabrics, has been adapted for plastics films in ASTM D 1922. Acting by gravity, a calibrated pendulum swings through an arc, tearing the specimen from a precut slit. The energy absorbed is indicated by a pointer and scale. Some other modes of measuring tear resistance are spelled out in ASTM D 1004, and D 2582.

Elongation \(▮)ē- ▮lóŋ- ▮gā-shən\ *n* (14c) In tensile testing, the fractional increase in length of a marked test length as the test specimen is stretched and stress rises. At any point during the test, percent nominal elongation = 100 × the increase in gage length/original length. *Ultimate elongation* is the elongation just prior to rupture of the sample and is the "elongation at break" reported in most properties tables. See also ▶ True Strain.

Elongational Flow *n* (extensional flow) Flow caused by stretching a material, usually a hot melt, as in fiber drawing, film blowing, parison drawdown, and biaxial stretching of sheet. This flow is always accompanied by a reduction in cross section. The *rate* of elongation, at any moment during one-dimensional, elongational flow, is given by

$$de/dt = (1/s)(ds/dt) = d(\ln s)/dt$$

where *e* = the true elongation and *s* = the strand length at time *t*.

Elongational Viscosity *n* (extensional viscosity, Trouton viscosity) The viscosity that characterizes an element undergoing ▶ Elongational Flow (above). It is equal to the tensile stress divided by the rate of elongation and for polymers it depends on the rate, but may increase with rate, unlike the usual reduction of shear viscosity with rate. Tensile viscosities are apt to be many times larger than shear viscosities for the same resin, temperature, and deformation rate. Values in the range of 10^4–10^7 Pa·s have been reported. For Newtonian liquids, the elongation viscosity is three times the shear viscosity (at the same temperature).

Elongation at Break *n* The increase in length when the last component of the specimen breaks.

Elongation at Rupture See ▶ Elongation, ▶ Elongation at Break.

Elongation at Yield *n* The increase in distance between two gauge marks, resulting from stressing the specimen in tension to the yield point. See also ▶ Elongation.

Eluent \ ▮el-yə-wənt\ *n* [L *eluent-*, *eluens*, pp of *eluere*] (1941) In gas and liquid chromatography, the fluid that carries the solute out of the column.

Elution *n* [L *elutus*, pp of *eluere* to wash out, fr. *e-* + *lavere* to wash] (1731) The removal, in chromatography, of the species adsorbed on the column matrix by a flowing liquid or gas.

Elutriate \ē- ▮lü-trē- ▮āt\ *vt* [L *elutriatus*, pp of *elutriare* to put in a vat, perhaps fr. *elutrum* vat, fr. Gk *elytron* reservoir, literally, covering] (ca. 1727) To purify, separate, or remove by washing.

Elutriation *n* The separation of less massive particles from a sample of distributed particle sizes or densities by upward flow of a liquid or gas.

Elvanol *n* Poly(vinyl alcohol), manufactured by DuPont, U.S.

EMA *n* Abbreviation for ▶ Ethylene–Acrylic Acid Copolymer.

EMAC See ▶ Ethylene Methyl Acrylate Copolymer.

Embedding *n* The process of encasing an article in a resinous mass, performed by placing the article in a simple mold, pouring a liquid resin into the mold to completely submerge the article, sometimes under vacuum so as to suck out hidden air bubbles, curing the resin, and removing the encased article from the mold. In the case of electrical components, the lead wires or terminals may protrude from the embedment. The main difference between embedding and potting is that in potting, the model is a container that remains fixed to the resinous mass. The liquid resin may contain microspheres to reduce the final mass of the embedment. See also ▶ Encapsulation, Impregnation, and Potting.

Embedment Decorating *n* A technique for decorating reinforced-plastics articles in which a mat or web of fibrous material printed with a design is embedded in the surface of the article and covered with a transparent gel coat. The technique can be adapted for use in hand lay-up, continuous laminating, pultrusion, matched-die molding.

Embossed Paper *n* Wallpaper run through rollers with raised areas, to provide a light relief effect.

Embossing *n* Any of several techniques used to crease depressed patterns in plastics films or sheeting. In the case of cast film, embossing can be accomplished directly by casting on an inverse-textured belt or roll. Calendered films are frequently embossed by rolls following the calender rolls. Other films or coated fabrics can be embossed subsequent to manufacture by reheating them and passing them through embossing rolls, or compressing them between plates. Extruded sheets, up to 3 mm or thicker, are commonly embossed as the sheets emerge from the extruder with an embossing roll on the takeoff.

Embrittlement *n* A reduction or loss of ductility or toughness in materials such as plastics resulting from chemical or physical damage.

Embroidery \im- ▮brói-d(ə)rē\ *n* (14c) Ornamental designs worked on a fabric with threads. Embroidery may be done either by hand or by machine.

Emerald Green \ˈem-rəld ˈgrēn\ *n* (1646) Name applied to two distinctly different green pigments: complex copper acetoarsenite (*also known as Paris Green and Schweinfurt Green*); and chromium hydroxide. See ▶ Hydrated Chromium Oxide.

Emerald Oxide of Chromium See ▶ Hydrated Chromium Oxide.

Emery \ˈem-rē, ˈe-mə-\ *n* [ME, fr. MF *emeri*, fr. OIt *smiriglio*, fr. ML *smiriglum*, fr. Gk *smyrid-*, *smyris*] (15c) A mixture of mostly corundum and some magnetite that, because of its great hardness, is widely used for grinding and polishing, including tumble-polishing of plastics moldings. Emery is available in many grades from coarse to extremely fine, as powder and bonded to paper and cloth.

Emery Abrasive *n* A natural composition of corundum and iron oxide, found in large deposits both in the United States and the Near East. It is magnetic and partly soluble in hydrochloric acid. The grains are blocky, cut slowly, and tend to polish the material being abraded.

Emery Cloth *n* Coated abrasive cloth used for light polishing of metal or for removing rust spots or similar light work; not recommended for large metal surfaces or tough metal, where aluminum oxide cloth is much more satisfactory.

EMI *n* Abbreviation for ▶ 2-Ethyl-4-Methylimidazole.

Emission \e-ˈmi-shən\ *n* (1607) (1) Discharges into the air by a pollution source as distinguished from effluents, which are discharged into water. (2) The emitting of radiation. Emission depends on the temperature to which a material is heated relative to the temperature of its surroundings, on the time, and on the nature of the surface. Basically, therefore, emission is a net rate at which the body emits radiation. As an adjective, it is used to describe the characteristic radiation emitted by elements in a spectroscope, e.g., emission spectrum.

Emission Factor *n* The average amount of pollutants emitted from a polluting source per unit of material produced.

Emissive Power *n* (or emissivity) Emissive Power is measured by the energy radiated from unit area of a surface in unit time for unit difference of temperature between the surface in question and surrounding bodies. For the cgs system the emissive power is given in ergs per second per square centimeter with the radiating surface at 1° absolute and the surroundings at absolute zero.

Empirical Formula \-i-kəl ˈfȯr-myə-lə\ *n* (1885) A formula expressing the simplest whole-number ratio of atoms of each element in a compound without providing information on the grouping of the atoms. For example, the empirical formula of oleic acid is $C_{18}H_{34}O_2$. Also called *Simplest Formula*.

Emulgator See ▶ Emulsifier.

Emulsification \i-ˌməl-sə-fə-ˈkā-shən\ *n* (1859) (1) In lithography, a condition resulting from the distribution of fountain solution in the ink. Excessive emulsification will produce poor printing. (2) The process of dispersing one liquid in another (the liquids being mutually insoluble or sparingly soluble in each other). When water is one of the liquids, two types of emulsions are possible: oil-in-water (water is the continuous state), and water-in-oil. The term "oil" describes any organic liquid sparingly soluble in water.

Emulsified Asphalt See ▶ Asphalt Emulsion.

Emulsifier *n* (1888) Substance that intimately mixes, modifies the surface tension of colloidal droplets, and disperses dissimilar materials ordinarily immiscible, such as oil and water, to produce a stable emulsion. The emulsifier has the double task of promoting the emulsification and of stabilizing the finished products. *Also known as Emulsifying Agent and Emuglator* {G Emulgator m, F émulsifiant m, S emulsionante m, I emulsionante m}.

Emulsifiers *n* A surface-active agent promoting the formation and stabilization of an emulsion.

Emulsifying Agent *n* A substance used to facilitate the formation of an ▶ Emulsion from two or more immiscible liquids, and/or to promote the stability of the emulsion. As surface-active agents, emulsifiers act to reduce interfacial tensions between the several phases. They also act as protective colloids to promote stability. See ▶ Emulsifier.

Emulsifying Agents *n* Materials used to facilitate the preparation of emulsions and to improve their stability.

Emulsion \i-ˈməl-shən\ *n* [NL *emulsion-*, *emulsio*, fr. L *emulgēre* to milk out, from e- + *mulgēre* to milk; akin to OE *melcan* to milk, Gk *amelgein*] (1612) Strictly, an emulsion is a two-phase, substantially permanent, intimate mixture of two incompletely miscible liquids, one of which is dispersed as finite globules in the other. However, in plastics and other industries the term is sometimes broadened to include colloidal suspensions of solids such as waxes and resins in liquids. In an emulsion, the liquid that forms globules is known as the *dispersed*, *discontinuous*, or *internal phase*. The surrounding liquid is called the *continuous* or *external phase*. The dispersed phase may be held in suspension by mechanical agitation or by the addition of small amounts of emulsifying agents {G Emulsion f, F émulsion f, S emulsión f, I emulsione f}.

Coating, emulsion polymerization 1000x mag SEM

Coating, emulsion, 2500x mag SEM

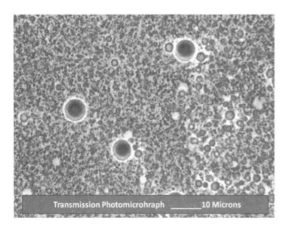

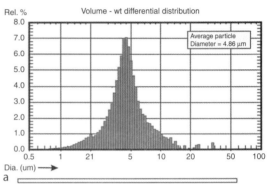

a

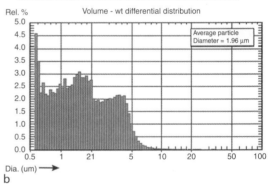

b

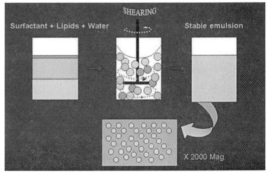

Preparation of Aqueous Emulsions.
(Gooch, J. W., Emulsification and Polymerization of Alkyd Resins, Kluwer Academic, 2002)

Emulsion Paint *n* Paint, the vehicle of which is an emulsion of binder in water. The binder may be oil, oleoresinous varnish, resin, or other emulsifiable binder. Not to be confused with a latex paint in which the vehicle is a latex.

Emulsion Polymerization *n* A polymerization process in which the monomer or mixture of monomers is emulsified in a low-viscosity aqueous medium by means of soaps or other surface-active, solubilizing, and emulsifying agents. The emulsion does not require intensive stirring as in suspension polymerization, and produces polymers of higher molecular weight than those produced by bulk or suspension processes. The polymers remain in emulsion, and must be recovered from the latex by freezing or chemical precipitation. The polymerization medium usually also contains a water-soluble initiator, catalyst, or chain-transfer agent. Examples of polymers produced from emulsions are polyvinyl acetate, acrylo-nitrile-styrene terpolymer, PVC, polyethylene, acrylics, and polystyrene.

Emulsion Spinning *n* Process by which a polymer matrix suspension of very fine particles is subjected to a high-temperature sintering and drawing process, during which the matrix polymer is burnt off and the dispersed particles are coalesced.

Emulsion Stabilizer See ▶ Protective Colloid.

Emulsoid \i-ˈməl-ˌsóid\ *n* (ca. 1909) Colloidal particle which takes up water. Syn: ▶ Hydrophile.

Enamel \i-ˈna-məl\ *vt* [ME, fr. MF *enamailler*, fr. *en-* + *esmail* enamel, of Germanic origin; akin to OH Gr *smelzan* to melt] (14c) (1) A dispersion of pigment in a liquid that forms a solid adherent film, on the surface to which it is applied, by means of oxidation polymerization, or other chemical reaction. The liquid vehicle of an enamel usually contains a thermosetting resin and a solvent. An initial soft film is formed by evaporation of the solvent, then the film hardens or cures at room temperature or during baking. (2) A class of substances having similar composition to glass with the addition of stannic oxide, SnO_2, and other infusible substances to render the enamel opaque.

Enamel Blue See ▶ Cobalt Blue.

Enantiomer \i-ˈnan-tē-ə-mər\ *n* [Gk *enangtios* + E – *mer*] (ca. 1929) Either of a pair of chemical compounds whose *molecular* structures have a mirror–image relationship to each other. An asymmetric molecule that is the mirror image of its stereoisomer. The two isomers are given the prefixes *dextro-* and *levo-*, e.g., *d-* and *l-*lactic acid. The physical properties of pure enantiomers are equal within experimental error, yet mixtures of the two, called *racemic* mixtures, may have different properties. For example, 50–50 *di-*lactic acid melts 20°C lower than its pure enantiomers.

Enantomeric \-ˌnan-t ē-ə-ˈmer-ik\ *adj.*

Enantiomeric Configurational Unit *n* Either of two stereoisomeric groups in a polymer that are mirror images at the plane containing the main-chain bonds.

Enantiomorph \i-ˈnan-tē-ə-ˌmórf\ *n*. [GK *enantios* opposite (fr. *enanti* facing, fr. *en* in + *anti* against) + ISR - morph] (1885) Enantiomer; also, either of a pair of crystals (as of quartz) that are structural mirror images.

Enantiotropic *n* Compound or an element able to exist in two distinct crystalline forms, depending upon its maintenance above or below a certain transition temperature.

Encapsulating \in-ˈkap-sə-ˌlāt-iŋ\ *v* (1876) The process of combining elements to create a new entity.

Encapsulation *n* The process of applying a fairly thick coating that conforms to the shape of the coated object. The coating, of either thermoplastic or thermosetting resin, may be applied by brushing, dipping, spraying, or thermoforming. The process is much used for the protection and insulation of electrical components and assemblies. See also ▶ Embedding, ▶ Impregnation, ▶ Potting, ▶ and Microencapsulation.

Encapsulization *n* The enclosure of adhesive particles with a protective film that prevents them from coalescing until such time as proper pressure or salvation is applied.

Encaustic \in-ˈkó-stik\ *n* [*encaustic*, adj, fr. L *encausticus*, fr. Gk *enkaustikos*, fr. *enkaiein* to burn in, fr. *en-* + *kaiein* to burn] (1601) (1) Painted with a mixture of a paint solution and wax which, after application, is set by heat. (2) Colors which have been applied to brick, glass, porcelain, and tile and set by the application of heat.

Encounter *n* The period of time during which two reactant particles are trapped by a solvent–molecule cage in a liquid–solution reaction.

End *n* (1) A strand of roving consisting of a given number of filaments gathered together. The group of filaments is considered to be an *end* or *strand* before twisting, and a *yarn* after the twist has been applied. (2) An individual warp yarn, thread, or fiber. (3) A short length or remnant of fabric.

End-Capping *n* Conversion, by chemical reaction, of the end groups of polymer chains to less reactive, more stable groups, thus preventing "unzipping" of chains and rendering the polymer itself more stable to processing. An example is the conversion of –OH end groups in polyoxymethylene to acetate groups.

End Grain *n* The surface of timber exposed when a tree is felled or when timber is cross-cut in any other way.

End Group *n* A chemical group or radical forming the end of a polymer chain. These are normally different from the repeating unit group or groups. Although end groups constitute a minute fraction of the polymer, they may vary considerably from the main-chain chemical structure and may exert effects on polymer properties that are stronger than one would expect from their numbers.

End-Group Analysis *n* The quantitative determination, by chemical or spectrographic methods, of the number of end groups present in a sample of polymer. With linear, unbranched polymers, the mass of the sample divided by half the measured number of end groups equals the ▶ Number-Average Molecular Weight.

Endo - {*combining form*} [F, fr. Gk, fr. *endon* within; akin to Gk *en* in, OL *indu*, Hittite *andan* within] (1) A chemical prefix denoting an inner position, for example in a ring rather than in a side chain, or attached as a bridge within a ring. (2) When prefixing "-thermic," denoting that the reaction so labeled takes heat from the surroundings when proceeding from left to right. In both sense, the opposite of EXO-.

Endothermic \ˌen-də-ˈthər-mik\ *adj* [ISV] (1884) Pertaining to a chemical reaction or an operation that is accompanied by the absorption of heat. Opposite of exothermic.

Endothermic Reaction *n* A reaction which absorbs heat.

End Out A void caused by a missing warp yarn.

End Point *n* (1899) (1) Maximum distillation temperature when a substance is distilled. (2) Stoichiometric point as shown by an indicator, potentiometer or other means, in a titration. (3) Required values of viscosity, acid values, etc., the attainment of which indicates the conclusion of a process in resin or varnish manufacturing.

End-to-End Distance *n* The square root of the average square of the distance from the one of the polymer chain to the other end of the polymer chain.

Endurance Limit *n* (fatigue limit) The stress level below which a specimen will withstand cyclic stress indefinitely without exhibiting fatigue failure. Rigid, elastic, low-damping materials such as thermosetting plastics and some crystalline thermoplastics do not exhibit endurance limits.

Endurance Ratio *n* The ratio of the ▶ Endurance Limit, under cyclic stress reversal, to the short-time, static strength of a material. If the mode of stress is not specified, tension/compression may be presumed.

Energy \ˈe-nər-jē\ *n* [LL *energia* fr. Gk *energeia* activity, fr. *energos* active, fr. *en-* in + *ergon* work] (1599) The capability of doing work. *Potential energy* is energy due to position of one body with respect to another or to the relative parts of the same body. *Kinetic energy* is energy due to motion. Cgs units, the erg, the energy expended when a force of 1 dyne acts through a distance of 1 cm; the joule is 1×10^7 ergs. The potential energy of a mass *m*, raised through a distance *h*, where *g* is the acceleration due to gravity is

$$E = mgh.$$

The kinetic energy of mass *m*, moving with a velocity *v*, is

$$E = \frac{1}{2}mv^2$$

Energy will be given in ergs if *m* is in grams, *g* in centimeter per square second, *h* in centimeter and *v* in centimeter per second.

Energy Absorption *n* The energy required to break or elongate a fiber to a certain point.

Energy *n*, *E* The ability to do work.

Energy Dispersive X-Ray Analysis (EDXRA) This method of elemental analysis (microanalysis) is often used in conjunction with scanning electron microscopy (SEM) (Staniforth M, Goldstein J, Echlin P, Lifshin E, Newbury DA (2003) Scanning electron microscopy and X-ray microanalysis. Springer, New York). An example of an EDXRA spectrograph is shown.

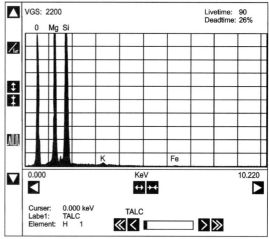

EDXRA spectrogram of talc mica particle

Energy *n*, **(Free Energy)** Usable power (as heat or electricity); also; the resources for producing such power.

Energy of a Charge *n* In ergs where Q is the change and V the potential in electrostatic units.

$$E = \frac{1}{2}QV.$$

Energy of Rotation *n* If a mass whose moment of inertia about an axis is I, rotates with angular velocity ω about this axis, the kinetic energy of rotation will be

$$E = \frac{1}{2}I\omega^2$$

Energy will be given in ergs if I is in gram-square centimeter and ω in radians per second.

Energy of the Electric Field *n* If H is the electric field intensity in electrostatic units and K the specific inductive capacity, the energy of the field in ergs per centimeter cube is

$$E = \frac{KH^2}{8\pi}$$

Energy-to-Break *n* The total energy required to rupture a yarn or cord.

Engineering Plastic *n* (1) A broad term covering those plastics, with or without fillers and reinforcements that have mechanical, chemical, electrical, and/or thermal properties suitable for industrial applications. R.B. Seymour, an outstanding authority, defined them as "….polymers….thermoplastic or thermosetting, that maintain their dimensional stability and major mechanical properties in the temperature range 0–100°C." He listed the "big five" (among neat resins) as nylons, polycarbonate, acetals, polyphenylene ether, and thermoplastic polyesters. Among many others are acrylics, fluorocarbons, phenoxy, acrylonitrile–butadiene–styrene terpolymer, polyaryl either, polybutylene, chlorinated polyether, polyether, and many polymers reinforced with ▶ Advanced Fibers. At the high end of the spectrum of performance, usable temperature range, and price, unreinforced engineering plastics are dubbed ▶ Advanced Resins.

Engineering Plastics *n* Plastics which are modified by using modifiers, additives, fillers, and reinforcements.

Engineering Polymers *n* Polymers referred to as "commodity thermoplastics," accounting for 85% of total plastics production. The major engineering polymers are ethylene, propylene, styrene, and vinyl chloride, which offer the optimum balance of easy processability and moderate properties at low cost.

English Gallon See ▶ Imperial Gallon.

English Vermillion See ▶ Mercuric Sulfide.

Engraved-Roll Coating See ▶ Gravure Coating.

Engraving *n* (1601) Machine-printing (of wallpaper) with etched-out rollers to obtain subtle and fine-effects not possible with surface printing.

Enhancement Ratio *n* In a filled or reinforced plastic, the ratio of the modulus of the filled material to that of the neat resin. The ratio is likely to be higher at low strains than high strains because of slippage between filler and matrix.

Enjay-Butyl *n* Isobutylene/isoprene copolymer, manufactured by Enjay, U.S.

Enkatherm *n* Poly(terephthaloyl oxamidrazone), manufactured by AKZO, The Netherlands.

Entangled Yarns See ▶ Compacted Yarns.

Entangling *n* (1) A method of forming a fabric by wrapping and knotting fibers in a web about each other, by mechanical means, or by the use of jets of pressurized water, so as to bond the fibers. (2) See ▶ Intermingling.

Entering *n* The process of threading each warp yarn on a loom beam through a separate drop wire, heddle, and reed space in preparation for weaving. This process may be done by hand or by a semiautomatic machine.

Enthalpy *n* \ ⁞en- ⁞thal-pē\ *n* [*en-* + Gk *thalpein* to heat] (ca. 1924) (heat content) Thermodynamically, the enthalpy of a system is the sum of the internal energy and the pressure–volume product multiplied by the pressure $H = E + pv$ where H is enthalpy or heat content, E is internal energy of the system, p is pressure, and v is volume. We are usually concerned about changes in enthalpy rather than absolute values. Enthalpies of polymers are usually stated per unit mass, e.g., kilojoule per kilogram, and are ordinarily referred to room temperature, 20–25°C, at which temperature enthalpy is arbitrarily set to zero. The rate of change of enthalpy with temperature is ▶ Heat Capacity.

Enthalpy *n*, *H* A thermodynamic quantity which is useful for describing heat exchanges taking place under constant-pressure conditions. The enthalpy of a system is defined as the sum of its energy, E, and its pressure–volume product: $H = E + PV$.

Enthalpy of Formulation *n*, ΔH_f The enthalpy change for a reaction in which compound is formed from its uncombined elements.

Entrance Angle *n* (entry angle) In an extrusion diem the total included angle, never more than 180°, of the main converging surfaces of the flow channel leading to the land area of the die.

Entropy, S \ˈen-tre-pe\ *n* [ISV ^{2}en- + Gk *tropē* change, literally, turn, fr. *trepein* to turn] (1875) (1) A measure of the unavailable energy in a thermodynamic system, commonly expressed in terms of its changes on an arbitrary scale, the entropy of water at 0°C being assigned the value of zero. (2) When a system (or sample of material) is heated or cooled from one temperature, T_1, to another T_2, the change in its entropy is given by the equation

$$\Delta S = \int_{T_1}^{T_2} \frac{dQ}{T}$$

(3) Also, entropy is a thermodynamic quantity which measures the degree of disorder or randomness in a system.

Entropy *n*, **of Elasticity** A measure of the unavailable energy in a closed thermodynamic system that is also usually considered to be a measure of the system's disorder and that is a property of the system's state and is related to it in such a manner that a reversible change in heat in the system produces a change in the measure which varies directly with the heat change and inversely with the absolute temperature at which the change takes place.

Environment \in-ˈvī-rə(n)-mənt\ *n* (1827) The sum of all external conditions and influences affecting the life, development, and survival of an organism.

Environmental Protection Agency *n* **(EPA)** Charged by the Clean Air Amendments of 1970 to the Air Quality Act of 1967 with establishing minimum air quality standards which must be incorporated in regional state standards. These standards are essentially similar to those first promulgated by the Los Angeles region of California in their Rule 66.

Environmental Stress Cracking *n* The formation of internal or external cracks in a plastic caused by tensile stresses well below its short-time strength, and induced by exposure to heat, solvent vapor, or chemically active solutions. ASTM Test F 1248 describes the measurement of the environmental stress-cracking resistance of polyethylene pipe in the presence of a surface-active agent. Other ASTM tests treating this subject are D 1693, D 2561, D 2951, and, in Section 15.03, F 484, F 791, and F 1164.

Enzyme \ˈen-ˌzīm\ *n* [Gr *Enzym*, fr. MGk *enzymos* leavened, fr. Gk *en-* + *zymē* leaven] (1881) Any of numerous proteins or conjugated proteins produced by, and functioning as, biochemical catalysts in living organisms.

EOS Abbreviation for ▶ Equation of State.

Eosine \ˈē-ə-sən\ *n* [ISV, fr. Gk *ēōs* dawn] (1866) $C_{20}H_8Br_4O_5$ Red acid dye used in the manufacture of lake pigments. Sodium salt of brominated fluorescein.

EP *n* (1) Abbreviation for ▶ Epoxy Resin or ▶ Epoxide. (2) (Usually E/P) Abbreviation sometimes used for copolymers of ethylene and propylene.

EPA See ▶ Environmental Protection Agency.

EPDM *n* Elastomer from ethylene, propylene, and a diene. Abbreviation for Elastomeric Terpolymer from ethylene, propylene, and a conjugated diene. See ▶ Ethylene–Propylene Rubber.

EPDM Rubber *n* Sulfur-vulcanizable thermosetting elastomers produced from ethylene, propylene, and a small amount of nonconjugated diene such as hexadiene. Have good weatherability and chemical and heat resistance. Used as impact modifiers and for weather stripping, auto parts, cable insulation, conveyor belts, hoses, and tubing. Also called EPDM.

Epi- (1) A prefix signifying a chemical compound or group differing from a parent compound or group by having a bridge connection. (2) EPI: See ▶ Epichlorohydrin.

Epichlorohydrin \ˌe-pi-ˌklōr-ə-ˈhī-drən\ *n* (ca. 1891) (chloropropylene oxide, EPI) An active solvent for cellulosic and other resins, and a key reactant for epoxy resins having the structure shown below

EPI is highly reactive with polyhydric phenols such as bisphenol A and forms glycidyls with many compounds containing active hydrogens. Epichlorohydrin (ECH) is used primarily in the manufacture of epoxy resins and synthetic glycerol. The main other uses include the production of epichlorohydrin elastomers, polyamide–epichlorohydrin resins, water treatment chemicals, and a variety of glycidyl derivatives.

Epichlorohydrin Rubber *n* (CO, CEO) Any of several elastomers comprising polymers and copolymers of epichlorohydrin, with good high-temperature resistance, low-temperature flexibility, resistance to fuels, oils, and ozone, and low gas permeability. The homopolymer (CO) is a saturated, aliphatic polyether with a chloromethyl side chain. The ECO type is an equimolar copolymer of epichlorohydrin and ethylene oxide.

Epikote *n* Epoxide resin, manufactured by Shell, The Netherlands.

Episulfide *n* The sulfur analog of epoxides in which the sulfur is part of a ring. The two most important members are ethylene sulfide and propylene sulfide. Episulfides are starting materials for ▶ Polysulfide Rubber.

Epitropic Fibers *n* Fibers with an altered surface property, e.g., electrically conducting, abrasive, etc.

EPM *n* Elastomer from ethylene and propylene. Abbreviation for ▶ Ethylene–Propylene Rubber.

Epon *n* Epoxide resin, manufactured by Shell, The Netherlands.

Epoxidation \(ˌ)e-ˌpäk-sə-ˈda-shən\ *n* (1944) A chemical reaction in which an oxygen atom is joined to an olefinically unsaturated molecule to form a cyclic, three-membered either. The products of epoxidation are known as *Oiranes* or ▶ Epoxides.

Epoxide *n* (1930) Any compound containing the oxirane structure, a three-membered ring containing two carbon atoms and one oxygen atom. The most important members are ethylene oxide and propylene oxide.

Epoxide Equivalent *n* The mass of resin in grams that contains one gram-equivalent of ▶ Epoxide.

Epoxidized Soybean Oil See ▶ Epoxy Plasticizer.

Epoxies See ▶ Epoxy Resins.

Epoxy *adj* [*epi-* + *oxy*] (1916) (epoxy group, oxirane group) In textiles, a compound used in durable-press applications for white fabrics. It provides chlorine resistance but causes loss of tensile strength. A label denoting an oxygen atom joined in a ring with two carbon atoms, as shown below.

Epoxy Adduct *n* Resin having all the required amine incorporated but requiring additional epoxy resin for curing.

Epoxy Compounds See ▶ Epoxides.

β-(3,4-Epoxycyclohexyl) Ethyltrimethoxy Silane *n* A coupling agent for reinforced polyester, epoxy, phenolic, melamine, and many thermoplastics.

Epoxy Ester *n* An epoxy resin partially esterified with fatty acids, rosin, etc.; single package epoxy.

Epoxy Foam *n* Two basic types of epoxy foams are in use, *chemical foams* and *syntactic foams*. Chemical-foam compositions contain the resin, curing agent, blowing agent, wetting agent, and a small percentage of an inert organic compound such as toluene to dissipate the exothermic heat of curing, and thus control the foaming action. Because foaming is rapid, the curing agent is withheld until all the other ingredients have been mixed, to be added just prior to casting. These systems may also contain amine-terminated polyamide resins to impart resiliency to the foam. In syntactic foams, the voids are provided by hollow phenolic microspheres and the resin does not foam but acts as a binder for the spheres. Epoxy foams are used in casting, in potting and encapsulating of electrical assemblies, in insulating coatings for chemical-storage tanks, and in cores of laminates for aircraft and boats.

Epoxy-Novolac Resin *n* A two-step resin made by reacting epichlorohydrin with a phenol–formaldehyde condensate. Such resins are also known as *thermoplastic*, B-stage phenolic resins that are in a state of partial cure. Whereas bisphenol-based epoxy resins contain up to two epoxy groups per molecule, the epoxy novolacs may have seven or more such groups, producing more tightly crosslinked structures in the cured resins. Thus they are stronger and superior in other properties.

Epoxy Number *n* The number of gram-equivalents of epoxy groups per 100 g of polymer, equal to 1/100 of the reciprocal of the epoxides equivalent.

Epoxy Paint *n* Paint based on an epoxy resin.

Epoxy Plasticizer *n* (epoxides plasticizer) Any of a large family of plasticizers obtained by the epoxidation of vegetable oils or fatty acids. The two main types are (a) epoxidized unsaturated triglycerides, e.g., soybean oil and linseed oil; and (b) epoxidized esters of unsaturated fatty acids e.g., oleic acid, or butyl-, octyl-, or decyl- esters. Most epoxy plasticizers have a heat-stabilizing effect and they are often used for stabilization in conjunction with other stabilizers. Epoxidized oils generally have good resistance to extrusion and migration and low volatility, but they cannot be used as sole plasticizers in unfilled vinyl compounds and hence are not considered to be primary plasticizers. Certain epoxidized soybean oils have been FDA-approved for food-contact use.

Epoxy Plastics *n* Plastics based on resins made by the reaction of epoxides or oxiranes with other materials such as amines, alcohols, phenols, carboxylic acids, acid anhydrides, and unsaturated compounds.

Epoxy Resin *n* (1950) Any of a family of thermosetting resins containing the oxirane group (See ▶ Epoxy for structure). Originally made by condensing epichlorohydrin and bisphenol A, epoxy resins are not more generally formed from low-molecular-weight diglycidyl ethers of bisphenol A and modifications thereof; or, another type, by the oxidation of olefins with peracetic acid. Depending on molecular weight, the resins range from liquids to solids. The liquids, used for casting, potting, coating, and adhesives, are cured with amines, polyamides, anhydrides, or other catalysts. The solid resins are often modified with other resins and unsaturated fatty acids. Epoxy resins are widely used in reinforced plastics, have strong adhesion to glass fibers. Epoxies based on epoxidized heterocyclic hydantoin are useful in electrical composites because their thermal expansion coefficient can be matched to that of copper. Their low viscosities are effective in wetting the various reinforcing materials used with them. Fast-curing epoxies are based on the diglycidyl ether of 4-methylol resorcinol (DGEMR). The methylol group appears to effectively catalyze the curing reactions. This resin is curable with all types of conventional epoxy hardeners including aliphatic and aromatic amines, anhydrides, and amidoamines. DGEMR cures approximately 30 times as fast as a conventional bisphenol A epoxy and two to five times as fast as older fast-gelling epoxies, and at lower temperatures. DGEMR may be formulated with flexibilizers and fillers without prolonging gel time. These same properties make the resin well suited for adhesives, coatings, and low-temperature applications. See also ▶ Epoxy-Novolac Resin.

Epoxy Resins *n* Plastic or resinous materials used for strong, fast-setting adhesives, as heat resistant coatings and binders, etc. Crosslinking resins based on the reactivity of the epoxides group. One common type is the resin made from epichlorohydrin and bisphenol A. Aliphatic polyols such as glycerol may be used instead of the aromatic bisphenol A or bisphenol F.

Epoxy Stabilizer *n* (epoxides stabilizer) Most ▶ Epoxy Plasticizers also serve as stabilizers because of the ability of the epoxides group to accept HCl, or to serve as an intermediate, in the presence of metallic salts, to convert HCl to a metallic chloride. Epoxy stabilizers are most often used in conjunction with barium–cadmium and other stabilizers, with which they have a synergistic effect.

EPR *n* Abbreviation for ▶ Ethylene–Propylene Rubber.

EPS *n* Anbbreviation for ▶ Expanded Polystyrene. See ▶ Polysty-Rene Foam.

Epsom Salts \ˈep-səm-\ *n* [*Epsom*, England] (1876) $MgSO_4 \cdot 7H_2O$. Magnesium sulfate. Water-soluble magnesium salt for the preparation of precipitated soaps. Rhombic, colored or white; sp gr, 1.68.

Equalizer Rod A metal rod wound with a fine wire around its axis so that an ink can be drawn down evenly and at a given thickness across a piece of paper. *Also called a Meyer Rod or Bar.* See ▶ Wire-Wound Rod.

Equant *n* A shape having nearly equal dimensions.

Equation of State *n* (EOS) For an ideal gas, if the pressure and temperature are constant, the volume of the gas depends on the mass, or amount of gas. Then, a single property called the gas density (ratio of mass/volume). If the mass and temperature are held constant, the product of pressure and volume are observed to be nearly constant for a real gas. The product of pressure and volume is exactly for an ideal gas. This relationship between pressure and volume is called Boyle's Law. Finally, if the mass and pressure are held constant, the volume is directly proportional to the temperature for an ideal gas. This relationship is called Charles and Gay-Lussac's Law. The gas laws of Boyle and Charles and Gay-Lussac can be combined into a single equation of state: PV = nRT, where p is pressure, V is volume, T is absolute temperature, n is number of moles, and R is the universal gas constant (Atkins PW, Atkins P, De Paula J (2001) Physical chemistry. W.H. Freeman, New York; Perry RH, Green DW (1997) Perry's chemical engineer's handbook, 7th edn. McGraw-Hill, New York). Anerodynamicists us a different form of the equation of state that is specialized for air. Regarding polymers and monomers, equation of state is an equation giving the specific volume (*v*) of a polymer from the known temperature and pressure and, sometimes, from its morphological form. An early example is the modified Van der Waals form, successfully tested on amorphous and molten polymers. The equation is:

$$v = b + RT/M(P + \pi)$$

where b = the "unfree" specific volume occupied by the polymer molecules, roughly 90 + percent of v, R = the universal molar-energy constant, T = the absolute temperature, P = pressure, M = an empirical molecular weight that for several thermoplastics has been closely equal to the mer weight, and π = an empirical internal pressure much larger than the highest injection-molding pressures. A number of more complex models have since been introduced and tested against experimental data (Carley JF (ed) (1993) Whittington's dictionary of plastics. Technomic Publishing).

Equilibrium \ˌē-kwə-ˈli-brē-əm\ *n*. [L *aequilbrium*, fr. *aequilibris* being in equilibrium, fr. *aequi-* + *libra*, weight, balance] (1608) A state of balance between opposing forces or actions that is either static (as in a body acted on by forces whose resultant is zero) or dynamic (as in a reversible chemical reaction when the velocities in both directions are equal) Chemical: A state of affairs in which a chemical reaction and its reverse reaction are taking place at equal velocities, so that the concentrations of reacting substances remain constant.

Equilibrium Condition *n* the condition, that the mass-action expression equals the value of the equilibrium constant for a reaction, which is satisfied when the reaction system is at equilibrium.

Equilibrium Constant *n* (1929) The product of the concentrations (or activities) of the substances produced at equilibrium in a chemical reaction divided by the product of concentrations of the reacting substances, which concentration raised to that power which is the coefficient of the substance in the chemical equation.

Equivalence Point \i- ˈkwiv-lən(t)s ˈpóint\ *n* The state in a titration at which equal numbers of equivalents of reactants and products have been mixed.

Equivalent, Acid-Base *n* The quantity of an acid (or base) which will furnish (or react with) 1 mole of H^+.

Equivalent Conductance *n* The equivalent conductance of an electrolyte is defined as the conductance of a volume of solution containing one equivalent weight of dissolved substance when placed between two parallel electrodes 1 cm apart, and large enough to contain between them all of the solution. λ is never determined directly, but is calculated from a specific conductance. If C is the concentration per cubic centimeter is C/1,000, and the volume containing one equivalent of the solute is, therefore, 1,000/C. Since L_s is the conductance of a centimeter cube of the solution, the conductance of 1,000/C cc, and hence λ will be

$$\Lambda = \frac{1000L_s}{C}$$

Equivalent, Redox *n* The quantity of an oxidizing agent (or a reducing agent) which will accept (or furnish) 1 mole of electrons.

Equivalent Single Yarn Number See ▶ Yarn Number.

Equivalent Weight *n* (1904) (combining weight, equivalent mass) The atomic or formula weight of a given element or ion divided by its valence in a reaction under consideration. Elements entering into combination always do so in quantities proportional to their equivalent weights. In oxidation–reduction reactions the equivalent weights of the reacting entities are dependent upon the change in oxidation numbers of the particular substances.

Erioglaucine Dye See ▶ Acid Dyes.

Erlangen Blue *n* Another name for Prussian blue.

Erode \i- ˈrōd\ *v* [L *erodere* to eat away, fr. *e-* + *rodere* to gnaw] (1612) See ▶ ETCH.

Erosion *n* (1541) (1) Wearing away of the top coating of a painted surface, e.g., by chalking or by the abrasive action of windborne particles of grit, which may result in exposure of the underlying surface. (2) Phenomenon manifested in paint films by the wearing away of the finish to expose the substrate or undercoat. The degree of failure is dependent on the amount of substrate or undercoat visible. Erosion occurs as the result of chalking or by the abrasive action of windborne particles of grit.

Erosion Breakdown *n* In an electrical-conductor insulation, deterioration caused by chemical attack of corrosive chemicals such as ozone and nitric acid that are formed by corona discharge from a high-voltage cable. This breakdown can occur even for the most chemically stable polymers, such as fluorocarbons, after long exposure to the condition.

Erosion Control Fabrics See ▶ Geotextiles.

Erosion Resistance *n* The ability of a coating to withstand being gradually worn away by chalking or by the abrasive action of water or windborne particles of grit. The degree of resistance is dependent upon the amount of coating retained. See ▶ Erosion.

Erucamide *n* $CH_3(CH_2)_7CH=CH(CH_2)_{11}COONH_2$. The amide of *cis*-13-docosenoic acid, used in fractional percentages as a lip agent in polyethylene film resins.

Erucyl Alcohol *n* $CH_3(CH_2)_7CH=CH(CH_2)_{11}CH_2OH$. A monounsaturated, fatty alcohol used as a mold lubricant. IUPAC Name: Doclos-13-en-1-0l.

Erucic Acid \i- ˈrü-sik-\ *n* [NL *Eruca*, genus of herbs, fr. L, colewort] (1869) $CH_3(CH_2)_7CH=CH(CH_2)_{11}COOH$. Unsaturated fatty acid found in many vegetable oils. Properties: mp, 33°C; bp, 281°C/30 mm Hg; sp gr, 0.860/55°C; and iodine value, 75.2. *Also known as cis-13-Docosenoic Acid.*

Erythrene *n* 1-3 butadiene. Old Syn: ▶ Butadiene.

Erythritol *n* $H(CHOH)_4H$. Tetrahydric alcohol. Properties: mp, 112°C; bp, 330°C. IUPAC Name: Butane-1, 2, 3, 4-tetraol.

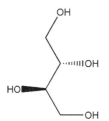

ESC *n* Abbreviation for ▶ Environmental Stress Cracking.

ESCA Energy Spectoscopy Chemical Analysis. An example of an ESCA spectrum for chloride is shown (Siegbahn K (1967) ESCA spectrogram of paint pigment, lead carbonate and calcium sulfate, atomic, molecular and solid state structure studied by means of electron spectroscopy. Almquist & Siks, Uppsala, Sweden).

Escaping Tendency *n* The tendency shown by a substance to escape from its phase to another.

ESO *n* Abbreviation for ▶ Epoxidized Soybean Oil. See ▶ Epoxy Plasticizers.

Esparto Wax \is- ˈpär-(ˌ)tō-\ Hard vegetable wax obtained from esparto grass, indigenous to North Africa and parts of Spain. It is used to some extent as a substitute for carnauba, candellia, and similar hard waxes. Properties: mp, approximately 71°C; acid value, 30, saponification value, 75.

ESR *n* Abbreviation for Electron-Spin-Resonance Spectros-Copy.

Essential Oils *n* Volatile oils or essences derived from vegetation and characterized by distinctive odors and a substantial measure of resistance to hydrolysis. Chemically, essential oils are often principally terpenes. Some essential oils are nearly pure single compounds. Some contain resins in solution and are then called oleoresins or balsams (Shahidi F, Bailey AE (eds) (2005) Bailey's industrial oil and fat products. Wiley, New York, New York; Langenheim JH (2003) Plant resins: chemistry, evolution ecology and ethnobotany. Timber Press, Portland, OR; (2001) Paint: pigment, drying oils, polymers, resins, naval stores, cellulosics esters, and ink vehicles, vol 3. American Society for Testing and Material, West Conshohocken, Pennsylvania).

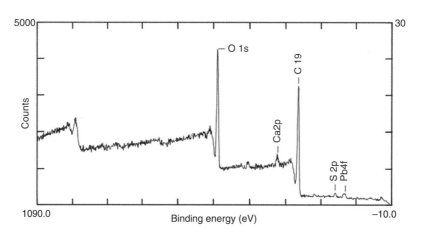

ESCA spectrogram of point pigment, lead carbonate and calcium sulfate

Estane *n* Polyurethane. Manufactured by Goodrich, U.S.

Ester \ˈes-tər\ *n* [Gr. fr. *Essigäther* ethyl acetate, fr. *Essig* vinegar + *Äther* ether] (ca. 1852) An organic compound with the general formula R–C–O–R' analogous to an inorganic salt. Esters are formed by reacting an acid with an alcohol, or by the exchange of a replaceable hydrogen atom of an acid for an organic alkyl radical. Esters of many monofunctional alcohols and organic acids are oily, fruity-smelling liquids, forming important families of solvents and plasticizers. When the alcohol selected is polyfunctional, that is, contains two or more –OH groups, and the acid is di- or polybasic, long chains of repeating units can be formed by their reaction. These are polyesters.

Ester Gum *n* Resin made from rosin or rosin acids and a polyhydric alcohol, such as glycerol or pentaerythritol. Used as an ingredient in certain printing ink varnishes (Whistler JN, BeMiller JN (eds) (1992) Industrial gums: polysaccharides and their derivatives. Elsevier Science and Technology Books).

Esterification \-ˌster-ə-tə-ˈkā-shən\ *n* (ca. 1905) The chemical process of combining an acid and an alcohol to form an ester. Cellulose acetate is an ester formed by the reaction of acetic acid and the hydroxyl groups of cellulose. Polyethylene terephthalate, the most common fiber-forming polyester, is a product of esterification of terephthalic acid with ethylene glycol (Morrison RT, Boyd RN (1992) Organic chemistry, 6th edn. Prentice-Hall, Englewood Cliffs, NJ).

Ester Interchange *n* (ester exchange) A reaction between an ester and another compound in which occurs an exchange of alkoxy or acyl groups, resulting in the formation of a different ester. When an ester is reacted with an alcohol, the process is called *alcoholysis*; reaction between an ester and an acid is called *acidolysis*. Tester interchanges are used in producing plasticizers, polyvinyl alcohol, acrylics, polyesters, and polycarbonates (Morrison RT, Boyd RN (1992) Organic chemistry, 6th edn. Prentice-Hall, Englewood Cliffs, NJ). See ▶ Transesterification.

Ester Number *n* Difference between the saponification value and the acid number. Therefore, it is the number of milligram of potassium hydroxide necessary to saponify the neutral esters in 1 g of fat, wax, or oil.

Esters *n* A class of organic liquids, used as solvents, which are the products of the reactions of organic acids and alcohols. Examples are ethyl acetate, dioctyl phthalate.

Esthetics See ▶ Aesthetics.

Eta (η) *n* Greek letter; symbol for absolute viscosity.

Etch \ˈech\ *v* [D *etsen*, fr. Gr. *ätzen* to etch, corrode, fr. OH Gr *azzen* to feed; akin to OH Gr *ezzan* to eat] (1634) Wear away or roughen a surface with, or as if with, an acid. Erode is sometimes used as a synonym.

Etching *v* (1634) (1) In lithography, the use of acidic substances to produce a surface in the nonprinting areas that is receptive to the fountain solution but not to the ink. (2) In engraving, a treatment with acid or by mechanical means to make certain areas considerably lower than the surface of the engraving.

Etching Primer See ▶ Wash Primer.

ETFE *n* Abbreviation for ▶ Ethylene–Tetrafluoroethylene Co-Polymer.

Ethanal Syn: ▶ Acetaldehyde.

Ethanite *n* Synthetic rubber made by the interaction of potassium polysulfide and ethylene dichloride.

Ethanol *n* \ˈe-thə-ˌnȯl, -ˌnōl, *British also* ˈē-\ *n* (1900) Syn: ▶ Ethyl Alcohol.

Ethanolamine \ˌe-thə-ˈnä-lə-ˌmēn\ *n* (1897) NH₂(CH₂)₂OH. Bp, 171°C/150 mm Hg. *Known also as Monoethanolamine, Colamine, 2-Amino Ethanol, and Betahyroxyamine.*

Ethanolurea *n* NH₂CONHCH₂CH₂OH. A white compound melting at 71–74°C. It condenses with formaldehyde to form permanently thermoplastic, water-soluble resins. Simple urea can be incorporated in the condensation reaction to give modified resins with any desired degree of water solubility and flexibility, both of which properties increase with urea content.

Ethene \ˈe-ˌthēn\ *n* (1873) IUPAC's name for ▶ Ethylene.

Ethenoid Plastics *n* (1) Plastics made from monomers containing the polymerizable double-bond group C=C, for example ethylene. Thermosetting ethenoid resins are made from monomers for linear polymers capable of giving crosslinked structures as a result of double-bond polymerization. (2) A British generic term that includes acrylic, vinyl, and styrene plastics.

Ether \ē-thər\ *n* [ME, fr. L *aether*, fr. Gk *aithēr*, fr. *aithein* to ignitem blaze; akin to OE *ād* pyre] (14c) (1) Any organic compound in which an oxygen atom is interposed between two carbon atoms or organic radicals in the molecular structure. Ethers are often derived from alcohols by elimination of one molecule of water from two molecules of alcohol. (2) Specifically, diethyl ether, $(C_2H_5)_2O$.

Etherified Urea Resins *n* Urea resins in which the methylol groups have been etherified with suitable alcohols. Alcohols commonly used include butyl and octyl alcohols. The etherified resins are characterized by improved hydrocarbon tolerance, some types permitting substantial additions of aliphatic hydrocarbons.

Ethocel *n* Cellulose ether, manufactured by Dow, U.S.

Ethyl \e-thəl\ *n* [Gr *Ethyl* (now *Äthyl*), fr. *Äther* ether + -*yl*] (1838) —C_2H_5. Monovalent alkyl radical.

•H_2C——

Ethyl Abietate *n* $C_{19}H_{29}COOC_2H_5$. Amber colored viscous liquid which hardens upon oxidation. Properties: bp, 350°C; sp gr. 1.03; flp, 171°C (340°F).

Ethyl Acetanilide *n* (ethyl phenylacetamide) A substitute for camphor in the manufacture of celluloid. IUPAC Name: N-Phenyl-butyramide.

n-Ethyl Acetanilide *n* $C_6H_5N(C_2H_5)COCH_3$. Properties: bp, 258°C; flp, 124°C (255°F). Substitute for camphor in the nitrocellulose industries. IUPAC Name: N-Ethyl-N-phenyl-acetamide.

Ethyl Acetate *n* (1874) $CH_3COOC_2H_5$. A colorless liquid made by heating acetic acid and ethyl alcohol in the presence of sulfuric acid, then distilling. It is a powerful solvent for ethyl cellulose, polyvinyl acetate, cellulose acetate–butyrate, acrylics, polystyrene, and coumarone–indene resins. It is also used in flexographic and rotogravure inks. Although it is highly flammable, it is the least toxic of common industrial solvents. Properties: bp, 77°C; sp gr, 0.−901/20°C; flp. 0.56°C (31°F); refractive index, 1.373; vp, 77 mm Hg/20°C. Also known as *Acetic Ether* and *Acetic Ester*.

Ethylacetic Acid *n* Syn: ▶ Butyric Acid.

Ethyl Acetyl Ricinoleate *n* $C_{17}H_{32}(OCOCH_3)COOC_2H_5$. Properties: bp, 400°C; sp gr, 0.931.

Ethyl Acrylate *n* $CH_2=CHCOOC_2H_5$. A colorless liquid, insoluble in water; miscible with most organic solvents. Used in the manufacture of synthetic resins. Sp gr, 0.9283; bp, 101°C. IUPAC Name: Acrylic acid ethyl ester.

Ethyl Alcohol *n* (1869) (alcohol, ethanol, grain alcohol) An alcohol used, in denatured form, as a solvent for ethyl cellulose, polyvinyl acetate, and polyvinyl butyrate. The industrial grade of undenatured alcohol usually contains 5 wt% water. Properties: bp, 78°C; sp gr. 0.7938/15°C; refractive index, 1.367. The pure compound is called ▶ Absolute Alcohol. *Also known as Alcohol, Grain Alcohol, Ethanol, Fermentation Alcohol, Spirit of Wind, Ethyl Hydroxide, Cologne Sprits, and EtOH.*

Ethyl Aldehyde *n* Syn: ▶ Acetaldehyde.

Ethyl Aluminum Dichloride *n* $C_2H_5AlCl_2$. A clear, yellow, flammable liquid, a catalyst for olefin polymerization.

Ethyl Aluminum Sesquichloride *n* $(C_2H_5)_3Al_2Cl_3$. A catalyst for olefin polymerization.

Ethylbenzene \ˌe-thil-ˈben-ˌzēn\ *n* [ISV] (1873) Colorless liquid with an aromatic odor. Used as an intermediate in styrene production and as a solvent. Properties: molecular weight, 106.16; bp, 136°C; sp gr, 0.8673.

Ethyl Benzoate *n* (benzoic ether) A colorless liquid derived by heating ethyl alcohol and benzoic acid in the presence of sulfuric acid. It is a solvent for cellulosics.

2-Ethylbutyl Acetate *n* A solvent for cellulose nitrate.

Ethyl Butyrate *n* (ethyl butanoate) A solvent for cellulosics.

Ethyl Carbamate *n* Syn: ▶ Urethane.

Ethyl Carbonate *n* (glycol carbonate, 1,3-dioxolan-2-one) A solvent for many polymers and resins. Syn: ▶ Diethyl Carbonate.

Ethyl Cellulose *n* (1936) (EC) An ethyl ether of cellulose formed by reacting cellulose steeped in alkali with ethyl chloride; it is a white granular thermoplastic resin. Since the repeating units are etheric, it is chemically different from other cellulosics, which are esters, and is therefore not compatible with them. EC resin can be injection molded, extruded, cast into film, or used as a coating material. It has the lowest density of all cellulosic plastics, good toughness and impact resistance, and is dimensionally stable over a wide temperature range. Syn: ▶ Carbitol.

Ethyl Chloride *n* (ca. 1891) (chloroethane) A colorless gas at ambient conditions, used in the production of ethyl cellulose by reaction with sodium cellulose.

Ethyl Citrate See ▶ Triethyl Citrate.

Ethylene *n* (Ethene) (bicarburetted hydrogen, ethene) A colorless, flammable gas derived by cracking petroleum and by distillation from natural gas. In addition to serving as the monomer for polyethylene, it has many uses in the plastics industry including the synthesis of ethylene oxide, ethyl alcohol, ethylene glycol (used in making alkyd and polyester resins), ethyl chloride, and other ethyl esters. Properties: molecular weight, 28; bp, $-1.025°C$; sp gr of liquid, $0.610/0°C$.

Ethylene–Acid Copolymer Resins *n* Resins that are flexible, specialty thermoplastics created by high-pressure copolymerization of ethylene(E) and methacrylic acid (MAA) or acrylic acid(AA).

Ethylene–Acrylic Acid Copolymer *n* (EAA) Either of the block or random copolymers of ethylene and acrylic acid whose ionic character gives strong adhesion to metals and other surfaces. Their toughness has created uses in multilayer packaging films and golf-ball covers. Two trade names are DuPont's Surlyn® (block) and Dow's EAA (random).

Ethylene Acrylic Rubber *n* Copolymers of ethylene and acrylic esters. Have good toughness, low temperature properties, and resistance to heat, oil, and water. Used in auto and heavy equipment parts.

N,N′-Ethylene Bis-Stearamide *n* (Acrawax C®) A lubricant used in acrylonitrile-butadiene–styrene resins, PVC, and polystyrenes.

Ethylene-bis Tris-(2-cyanoethyl) Phosphonium Bromide *n* (ECPB) A flame retardant for thermoplastics. In polymethyl methacrylate, 20% ECPB caused the resin to become opaque and reduced its burning rate to zero.

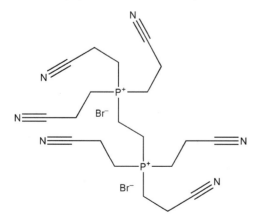

Ethylene Carboxylic Acid *n* A little used Syn: ▶ Acrylic Acid.

Ethylene Chloride *n* Syn: ▶ Ethylene Dichloride.

Ethylene Chlorotrifluoroethylene Copolymer *n* (ECTFE, E/CTFE) A fluoroplastic with good mechanical, thermal, electrical, processing, and resistance properties.

Ethylene Copolymers See ▶ Ethylene Polymers.

Ethylene Dichloride *n* $ClCH_2CH_2Cl$. Colorless, oily liquid used for metal degreasing; paint and varnish remover. Properties: bp, 83°C; sp gr, $1.-252/20°C$; vp, 65 mm Hg/20°C; mol wt. 99. *Also known as 1,2-Dichlorethane.*

Ethylene-Ethyl Acrylate Copolymer *n* (EEA, E/EA) A family of Elastomeric resins similar in appearance to polyethylene, but possessing properties like those of rubber and flexible vinyls.

Ethylene–Ethyl Acrylate Resins Ethylene Glycol *n* CH_2OHCH_2OH. A type of alcohol, completely miscible in water, used in latex and water-based paints to provide stability when frozen; used in heating and cooling systems as a fluid for transferring heat. Bp; 197°C; sp gr, 1.12/4°C; flp, 111°C (232°F). Used as a diol in manufacturing resins. *Also known as Ethylene Alcohol.*

Ethylene Glycol *n* (1901) (ethanediol, ethylene alcohol, glycol) $HOCH_2CH_2-OH$. A clear, syrupy liquid used as a solvent for cellulosics, particularly cellophane, and in the production of alkyd resins and polyethylene terephthalate.

Ethylene Glycol Diacetate *n* A very slowly evaporating solvent for cellulosic and acrylic resins, sometimes used as a fugitive plasticizer for vinyls and acrylics.

Ethylene Glycol Dibenzoate *n* A plasticizer for cellulosic resins, having limited compatibility with some vinyl resins.

Ethylene Glycol Dibutyrate (Glycol Dibutyrate) A plasticizer for cellulosic plastics.

Ethylene Glycol Dipropionate *n* (glycol propionate) A plasticizer for cellulosic resins.

Ethylene Glycol Monoacetate (glycol monoacetate) A solvent for cellulose nitrate and cellulose acetate.

Ethylene Glycol Monobenzyl Ether *n* (benzyl cellosolve) A solvent for cellulose acetate.

Ethylene Glycol Monobutyl Ether *n* (2-butoxyethanol, butyl cellosolve) A colorless liquid used as a solvent for cellulosic, phenolic, alkyd, and epoxy resins, especially in varnish and other coating formulations.

Ethylene Glycol Monobutyl Ether Acetate *n* A colorless liquid with a fruity aroma, used as a high-boiling solvent for cellulose nitrate, epoxy resins, and as a film-coalescing aid for polyvinyl-acetate latex.

Ethylene Glycol Monobutyl Ether Laurate *n* A plasticizer for cellulosics, polystyrene, and vinyls.

Ethylene Glycol Monobutyl Ether Oleate *n* A plasticizer for cellulose nitrate, ethyl cellulose, and PVC.

Ethylene Glycol Monobutyl Ether Stearate *n* A plasticizer for cellulose nitrate, ethyl cellulose, polystyrene, and polyvinyl butyral.

Ethylene Glycol Monoethyl Ether *n* (cellosolve, ethyl cellosolve) A solvent for cellulose nitrate, phenolic, alkyd, and epoxy resins. It is colorless, nearly odorless, has a low evaporation rate, and imparts good flow properties to coatings.

Ethylene Glycol Monoethyl Ether Acetate *n* (Cellosolve acetate) $C_2H_5OC_2H_4OOCCH_3$. A solvent for cellulose nitrate, ethyl cellulose, vinyl polymers and copolymers, polymethyl methacrylate, polystyrene, epoxy, coumarone–indene, and alkyd resins.

Ethylene Glycol Monoethyl Ether Laurate *n* A plasticizer for cellulosic and vinyl resins, and polystyrene.

Ethylene Glycol Monoethyl Ether Ricinoleate *n* A plasticizer.

Ethylene Glycol Monomethyl Ester *n* $CH_3OC_2H_4OOCC_{17}$–H_{35}. A plasticizer for cellulosics and polystyrene, having limited compatibility with other thermoplastics.

Ethylene Glycol Monomethyl Ether *n* (2-methoxyethanol, methyl cellosolve) A solvent for cellulose esters.

Ethylene Glycol Monomethyl Ether Myristate *n* A plasticizer for cellulosic plastics, PVC, and polyvinyl butyral.

Ethylene Glycol Monomethyl Ether Oleate *n* A plasticizer for cellulosic and vinyl resins.

Ethylene Glycol Monooctyl Ether *n* A solvent for cellulose esters, and a plasticizer.

Ethylene Glycol Monophenyl Ether *n* A solvent for cellulosics, vinyls, phenolics, and alkyd resins.

Ethylene Glycol Monoricinoleate *n* (ethylene glycol ricinoleate) plasticizer and an intermediate for urethane polymers.

Ethylene–Methyl Acrylate Copolymer *n* (EA, E/EA) An elastomer vulcanizable with peroxides or diamines. It resists attack by oils and temperatures to 175°C.

Ethylene Oxide *n* (1898) (epoxyethane) A three-membered ring compound with the formula H_2COCH_2, Colorless, flammable gas at ordinary room temperature and pressure. It is used in organic synthesis, especially in the production of ethylene glycol and it is the starting material for the manufacture of acrylonitrile and nonionic surfactants. Bp, 10.7°C; sp gr, 0.869.

Ethylene Plastic See ▶ Polyethylene.

Ethylene, Polymerization *n* A chemical reaction in which two or more molecules combine to form larger molecules that contain repeating structural units.

Ethylene Polymers *n* Ethylene polymers include ethylene homopolymers and copolymers with other unsaturated monomers, most importantly olefins such as propylene and polar substances such as vinyl acetate. The properties and uses of ethylene polymers depend on the molecular structure and weight. Also called ethylene copolymers.

Ethylene Propene Rubber *n* Stereospecific copolymers of ethylene with propylene. Used as impact modifiers for plastics. Also called EPR.

Ethylene–Propylene Rubber *n* (E/P, EPDM, SPM, EPR) Any of a group of elastomers obtained by the stereospecific copolymerization of ethylene and propylene (EOM), or of these two monomers and a third monomer such as an unconjugated diene (EPDM). Their properties are similar to those of natural rubber in many respects, and they have been proposed as potential substitutes for natural rubber in tires.

Ethylene–Tetrafluoroethylene Copolymer *n* A copolymer of ethylene and tetrafluoroethylene (DuPont Tefzel®), ETFE is readily processed by extrusion and injection molding. It has excellent resistance to heat, abrasion, chemicals, and impact, with good electrical properties.

Ethylene–Urea Resin *n* A type of AMINO RESIN.

Ethylene–Vinyl Acetate Copolymer *n* (EVA, E/VAC) Any copolymer containing mainly ethylene with minor proportions of vinyl acetate. They retain many of the properties of polyethylene but have considerably increased flexibility, elongation, and impact resistance. They resemble elastomers in many ways, but can be processed as thermoplastics.

Ethylene–Vinyl Acrylate Resins *n* Co-polymers of the polyolefins family derived from random co-polymerization of acetate and ethylene.

Ethylene–Vinyl Alcohol Copolymer *n* (EVAL, E/VAL, EVOH) A family of copolymers made by hydrolyzing ethylene–vinyl acetate copolymers with high VA content. Those containing about 20–35% ethylene are useful as barriers to many vapors and gases, though not to water. Because of their water sensitivity, they are usually sandwiched between layers of other polymers.

Ethyl Formate *n* solvent for cellulose acetate.

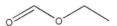

2-Ethyl-1,3-Hexanediol *n* A stable, colorless, nearly odorless, a high-boiling liquid with weak solvent action. In two-part urethane systems, the material acts as a viscosity reducer at room temperature. When the urethane mixture is heated to cure it, the diol reacts into the urethane matrix to eliminate solvent emissions.

2-Ethylhexyl- *n* An 8-carbon branched-chain radical of the formula $C_4H_9CH(C_2H_5)CH_2$–, often called *octyl* in the plastics industry. For example, the common plasticizer di-2-ethylhexyl phthalate is commonly referred to as dioctyl phthalate and by its abbreviation, DOP.

2-Ethylhexyl Acetate *n* (octyl acetate) A high-boiling retarder solvent with low evaporation rate and limited water solubility, used primarily in coating formulations based on cellulose nitrate. It is also used as a dispersant in vinyl organosols.

2-Ethylhexyl Acrylate *n* One of the monomers for acrylic resins, especially for those used in water-based paints.

2-Ethylhexyl Alcohol *n* (2-ethylhexanol, octyl alcohol) $C_4H_9CH(C_2-H_5)CH_2OH$. An involatile solvent with many uses in the plastics industry. As a solvent, it is used in coatings for stenciling, silk screening, and dipping. As an intermediate, the alcohol is an important raw material for the production of the 2-ethylhexyl esters of dibasic acids used as plasticizers, such as dioctyl phthalate, adipate, and azelate.

2-Ethylhexyl Decyl Phthalate *n* A mixed diester plasticizer for cellulosics, polystyrene, PVC, and polyvinyl acetate.

2-Ethylhexyl Epoxytallate *n* An epoxy ester used mainly as a combined plasticizer and stabilizer in vinyl compounds. At concentrations as low as 5 phr it reacts synergistically with many metallic stabilizers to provide stability comparable to similar combinations based on epoxidized soybean oils. As a partial replacement for other plasticizers, it imparts good low-temperature flexibility. It is also compatible with vinyl chloride–vinyl acetate copolymers, high-butyral cellulose acetate–butyrate resins, ethyl cellulose, polystyrene, and chlorinated rubbers.

2-Ethylhexyl Isodecyl Phthalate *n* (octyl isodecyl phthalate) $C_8H_{17}-OOCC_6H_4COOC_{10}H_{21}$. A mixed ester compatible with PVC, vinyl chloride–acetate copolymers, cellulose acetate–butyrates with higher butyrate contents, cellulose nitrate, and, in lower concentrations, with polyvinyl butyral. In vinyls, it is somewhat less volatile than dioctyl phthalate and has equivalent low-temperature properties.

2-Ethylhexyl-*p*-Hydroxybenzoate *n* A plasticizer for polyamides.

Ethyl Hydroxide See ▶ Ethyl Alcohol.
Ethyl-α-Hydroxyisobutyrate *n* $C_2H_5OOC(OH)C(CH_3)_2$. A solvent for cellulose nitrate and cellulose acetate.
Ethylidene Acetobenzoate *n* (ethylidene benzoacetate) $C_6H_5COO(CH_3-CH)COCH_3$. A solvent for cellulosics and synthetic resins.

Ethyl Lactate *n* CH$_3$CH(OH)COOC$_2$H$_5$. Used as a solvent for nitrocellulose and cellulose acetate. Colorless liquid with a mold odor. Properties: mol wt. 118; bp, 154°C; sp gr, 1.024/20°C; flp, 45°C (113°F); refractive index, 1.4111; vp. 5 min Hg/233°C.

Ethyl Levulinate *n* C$_2$H$_5$OOC(CH$_2$)$_2$COCH$_3$. A solvent for cellulose acetate.

Ethyl Methacrylate *n* A readily polymerizable monomer used for certain types of acrylic resins.

2-Ethyl-4-Methylimidazole *n* (EMI) An epoxy resin curing agent with the heterocyclic structure.

EMI is used with epoxies formed from epichlorohydrin and bisphenol. A or –F, and for novolac epoxy resins. It provides ease of compounding, long pot life, low viscosity, and nonstaining characteristics, and yields castings with excellent mechanical and electrical properties.

Ethyl-*n*-Butyl Ketone *n* (2-heptanone) A stable, colorless liquid with medium volatility, used in solvent mixtures for cellulosic and vinyl resins. When used in vinyl organosols it imparts good lone-time viscosity stability.

ethyl-n-butyl ketone

Ethyl Oleate *n* C$_2$H$_5$OOCC$_{17}$H$_{33}$. This monounsaturated fatty ester is a solvent, lubricant, and plasticizer.

Ethyl Oxalate *n* (Oxalic acid diethyl ester) Colorless, unstable, oily, aromatic liquid. Solvent for cellulose esters and ethers, many natural and synthetic resins. Sp gr., 109 20/20°C; bp, 186°C; mp, −40.6°C; fp, 75.6°C (168°F). *Also known as Diethyl Oxalate.*

Ethyl Phthalate See ▶ Diethyl Phthalate.

Ethyl Phthalyl Ethyl Glycolate (2-Ethoxy-1,3-dioxoindan-2-carboxylic acid ester) A plasticizer compatible with PVC and most common thermoplastics. It has been approved by the FDA for use in food packaging.

Ethyl Propionate *n* (Propionic acid ethyl ester) Colorless liquid, fruity odor. Medium boiling solvent for nitrocellulose. Properties: bp, 98°C; sp gr, 0.−912/0°C; flp, 12°C (43.6°F); vp, 43 mm Hg/30°C.

N-Ethyl-*p*-Toluene Sulfonamide *n* (2-Ethyl-5-methylbenzenesulfonamide) A solid plasticizer for rigid PVC.

Ethyl Succinate *n* (Succinic acid monoethyl ester) Saturated plasticizer. Properties: bp, 216°C; sp gr, 1.049. Also known as *Diethyl Succinate*.

EtOH See ▶ Ethyl Alcohol.

Ettinghausen's Effect *n* [*Von Ettinghausen's*] When an electric current flows across the lines of force of a magnetic field an electromotive force is observed which is at right angles to both the primary current and the magnetic field: a temperature gradient is observed which has the opposite direction to the Hall electromotive force (Lide DR (ed) (2004) CRC handbook of chemistry and physics. CRC Press, Boca Raton, FL).

EU Abbreviation for Polyether Type of Polyurethane Rubber.

Euhedral *adj* Euhedral crystals are those that are bounded by plane faces, cf., anhedral.

Eupolymer *n* Polymer with a molecular weight over 10,000.

Eutectic \yú-ˈtek-tik\ *adj* [Gk *eutēktos* easily melted, fr. *eu-* +*tēktos* melted, fr. *tēkein* to melt] (1884) A term applied to the specific mixture of two or more substances that has a lower melting point than that of any of its constituents alone or any other percentage composition of the constituents.

EVA *n* (E/VAC) Copolymer from ethylene and vinyl acetate. See ▶ Ethylene–Vinyl Acetate Copolymer.

Evaporation \i-ˈva-p(ə-)ˌrā-shən\ *v* [ME, fr. L *evaporatus*, pp of *evaporare*, from *e-* + *vapor* steam, vapor] (15c) The changing from the liquid to the gaseous or vapor state, as when the solvent leaves the printed ink film.

Evaporation Rate *n* A measure of the length of time required for a given amount of a substance to evaporate, compared with the time required for an equal amount of ethyl ether or butyl acetate (rated at 100) to evaporate.

Evaporation Rate *n*, **Final** Time interval for complete evaporation of all solvent.

Evaporation Rate *n*, **Initial** Time internal during which low boiling solvent evaporates completely.

Evaporometer *n* Instrument for measuring the evaporation rate of a liquid.

EVE See ▶ Vinylethyl Ether.

Evenness Testing *n* Determination of the variation in weight per unit length and thickness of yarns or fibers aggregates such as roving, sliver, or top.

EVOH See ▶ Ethylene–Vinyl Alcohol Copolymer.

Exa- The SI prefix meaning $\times 10^{18}$.

Excessive Clearer Waste *n* higher that normal amount of short and regular fibers that become attached to the drafting rolls and are transferred to the clearer brushes to accumulate in abnormal amounts until they are removed manually.

Excitation *n* \ˌek-ˌsī-ˈtā-shən\ (14c) In ultraviolet curing, the first state of the polymerization process, in which the photo-initiator, such as benzophenone amine, is stimulated by UV into a singlet or triplet state, with subsequent formation of free radicals.

Excitation Purity See ▶ Excitation.

Excited State *n* Any state higher in energy then the ground state.

Excluded Volume *n* The volume surrounding and including a given object, which is excluded to another object. This terminology comes from the statistical mechanics of gases, where this function arises in the leading order concentration expression (virial coefficient) for the pressure in the case of gas particles that repeal each other with a hard-core volume exclusion.

Exempt Solvents *n* Any solvent that has not been declared photochemically reactive by any of several regulatory agencies, most notably, the Los Angeles Air Pollution Control District. Many alcohols, many esters, some ketones, and mineral spirits are exempt under Rule 66.

Exfoliate \(ˌ)eks-ˈfō-lē-ˌāt\ *v* [LL *exfoliatus*, pp of *exfoliare* to strip of leaves, fr. L *ex-* + *folium* leaf] (1612) Sealing from a surface in flakes or layers.

Exhaustion *n* During wet processing, the ratio at any time between the amount of dye or substance taken up by the substrate and the amount originally available.

Exo- A chemical-structure prefix meaning attachment to a side chain rather than to a ring. Compare ▶ Endo-.

Exotherm \ˌek-sō-ˈthərm\ *n* (1) The temperature/time curve of a chemical reaction giving off heat, particularly the polymerization of a casting resin. ASTM Test D 2471 delineates a procedure for measuring this curve, which in practice is strongly dependent on the amount of material present and the geometry of the casting. (2) The amount of heat given off per unit mass of the principal reactant. As yet (1992) no standard method for determining this heat in reacting plastics has been adopted.

Exothermic *adj* [ISV] (1884) Denoting a chemical reaction that is accompanied by the evolution of heat. An example in the plastics industry is the curing reaction of an epoxy resin with an amine hardener. Opposite of endothermic.

Exothermic Reaction *n* A reaction which liberates heat.

Expandable Plastic *n* A plastic formulated so as to be transformable into a cellular plastic by thermal, chemical, or mechanical means. See ▶ Cellular Plastic.

Expanded Polystyrene (XPS) See ▶ Polystyrene Foam.

Expanding Agent See ▶ Blowing Agent.

Expansion Coefficient The fractional change in length or volume of a material for a unit change in temperature.

Expansion of Gases *n* (Charles' Law or Gay-Lussac's Law) The volume of a gas at constant pressure increases proportionately to the absolute temperature. If V_1 and V_2 are volumes of the same mass of gas at absolute temperatures, T_1 and T_2,

$$\frac{V_1}{V_2} = \frac{T_1}{T_2}$$

For an original volume V_o at 0°C the volume at t°C (at constant pressure) is

$$V_1 = V_o(1 + 0.00367t).$$

Expansion of Gases *n* (*General Law for Gases*)

$$p_t v_t = p_o v_o \left(1 + \frac{t}{273}\right)$$

where p_o, v_o, p_t, v_t represent the pressure and value at 0° and t°C or

$$\frac{p_1 v_1}{T_1} = \frac{p_2 v}{T_2}$$

where p_1, v_1 and T_1 represent pressure, volume and absolute temperature in one case and p_2, v_2, and T_2 the same quantities for the same mass of gas in another. The law may also be expressed as $pv = RmT$, where m is the mass of gas at absolute temperature T. R is the *gas constant* which depends on the units used. *Boltzmann's molecular gas constant* is obtained by expressing m in terms of the number of molecules. For volume in centimeter cube, pressure in dynes per square centimeter, and temperature in Centigrade degrees on the absolute scale R = 8.3136 × 10^7 (Watson P (1997) Physical chemistry. Wiley, New York).

Expansivity See ▶ Coefficient of Thermal Expansion.

Explosive Limits *n* When combustible vapor is mixed with air in the proper proportions, ignition will produce an explosion. This proper proportion is called the explosive range. The explosive range includes all concentrations of a mixture of flammable vapor or gas in air, in which a flash will occur or a flame will travel if the mixture is ignited. The lowest percentage at which this occurs is the lower explosive limit; and the highest percentage, the upper explosive limit. Explosive limits are express in percent by volume of vapor in air and, unless otherwise specified, under normal conditions of temperature and pressure.

Exposure Meter *n* (1891) A device for indicating correct photographic exposure under varying conditions of illumination.

Exposure Rack *n* Term given to a frame on which test panels are exposed for durability tests.

Exposure Tests *n* Tests which are conducted to evaluate the durability of a coating or film. They include exposure to ultraviolet light, moisture, cold, heat, salt water, mildew, etc. They can be generated either naturally or artificially.

Expression *n* Removal of a liquid from a solid by pressing, as in the manufacture of vegetable oils from meal cakes.

Extended Length *n* The length of a face pile yarn required to produce 1 in. of tufted carpet.

Extender \ik-ˈsten-dər\ *n* (1611) (1) See ▶ Extender Pigment. (2) A transparent or semi-transparent white pigment or a varnish that is used to alter the color strength and working properties of an ink, without affecting its hue.

Extender Pigment *n* A specific group of achromatic pigments of low refractive index (between 1.45 and 1.70) incorporated into a vehicle system whose refractive index is in a range of 1.5–1.6. Consequently, they do not contribute significantly to the hiding power of paint. They are used in paint to: reduce cost, achieve durability, alter appearance (e.g., decrease in gloss), control rheology, and influence other desirable properties. If used at sufficiently high concentration, an extender may contribute dry hiding and increase reflectance.

Extensibility *n* (1611) (1) The ability of a material to stretch or elongate upon application of sufficient tensile stress. It is expressed as a percentage of the original length. See ▶ Elongation. (2) The value of said ability just prior to rupture of the specimen; ultimate elongation.

Extensiometer *n* A rheometer for measuring the extensional flow properties of molten polymers. In one early form, the *Cogswell rheometer*, useful at tensile viscosities over 10^5 Pa·s, unidirectional tensile force was exerted on a polymer rod by a dead-weight acting through a cam and pulley. As the cam rotated, the moment arm exerted by the weight on the rod decreased in proportion to the rod cross section so as to maintain constant stress.

Extensional Strain Rate See ▶ Elongational Flow.

Extensional Viscosity See ▶ Elongational Viscosity.

Exterior Finishes *n* Coatings which are expected to possess reasonable durability when exposed to natural weathering. See ▶ Exterior Paints and Varnishes.

Exterior Paints and Varnishes *n* Material formulated for use in conditions exposed to the weather.

Exterior Type Plywood *n* Plywood bonded with a fully waterproof glueline.

External Mix Spray equipment in which fluid and air join outside of aircap.

External Phase of an Emulsion Another term for the continuous phase.

External Plasticizer Post added plasticizer as opposed to plasticization by means of internally combined groups, such as copolymerization. See ▶ Internal Plasticizers.

External Undercut Any recess or projection on the outer surface of a molded part that prevents its direct removal from its mold cavity. Parts with such undercuts may be molded by splitting the mold vertically and opening the split to withdraw the part; or by providing *side draws*, i.e., mold parts that are withdrawn to the sides to relieve the undercuts.

Extinction \ik-ˈstiŋ(k)-shən\ *n* (15c) If the orientation of the crystals that appear white or colored between crossed polars is changed by rotating the stage, all single crystals will be observed to disappear (become black) four times during complete rotation of the stage. The positions are 90° apart; they reveal the vibration directions of each crystal. These directions will parallel the vibration directions of the two polars when the crystal is extinct.

Extinction Angle *n* The angle between the nearer vibration direction and a prominent direction of the crystal. It never exceeds 45°.

Extinction Coefficient *n* (1902) An older term synonymous with absorption coefficient. See ▶ Absorption Coefficient.

Extinction, Oblique *n* Extinction is oblique if the vibration directions are oblique to the long direction of the crystal.

Extinction, Parallel *n* Extinction is parallel if the vibration directions are parallel and perpendicular to the long direction of the crystal.

Extinction, Symmetrical *n* Extinction is symmetrical if the vibration directions bisect a prominent angle.

Extractable \ik-ˈstrak-tə-bəl\ *n* [ME, fr. L *extractus*, pp of *extrahere*, fr. *ex-* + *trahere* to draw] (15c) The amount of soluble material extracted from a polymer specimen when it has been exposed to a solvent under specified conditions. In ASTM Test D 4754, disks of the polymer and glass separator beads are alternately threaded onto a stainless-steel wire and slipped into a vial containing the test solvent. The tightly closed vial is placed in an oven, and the solvent is sampled and analyzed periodically.

Extraction *n* The transfer of a constituent of a plastic mass to a liquid with which the mass is in contact. The process is generally performed with a solvent selected to dissolve one or more specific constituents; or it may occur as a result of environmental exposure to a solvent.

Extraction Extrusion *n* An extrusion operation in which a volatile component present in the feedstock is removed by flash vaporization through a vent connected to a vacuum pump. The volatile component is typically a small amount of water, but may be monomer or solvent. In a two-stage, single-screw extruder, the vent is located over the deep extraction section that begins the second stage of the screw. A few double-vented (three-stage) machines have been made. Some twin-screw machines have greater capacity for removal of volatiles.

Extract Printing See ▶ Printing, ▶ Discharge Printing.

Extranuclear Region All of an atom except the nucleus.

Extrudate (1) The product or material delivered from an extruder, for example film, pipe, profiles, and wire coatings. (2) The extruded melt just as it emerges from the die.

Extrudate Roughness See ▶ Melt Fracture.

Extrudate Swelling *n* (Barus effect) The increase in thickness or diameter, due mainly to the release of stored elastic energy, as a hot melt emerges *freely* from an extrusion die. In many commercial operations, because the extrudate is drawn away at speeds higher than the mean flow velocity in the die, swelling is more than offset by draw down and is not actually observed. Swelling tends to increase with extrusion rate, abruptness of the approach to the die land, and inversely with land length and melt temperature (Strong AB (2000) Plastics materials and processing. Prentice-Hall, Columbus, OH).

Extruded-Bead Sealing *n* (melt-bead sealing) A method of welding or sealing continuous lengths of thermoplastic sheeting or thicker sections by extruding a bead of the same material between two sections and immediately pressing the sections together. The sensible heat in the bead is sufficient to achieve the weld to the adjacent surfaces.

Extruded Foam *n* Cellular plastic produced by extrusion with the aid of a blowing agent – a decomposable, gas-generating chemical – or by injection into the extruder of a gas such as nitrogen or carbon dioxide, or a highly volatile liquid, such as a pentane.

Extruded Shape *n* (profile) Any of a huge variety of cross-sectional shapes produced in continuous or cut lengths by extrusion through profile dies. Some complex shapes reach their final form with the aid of post forming jigs that alter the shape before the extrudate has cooled and become firm.

Extruder \-ˈstrü-dər\ *n* [L *extrudere*, fr. *es-* + *trudere* to thrust] (1566) (1) An extrusion machine. (2) A machine in which molten or semisoft materials are forced under pressure through a die to form continuous tubes, sheets, or fibers. It may consist of a barrel, heating elements, a screw, ram or plunger, and a die through which the material is pushed to give it shape. (3) In fiber manufacture the machine that feeds molten polymer to an extrusion manifold or that first melts the polymer in a uniform manner then feeds it to a manifold and associated equipment for extrusion. (4) In plastics compounding, a material added to the mixture to reduce its cost per unit volume. The material may be a resin, plasticizer, or filler. See also ▶ Blending Resin. (5) A substance, generally having some adhesive capacity, added to an adhesive formulation to reduce the amount of the primary (i.e., more costly) binder required per unit of bonding area. See ▶ Binder and Filler (Strong AB (2000) Plastics materials and processing. Prentice-Hall, Columbus, OH). Also see ▶ Screw Melter.

Extruder Barrel *n* (extruder cylinder) A thick-walled, cylindrical steel tube, lined with a special hard alloy to resist wear that forms the housing for the extruder screw and contains, between itself and the screw, the plastic material as it is conveyed from feed hopper to die. Barrels are usually surrounded by heating and cooling media, such as electrical heater bands, cast-in-aluminum calrods and tubing for coolant, induction heaters, or, rarely, by a compartmented jacket for the circulation of hot oil or steam. Electrically heated barrels are usually furnished with some means of cooling, such as air blowers or coils through which fluid may be circulated. Small holes drilled radically into (but not through) the barrel accommodate temperature sensors whose signals are used to control the means of heating and cooling.

Extruder Burn *n* The local decomposition of the plastic resin inside an extruder caused by the excessive mechanical heating (Strong AB (2000) Plastics materials and processing. Prentice-Hall, Columbus, OH).

Extruder Breaker Plate See Breaker Plate.

Extruder Drive *n* The system comprising an AC or DC motor, speed reducer (gearbox), screw-shaft bearings, coupling, and controls that supply power to the screw and regulate its speed. To facilitate startup and accommodate various operating conditions, modern extruders are always equipped with some means of varying the screw speed, with close control, over a wide range.

Extruder, Dual-Ram *n* A modification of the original ram extruder employing two identical units, stroking alternately, and delivering to the same die. Aided by values, the result is the conversion of a batch operation into one that makes continuous extrusions from sinterable resins such as polytetrafluoroethylene and ultra-high-molecular-weight polyethylene.

Extruder, Elastic-Melt *n* (elastodynamic extruder) A type of extruder in which the material is fed into a fixed gap between stationary and rotating, vertical disks, is melted by frictional heat, and flows in a spiral path toward the center of rotation, from which it is discharged into a secondary device that can develop the high pressure required for extrudate shaping. Only rubbery polymers with certain viscoelastic properties are suitable for the process.

Extruder, Hydrodynamic *n* A device similar to the elastic-melt extruder (see preceding entry) in that the plastic pellets are sheared between relatively rotating disks. However, the disks in a hydrodynamic extruder are shaped to provide positive driving force, whatever the properties of the melt. It, too, can provide efficient melting while developing little pressure.

Extruder, Piggy-Back *n* A system in which the two chief functions of a plasticating extruder – melting and pressure development – are made independently controllable by using two extruders in tandem. The first receives the cold feed, melts it, and delivers the melt to the second extruder, which is essentially a melt pump with possible mixing and/or extraction zones, and which develops the die pressure. Though the piggy-back principle is sound, the market has favored the combination of extruder with gear pump. However, see ▶ Extruder.

Extruder, Planetary Screw *n* A multi-screw device in which a number of satellite screws, generally six, are arranged around one longer central screw. The portion of the central screw extending beyond the satellite screws serves as the final pumping screw as in a single-screw extruder, while the planetary screws aid in plastication and permit the discharge of volatiles toward the hopper. A few of these have served in processing fine powders such as PVC dry-blend.

Extruder, Ram *n* An extruder in which the material is advanced through the barrel and die by means of a ram or plunger rather than by a screw. Melting is accomplished either by preheating the feedstock close to the fusing temperature and by conductive heating from the barrel wall, or both. The ram extruder was the earliest type to be used in the plastics industry, dating back to 1870 when cellulose nitrate was extruded into rods. Among plastics today, polytetrafluoroethylene, which softens to a gel but does not achieve a true melt state, is the one mainly processed with ram extruders.

Extruder Screen Pack *n* A layered group of woven metal-wire screens placed at the end of the screw and supported by the Breaker Plate, to prevent contaminants from obstructing or passing through the die. Screen packs have also been used to create additional resistance to flow, thus raising the head pressure and increasing the level of viscous working and mixing. Today the second function is more conveniently accomplished with an adjustable valve in the adapter between screw and die. The pack usually contains several screens of different meshes, the finest one facing the incoming melt and supported by successively coarser ones with their thicker wires. Screens gradually become loaded with the foreign material they retain, and, with fewer openings remaining clear, the head pressure gradually rises, requiring increases in screw speed to maintain the rate. Eventually clean screens must be substituted for dirty ones. A device that permits making this change with minimal disturbance of the extrusion operation is a ▶ Screen Changer. Frequency of changing can range from once a week in plants processing mostly virgin resin and carefully handled regrind, to every half hour in plants recovering scrap resins.

Extruder Screw *n* A solid or cored shaft with a continuous helical channel (sometimes two channels) cut into it, usually extending from the feed throat of the extruder barrel to the die end of the barrel. In most screws, the channel varies in its volume per turn of the helix, being larger at the end receiving the feed with its low bulk density, to much less – roughly one third – in the pumping section at the delivery end. The reduction in volume is usually accomplished by reducing the channel depth but can be done by reducing the helical lead; sometimes a combination of both has been used. The reduction in volume serves several purposes: feeding, compressing the particles and forcing the interstitial air back out the feed hopper, melting the polymer, and developing pressure to overcome resistance at the die. Extruder screws are made of tough steel – SAE 4140 is common – are usually chrome plated, and have flight tips hardened by one of several techniques. Many designs have been, and continue to be, developed and marketed for extruder screws. Some terms used in describing extruder screws are defined below.

- **Barrier Screw** See ▶ Solids-Draining Screw.
- **Constant-Lead Screw** *n* (uniform-pitch screw) A screw with a flight of constant helix angle over its whole length.
- **Constant-Taper Screw** *n* A screw of constant lead and uniformly.
- **Cored Screw** A screw with a hole bored along its axis for circulation of heat-transfer medium or insertion of a heater. The core may extend only through the feed section or further, even to the screw tip.
- **Decreasing-Lead Screw** *n* A screw in which the helix angle decreases steadily over the length of the screw. Channel depth is usually constant.
- **Metering-Type Screw** *n* A screw whose final section is of constant lead and relatively shallow depth.
- **Multiple-Flighted Screw** *n* (multi-flight screw) A screw having more than one flight, thus having two or more parallel channels.
- **Single-Flighted Screw** *n* A screw having just one flight – the usual case, presumed if not otherwise stated.
- **Two-Stage Screw** *n* Essentially two metering-type screws in series, typically used for vented operation. The first stage consists of a feed section, compression zone, and metering zone. The second stage consists of a deep, constant-depth section (decompression zone), usually running only fractionally full to permit expansion of bubbles and release of volatiles, followed by a short, steep compression zone and a metering zone of about one third greater capacity than that of the first stage. There may be a restriction at the end of the first stage.

- **Vented Screw** *n* A two-stage screw with a screw vent in the decompression zone, permitting volatiles to escape through the screw core.
- **Water-Cooled Screw** *n* A screw cored in its feed section to permit circulation of water there.
- **Extruder, Single-Screw** *n* An extruder with one tubular barrel within which a solid or cored screw rotates.

(Strong AB (2000) Plastics: materials and processing. Prentice-Hall, New York; Pittance JC (ed) (1990) Engineering plastics and composites. SAM International, Materials Park, OH; Carley JF (ed) (1993) Whittington's dictionary of plastics. Technomic Publishing).

Extruder Size *n* Traditionally, the nominal inside diameter of the extruder barrel, usually stated in inches or millimeters. However, the output power rating of the drive is more directly related to output capability than is the diameter. See also ▶ L/D Ratio.

Extruder, Tandem *n* A pair of extruders used sequentially for the production of foamed-polystyrene board. The first extruder, usually fitted with a two-stage screw operating at high speed, melts the resin and intimately mixes it with the nucleating agent and the blowing agent injected at the second stage. It feeds directly into a larger, slowly turning extruder where the foamable melt is cooled to a lower temperature, higher viscosity, and higher melt strength, permitting the extruded foam to be controllably expanded to the desired density as it emerges from the die onto a long conveyor.

Extruder, Twin-Screw *n* (double-screw extruder) An extruder with a barrel consisting of two side-by-side intersecting cylinders internally open to each other along their intersection. There are two basic types. If the two internal cylinders are tangentially joined, the two screws are also nearly tangential and are normally of opposite "hands" and counter-rotating, turning downward at their juncture (Welding Engineers). This design permits the use of long vented sections and that feature, together with the milling action of the screws, makes it possible to remove large percentages of volatiles from the feedstock. One of the two screws is extended to become the metering section that provides die pressure. In the second basic type, the two cylinders of the barrel intersect more deeply and the screws intermesh with each other. Rotation may be co- or counter-, depending on whether the screws have the same or opposite "hands." In these machines the intermeshing of the screws traps the plastic and moves it much more positively than in the tangential type, with virtually no back flow, and shear working is less severe. The screws of both types of assembled from segments, each segment designed to perform a particular function. An important application for the second type has been extrusion of rigid PVC compounds from powder, heating gently by conduction and fluxing near the end of the screws and into the die entry and extruding at die pressures up to 100 MPa. Other applications are compounding fluffy polyolefin powders, volatiles extraction, and compounding. Twin-screw machines have also proved useful in reactive extrusion, in which chemical reactions, including polymerization, are performed with the extruder.

Extruder, Vented *n* An extruder provided with a vent opening, most often equipped with a vacuum pump to draw off water vapor or other volatiles. In nonintermeshing twin-screw machines, because of the mutual forwarding action developed by the screws, large vent openings are feasible. In single-screw machines, a two-stage screw is required and a circular vent hole through the top center of the barrel is located above the deep-flighted decompression section following Stage 1. Some vented extruders have valved first stages to help control flow from State 1 into the vented zone. In others, the first metering section is designed to deliver considerably less than the capacity of the second metering section to discharge against the total resistance of the screen pack (if any), adapter and die. A resistance-altering valve may also be used at the head end to adjust the balance between stages for the most uniform rate of discharge.

Extrusion *n* [ML *extrusion-*, *extrusio*, fr. L *extrudere*] (1540) (1) A shaping/molding process use in plastics processing in which the material is melted and then pushed out the end of the machine, usually through a forming die (Strong AB (2000) Plastics materials and processing. Prentice-Hall, Columbus, OH). (2) Any process by which lengths of constant cross section are formed by forcing a material, e.g., a molten plastic, through a die. Typical shapes extruded are hose, tubing, flat films and sheets, wire and cable coatings, parisons for blow molding, filaments and fibers, strands cut hot or cold to make pellets for further processing, webs for coating and laminating, and many of the above in multiple layers by ▶ Coextrusion (Carley JF (ed) (1993) Whittington's dictionary of lastics. Technomic Publishing).

Extrusion, Autothermal *n* (autogenous extrusion, "adiabatic" extrusion) An extrusion operation in which the entire increase in enthalpy of the plastic, from feed throat to die, or very nearly all of it, is generated by the frictional action of the screw. In such an operation, which most commercial single-screw extrusions approach closely, the functions of the barrel heaters are to preheat the machine at startup and, during steady operation, to prevent heat loss from the plastic through the barrel to the surroundings.

Extrusion Blow Molding *n* The most common process of ▶ Blow Molding in which the parison is formed by extrusion.

Extrusion Casting A term sometimes employed in the industry for the process of extruding unsupported film, especially a composite of two or more integral resin layers formed by ▶ Coextrusion. Such extrusion-cast composite films possess desired properties on each of the respective sides, e.g., heat-sealability on one side and stiffness on the other, or different levels of slip, or different colors.

Extrusion Coating *n* The process of coating a substrate by extruding a layer or molten resin onto the substrate with sufficient pressure to bond the two together without the use of an adhesive. A common application of the process is the coating of foil, paper, or fabric with polyethylene, by extruding a web directly into the nip of a pair of rolls through which the substrate is passing.

Extrusion Coloring *n* the method of adding colorants to a plastic compound by dry-blending the colorant with the solid granular resin, extruding the mixture into strands, and cutting these strands into pellets for use in subsequent processing operations.

Extrusion Die *n* The orifice-containing element, mounted at the delivery end of an extruder that shapes the extrudate. Elements of the die assembly are (1) the die block, (2) an adapter connecting the die to the extruder, (3) a manifold within the die that distributes the melt to the orifice, (4) in the case of dies for hollow sections, a mandrel inserted in the flow channel to form the interior surface of the extrudate, (5) a spider that holds the mandrel in position, and (6) the land section, i.e., the orifice that gives the extrudate its emergent shape. Extrusion dies are classified in four ways according to the relation between the screw axis and the direction of flow of the extrudate: straight (*in-line*), *offset*, *angle*, and *crosshead*. In an in-line die, the die-discharge channel is coaxial with the screw. In an offset die, those directions are parallel but not coaxial. In an angle die the axis of the die-discharge channel is at an angle, typically 45°, to that of the screw. A crosshead die is an angle die in which the two axes are perpendicular. Sheet and film dies are also classified as to the type of feed. In a *center-feed die*, the melt enters at the lateral center of the manifold and divides into two equal streams that move in opposite directions to the ends of the die. In an end-fed die, the melt enters one end of the manifold and flows toward the other. Theory tells that it is easier to maintain thickness uniformity over the width of the extruded film or sheet is the die is center-fed, but end-fed dies are sometimes used for logistical reasons, especially in extrusion coating.

Extrusion Laminating *n* A process in which a plastic layer is extruded between two layers of substrate(s) in a nip of a rubber roll and a chill roll. See ▶ Extrusion Coating.

Extrusion Mark *n* In extruded items, a cleft, gash, slit, or notch.

Extrusion Moldings *n* Moldings which are made from plastic material by forcing the melted resin through a shaped orifice by means of pressure.

Extrusion Plastometer *n* (melt-indexer) A simple viscometer consisting of a heated vertical cylinder with two bores, a central one that contains a close-fitting piston and a recess for an orifice block, the other, nearby, for a thermoswitch. The orifice is 2.1 mm in diameter and 8 mm long. Plastic particles are loaded into the bore, allowed to heat for 6–8 min, then the weighted piston is released, and the extrudate is collected for a measured time internal. The melt-flow index (MFI) is stated as the rate of extrusion in grams per 10 min. The instrument and its use are described in ASTM D 1238. Originally developed in 1953 for low-density polyethylene, the melt-indexer is now used with many other polymers with specific temperatures and piston weights. It is *essential* to state the condition (A through X) at which an MFI was measured. Most of the conditions result in shear rates far below those typical of commercial processing (MFI) = 1 g/10 min corresponds to about 2 s^{-1}. Thus, while the measurement is useful for product identification and quality control, it is a poor indicator of processability. An estimate of a resin's pseudoplasticity may be obtained by running two or more tests with substantially different piston weights.

Extrusion Pressure *n* (1) Broadly, pressure indicated anywhere within an extruder. (2) The pressure at the delivery, or head end of the screw in a screw extruder, immediately upstream of the screen pack if one is present. (3) In a ram extruder, the pressure at the face of the ram in contact withy the plastic.

Exudation \ˌek-s(y)ú-ˈdā-shən\ *n* (1612) The undesirable appearance on the surface of an article of one or more of its constituents that have migrated or exuded to the surface. In vinyls, such constituents may be residual emulsifier from the resin, stabilizer, lubricant, or plasticizer. Secondary plasticizers in particular have a tendency to exude when used in excessive percentages. Exudation may appear on a product shortly after it has been made, but more often it is delayed for periods ranging from several weeks to years. Products that do not exude for long periods under ideal storage conditions can be caused to exude by exposure to pressure, heat, high humidity, light, and other environmental agents.

Eyelet \ ī-lət\ *n* [alter. of ME *oilet*, fr. MF *oillet*, dim. of *oil* eye, fr. L *oculus*] (14c) (1) A series of small holes made to receive a string or tape. A buttonhole stitch is worked around the holes. (2) A type of yarn guide used on a creel. (3) A fabric style with areas of cut-outs surrounded by stitching (Complete textile glossary, Celanese Corporation, Three Park Avenue, New York, NY).

Eyring Model *n* (Prandtl-Eyring model) A rheological model, proposed for plastic melts that contains two constants that must be evaluated from experimental flow measurements on the material and at the temperature of interest. The model calls for Newtonian behavior at very low shear rates, with a gradual transition into every stronger pseudoplasticity as the shear rate increases (Elias H-G (2003) An introduction to plastics. Wiley, New York). It has the form

$$\tau_{xz} = A \sinh^{-1}\left(\frac{1}{B}\frac{dv_z}{dx}\right)$$

dv_z is the velocity in the z-direction and in the x-surface, dv_z/dx is the shear rate, and A and B are the two constants. The inverse hyperbolic sine function above is shorthand for

$$\ln\left[\frac{1}{B}\frac{dv_z}{dx} + \sqrt{\left\{\left(\frac{1}{B}\frac{dv_z}{dx}\right)^2 + 1\right\}}\right]$$

It is clear that this model is difficult to apply even to simple geometries. Also, the great body of data on polymer melts shows that, once clearly above the Newtonian, low-shear region, they are well represented by the ▶ Power Law and do not become increasingly more pseudoplastic (Harper CA (ed) (2002) Handbook of plastics, elastomers and composites, 4th edn. McGraw-Hill, New York; Strong AB (2000) Plastics materials and processing. Prentice-Hall, Columbus, OH).

F

f \\ˈef\ *n* (1) SI abbreviation for ▶ Femto-. (2) Symbol for frequency of an oscillating system.

F (1) *n* Chemical symbol for the element fluorine. (2) SI abbreviation for ▶ Farad.

Fabric \\ˈfa-brik\ *n* [MF *fabrique*, fr. L *fabrica* workshop structure] (15c) (cloth) A flexible structure, usually thin relative to its width and length, made up o intermingled yarns, fibers, filaments, or wires. In woven fabrics, the elements are alternately crossed over and under one or more of those oriented in the other directions, typically two perpendicular directions. In Nonwoven fabrics, such as felts, fibers are randomly oriented (Complete textile glossary, Celanese corporation. Three Park Avenue, New York).

Fabricate \-ˈkāt\ *vt* [ME, fr. L *fabricatus*, pp of *fabricari*, fr. *fabrica*] (15c) In the broadest sense, this term means to manufacture, devise, or to make an assembly of parts and sections. In the plastics industry it refers to the assembly or modification of preformed plastics articles by processes such as welding, heat sealing, adhesive joining, machining, and fastening. "Fabrication" is *not* generally used to mean basic manufacturing processes such as extrusion, calendering, molding and the like.

Fabric Construction *n* The details of structure of fabric. Includes such information as style, width, type of knit of weave, threads per inch in warp and fill, and weight of goods (Complete textile glossary, Celease acetate. Three Park Avenue, New York).

Fabric Crimp *n* The angulation induced between a yarn and woven fabric via the weaving or braiding process (Vincenti R (ed) (1994) Elsevier's textile dictionary. Elsevier Science and Technology Books, New York).

Fabric Crimp Angle *n* The maximum acute angle of a single weaving yarn's direction measured from a plane parallel to the surface of the fabric.

Fabric Set The number of warp threads per inch, or other convenient unit.

Fabric Stabilizer *n* Resin or latex treatment for scrims used in coated fabric manufacture to stabilize the scrim for further processing (Tortora PG (ed) (1997) Fairchild's dictionary of textiles. Fairchild Books, New York).

Fabrics, Asphalt Overlay *n* Fabric systems installed between the old and new asphalt layers during pavement resurfacing. The fabric absorbs the tack coat sprayed on the old surface thus forming a permanent moisture barrier to protect the subgrade from strength loss due to water intrusion. The fabric system also helps retard reflective cracking by serving as a flexible layer to diffuse stress 0.

Fabrics, Drainage *n* Fabrics used as filter media or separators in subsurface drainage systems. The fabric is installed around the drainage pipe, or coarse stone in the drain; it allows water to pass freely from the soil to the drain, but prevents soil particles from migrating into the drain system. These fabrics are also used as outer coverings in prefabricated drainage composites and serve the same function as a filtering medium.

Fabrics, Erosion Control *n* Fabrics used in the stabilization of embankments and the containment of silt run-off from erodible slopes. In embankment stabilization, the fabric functions as a filter medium behind stabilizing rip-rap revetments. In siltation control, the fabric acts as a filter to contain silt while allowing excess water to drain freely. In turf reinforcement, the mat is used to retain soil while allowing roots and stems to grow through. In fabric-forming systems for the construction of revetments, a double-layer, water-permeable fabric is positioned, then pumped full of structural grout. These systems are alternatives to rip-rap.

Fabrics, Reinforcement *n* Fabric system used in the construction of steep slopes and retaining walls. By stabilizing the soil mass, they reduce the stress on the retaining wall with corresponding decrease in load-bearing requirements for the wall design. In slope reinforcement, the stabilization permits steeper slope construction than would be possible based on soil properties.

Fabrics, Subsurface Stabilization *n* Fabrics used in the construction on access roads, railroads, parking, and storage areas over soft, unstable soil. The fabric is placed between the subgrade and the stabilizing fill material where it provides separation of subsoil and fill, filtration of moisture at the subsoil/fill interface, and added tensile reinforcement of the compacted fill.

Façade \fə-ˈsäd\ *n* [F *façade*, fr. It *facciata*, fr. *faccia* face, fr. (ass.) VL *facia*] (ca. 1681) One of the faces of a building, usually that containing the main entrance.

Face \\ˈfās\ *n* [ME, fr. OF, fr. (assumed) VL *facia*, fr. L *facies*, fr. *facere*] (13c) The correct or better-looking side of a fabric.

Face Color *n* Term used to describe the color observed on paints (particularly, metallic paints) when they are viewed near to the normal (near to the perpendicular). It may also be referred to as the "top color." It is generally used in contrast to the term "flop color," which is the color observed at an angle near to the grazing angle (parallel to the surface).

Jan W. Gooch, *Encyclopedic Dictionary of Polymers*, DOI 10.1007/978-1-4419-6247-8,
© Springer Science+Business Media LLC 2011

Face Putty *n* Triangular fillet of glazier's putty on the exposed surface of glass.

Facial (fac-) Isomer *n* An isomer of an octahedral complex in which three adjacent octahedral positions are occupied by one kind of ligand.

Faciated Yarn *n* Yarns consisting of a core of discontinuous fibers with little or no twist and surface fibers wrapped around the core bundle.

Facing *v* (15c) A lining or trim that protects the edges of a garment especially at collars, cuffs, and front closings.

Faconné *n* A broad term for fabrics with a fancy-type weave made on a Jacquard or dobby loom.

Facsimile \fak-ˈsi-mə-lē\ *n* [L *fac simile* make similar] (1691) The exact reproduction of a letter, document, or signature.

Factice *n* Elastomeric products made by reaction of sulfur or sulfuric chloride with vegetable oils.

Factitious Ultramarine Synthetic ultramarine blue. See ▶ Ultramarine Blue.

Fadding *n* Application of shellac lacquers by means of a pad known as a "fad."

Fade-Ometer *n* (1) Laboratory device used to determine the fastness of a colored fabric to exposure to light. The test pieces are rotated around a light source simulating the sun's rays at 45°N latitude in July between the hours of 9 a.m. and 3 p.m. Fabrics are rated by visual comparison with a gray scale according to degree of fading. (2) An apparatus for determining the resistance of resins and other materials to fading. It accelerates the fading by subjecting the article to high-intensity ultraviolet wavelengths similar to those found in sunlight. Also known as *Fugitometer*.

Fading *v* [ME, fr. MF *fader*, fr. *fade* feeble, insipid, fr. (ass.) VL *fatidus*, alter. of L *fatuus* fatuous, insipid] (14c) Subjective term used to describe the lightening of the color of a pigmented paint following exposure to the effects of light, heat, time, temperature, chemicals, etc. The observed fading may result from deterioration of the pigment, from deterioration of the vehicle, or from a decrease in gloss. A separation of the vehicle from the pigment particle in the interior of the film, with the subsequent introduction of Microvoids which scatter light, may also be interpreted visually as fading (Koleske JV (ed) (1995) Paint and coating testing manual. American Society for Testing and Materials; Paint and coating testing manual (Gardner-Sward handbook) MNL 17, 14th edn. ASTM, Conshohocken, 1995). See ▶ Light Resistance.

Fahrenheit Scale \ˈfar-ən-ˌhīt, ˈfer-\ *adj* [Gabriel D. *Fahrenheit*] (1753) Temperature scale on which the freezing point of water is represented by 32°F (0°C) and the bp by 212°F (100°C) under normal pressure at sea level (Whitten KW, Davis RE, Davis E, Peck LM, Stanley GG (2003) General chemistry. Brookes/Cole, New York).

Faille \ˈfī(ə)l\ *n* [F, fr. OF] (1869) A soft, slightly glossy woven fabric made of silk, rayon, cotton, wool, or manufactured fibers or combinations of these fibers and having a light, flat crossgrain rib or cord made by using heavier yarns in the filling than in the warp.

Failure, Adhesive *n* The rupture of an adhesive bond, such that the separation appears to be at the adhesive adherend interface. NOTE – Sometimes termed failure in adhesion (Skeist I (ed) (1990) Handbook of adhesives. Van Nostrand Reinhold, New York).

Falling-Ball Viscometer *n* (falling-sphere viscometer) An instrument well suited to determining polymer-melt viscosity at extremely low shear rates, i.e., the limiting Newtonian viscosity. A sphere more dense than the melt is placed between two premolded slugs of the test polymer within a steel cylinder, which is then kept for a preset time in a temperature-controlled oven. From the initial and final positions of the sphere the viscosity can be calculated by Stokes' law (with corrections). By repeating the test with spheres of different densities, a range of low shear rates can be explored (Van Wazer JR, Lyons JW, Kim KY, Colwell RE (1963) Viscosity and flow measurement. Interscience, New York).

Falling Bodies *n* For bodies falling from rest conditions are as for uniformly accelerated motion except that $v_o = Q$ and g is the acceleration due to gravity. The formulate becomes – air resistance neglected,

$$v_t = gt, s = \frac{1}{2}gt^2, v_s = \sqrt{2gs}$$

For bodies projected vertically upward, if v is the velocity of projection, the time to reach greatest height, neglecting the resistance of the air,

$$t = \frac{v}{g}$$

Greatest height

$$h = \frac{v^2}{2g}$$

(Giambattista A, Richardson R, Richardson RC, Richardson B (2003) College physics. McGraw Hill, New York). See ▶ Projectiles.

Falling-Dart Impact Test *n* In addition to the ASTM tests mentioned at ▶ Free-Falling-Dart Test, several similar tests exist for products such as pipe and bottles as well as sheeting. One procedure is the *staircase* *method*, also known as the *Up-and-Down Method* which for a given quantity of testing, provides a good estimate of the impact energy at which 50% of such samples may be expected to break. In the *probit method*, groups of samples are tested at preselected drop heights ranging from that at which most or all of the samples fail to that at which very few or none fail. This method also provides an estimate of the 50% point but, in addition, provides a better estimate of the standard deviation than does the staircase method. Combinations of the two methods have been used to optimize the amount of information per test specimen (Brown R (1999) Handbook of physical polymer testing, vol 50. Marcel Dekker, New York; Shah V (1998) Handbook of plastics testing technology. Wiley, New York).

Falling Sand Abrasion Test *n* Abrasion resistance of coatings of paint, varnish, lacquer and related products is determined by the amount of abrasive (Ottawa) sand required to wear through a unit thickness of the coating, when the sand falls against it at a specified angle from a specified height through a guide tube (Koleske JV (ed) (1995) Paint and coating testing manual. American Society for Testing and Materials).

False Becke Line *n* A second bright line which moves in the direction opposite to the Becke line. It is usually observed with thick particles or when the refractive index difference between particles and mountant is large. It appears as a concentration of reflected light at the low index side of the interface.

False Body *n* Thixotropic flow property of a suspension or dispersion. When a composition "thins down" on stirring and "builds up" on standing it is said to exhibit false body. See ▶ Thixotropy. The term "false body" is also used in practice for "buttery" materials which are characterized by a relatively low viscosity and high yield value.

False Neck *n* In blow molding of containers, a neck construction that is additional to the neck finish of the container and that is only intended to facilitate the blow-molding operation. Afterwards the false neck is removed from the container.

False Twisting See ▶ Texturing.

Family Mold (composite mold) A multicavity mold containing a variously shaped cavities, each of which produces a component of an item that is assembled from the components. For example, a family mold for a model-airplane kit would contain a cavity for each part, and components of a complete kit would be produced in one shot.

Fancy Yarn See ▶ Novelty Yarn.

Fan Gate A shallow gate becoming wider (and usually thinner) as it extends from the runner to the cavity.

Fantail Die (Fishtail die) An extrusion die, usually one making a wide strip or sheet, in which the flow passage diverges from the adapter to the die lip.

Farad \far-ˌad\ [Michael *Faraday*] (1873) (F) The SI unit of electrical capacitance. A capacitor with a I–V potential between its plates and holding a charge of 1 C has a capacitance of 1 F. Thus 1 F = 1 C/V (Serway RA, Faugh JS, Bennett CV (2005) College physics. Thomas, New York).

Faraday Effect \ˈfar-ə-ˌdā, -ə-dē-\ *n* [Michael *Faraday*] (1904) The rotation of the plane of polarization produced when plane-polarized light is passed through a substance in a magnetic field, the light traveling in a direction parallel to the lines of force. For a given substance, the rotation is proportional to the thickness traversed by the light and to the magnetic field strength (Serway RA, Faugh JS, Bennett CV (2005) College physics. Thomas, New York).

Faraday (F) A unit of electrical charge: 1 F equals 9.65×10^4 C (Serway RA, Faugh JS, Bennett CV (2005) College physics. Thomas, New York).

Faraday's Laws In the process of electrolytic changes equal quantities of electricity charge or discharge equivalent quantities of ions at each electrode. One gram equivalent weight of matter is chemically altered at each electrode for 96,501 int. coulombs, or 1 F, of electricity passed through the electrolyte (Serway RA, Faugh JS, Bennett CV, College physics. Thomas, New York).

Farnsworth–Munsell 100 Hue Test Test devised to check for both defective color vision and color discrimination ability. The test requires the subject to arrange four sets of randomized colors, contained in plastic caps, in order of hue change. The test can be administered in about 15 min. It may be purchased from Munsell Color Co. (McDonald R (1997) Colour physics for industry, 2nd edn. Society of Dyers and Colourists, West Yorkshire).

Fashioning \ˈfash-niŋ\ *vt* (15c) The process of shaping a fabric during knitting by increasing or decreasing the number of needles in action. Fashioning is used in manufacturing hosiery, underwear, and sweaters.

Fastness See ▶ Colorfastness.

Fastness, Fast The ability of a pigmented or dyed material to resist color change following exposure to the deleterious elements. *Also known as Fast to Light*. See ▶ Lightfastness and ▶ Colorfastness.

Fastopake Inks Compositions for printing on waxed surfaces.

Fast Ray The fast ray or fast component for a crystal corresponds to the lower refractive index.

Fast Solvent Solvent that evaporates rapidly under atmospheric conditions.

Fast to Light See ▶ Fastness.

Fat *n* (14c) A solid or semisolid ester of the troll glycerol and fatty acids.

Fat Edge Accumulation of paint in the form of a ridge at the edge of a painted surface.

Fatice Sometimes called "artificial rubber" or a "rubber substitute," fatice is made by vulcanizing with sulfur a vegetable oils such as soybean, rapeseed, or castor oil. It is used as a processing aid and extender in natural-rubber compounds and synthetic elastomers.

Fatigue \fə-ˈtēg\ *n* [F, fr. MF, fr. *fatiguer* to fatigue, fr. L *fatigare*; akin to L af*fatim* sufficiently] (1669) (1) Fatigue refers to the resistance of a material to weakening or failure during alternate tension-compression cycles, i.e., in stretch yarns, the loss of ability to recover after having been stretched. (2) A plot of the maximum cyclic stress applied to a fatigue specimen versus the number of cycles to failure, the abscissa being a logarithmic scale (S/N curve). Typically the S/N curve is linear or slightly concave upward, sloping gently downward, sometimes flattening at the low-stress (right) end, suggesting that there may be an ▶ Endurance Limit (Shah V (1998) Handbook of plastics testing technology. Wiley, New York; ASTM testing methods, www.astm.org).

Fatigue Failure The cracking or rupture of a plastic article under repeated cyclic stress, at a stress well below the normal short-time breaking strength as measured in a static (0.5-cycle) test (Shah V (1998) Handbook of plastics testing technology. Wiley, New York).

Fatigue Life The number of cycles of specific alternating stress required to bring about the failure of a test specimen.

Fatigue Limit Syn: ▶ Endurance Limit.

Fatigue Notch Factor The ratio of the fatigue strength of a specimen with no site of stress concentration (notch) to that of a duplicate specimen having a notch.

Fatigue Ratio The ratio of fatigue strength at a given number of cycles of stated alternating tensile stress to the static tensile strength.

Fatigue Strength The maximum-stress level at which a material subjected to cyclic alternating stress fails after a given number of cycles. This is a number read off a fatigue curve that is derived by measuring the cycles to failure of numerous specimens subjected to various known maximum alternating stresses (Shah V (1998) Handbook of plastics testing technology. Wiley, New York).

Fatigue Tests During testing, specimens are subjected to periodic varying of stresses by means of a mechanically operated device.

FATIPEC Acronym for Federation d'Associations de Techniciens des Industries des Peintures, Vernis, Emaux et Encres d'Imprimerie de l'Europe Continentale (Federation of Associations of Technicians in the Paint, Varnish, Lacquer, and Printing Ink Industries of Continental Europe).

Fats Fats are triglycerides(*oils*) or glyceryl esters of higher fatty acids, such as stearic and palmitic. Oils (especially drying oils) are used for the production of alkyd resins, emollients and cooking oils (Gooch JW (2002) Emulsification and polymerization of alkyd resins. Kluwer, New York; Paint: pigment, drying oils, polymers, resins, naval stores, cellulosics esters, and ink vehicles, vol. 3. American Society for Testing and Material, 2001; Morrison RT, Boyd RN (1992) Organic chemistry, 6th edn. Prentice Hall, Englewood Cliffs; Martens CR (1961) Alkyd resins. Reinhold, New York).

Fat Turpentine Oxidized turpentine. Turpentine thickened by exposure to air at elevated temperature or for long periods at ambient temperature.

Fatty Acid *n* (ca. 1872) Organic acid of the general formula $C_nH_{2n}O_2$, e.g., butyric acid and stearic acid; organic acids of aliphatic or open chain structure. These are generally classified as saturated or unsaturated. The saturated fatty acids range from C_4 to C_{26} (Gooch JW (2002) Emulsification and polymerization of alkyd resins. Kluwer, New York; Morrison RT, Boyd RN (1992) Organic chemistry, 6th edn. Prentice Hall, Englewood Cliffs).

Fatty acids, IUPAC name and sources (Gooch, 1997)

Common name	Carbon atoms	Double bonds	Scientific name	Oil/fat sources (triglyceride)
Butyric acid	4	0	Butanoic acid	Butterfat
Caproic acid	6	0	Hexanoic acid	Butterfat
Caprylic acid	8	0	Octanoic acid	Coconut oil

Fatty acids, IUPAC name and sources (Gooch, 1997) (Continued)

Common name	Carbon atoms	Double bonds	Scientific name	Oil/fat sources (triglyceride)
Capric acid	10	0	Decanoic acid	Coconut oil
Lauric acid	12	0	Dodecanoic acid	Coconut oil
Myristic acid	14	0	Tetradecanoic acid	Palm kernel oil
Palmitic acid	16	0	Hexadecanoic acid	Palm oil
Palmitoleic acid	16	1	9-hexadecenoic acid	Animal fats
Stearic acid	18	0	Octadecanoic acid	Animal fats
Oleic acid	18	1	cis-9-Octadecenoic acid	Olive oil
Ricinoleic acid	18	1	Hydroxy-9-cis-octadecenoic acid	Castor oil
Vaccenic acid	18	1	Octadec-trans-11-enoic acid	Butterfat
Linoleic acid	18	2	cis, cis-9,12-octadecadienoic acid	Grape seed oil
Linolenic acid (alpha)	18	3	cis,cis,cis-9,12,15-octadecatrienoic	Flaxseed (linseed) oil
Linolenic acid (Gamma)	18	3	cis,cis,cis-6,9,12-octadecatrienoic acid	Borage oil
Arachidic acid	20	0	Icosanoic acid	Peanut oil, fish oil
Gadoleic acid	20	1	cis-9-eicosenoic acid	Fish oil
Arachidonic acid (AA)	20	4	cis,cis,cis,cis-5,8,11,14-eicosatetraenoic acid	Liver fats
Eicosapentaenoic acid	20	5	cis,cis,cis,cis-5,8,11,14,17-eicosapentaenoic acid	Fish oil
Behenic acid	22	0	Docosanoic acid	Rapeseed oil
Erucic acid	22	1	cis-Docos-13-enoic acid	Rapeseed oil
Docosahexaenoic acid	22	6	cis,cis,cis,cis,cis-7,10,13,16,19-docosapentaenoic acid	Fish oil
Lignoceric acid	24	0	Tetracosanoic acid	Small amounts in most fats

Fatty Acid Pitches Pitch-like residues obtained from the distillation of fats and oils of animal or vegetable origin, or of their fatty acids. Fatty acid pitches include cottonseed pitch, palm oil pitch, wool grease pitch, etc. They have comparatively high acid and saponification values.

Fatty Nitrogen Products Nitrogen-containing compounds derived from vegetable and animal fatty acids.

Fatty Paint Paint which has thickened because of oxidation and polymerization of the drying oil vehicle during storage.

Fatty Polyamide (Versamid®, oldest of many trade names) A polymer formed by the condensation of a dibasic acid having a bulky side group and from 13- to 21-carbon chains, or the dimmer acids, C-36, with di-or polyamines. The commercially important dimmer acids are addition products of unsaturated C-18 fatty acids and can take several forms, giving different structures to the polyamides. They are used in hot-melt adhesives, inks, as epoxy flexibilizers and, in amine form, as curing agents for epoxies (Skeist I (ed) (1990, 1977, 1962) Handbook of adhesives. Van Nostrand Reinhold, New York).

Faux Hois Painted decoration simulating marble, tortoise shell or wood grain.

Fay \ fā\ v [ME feien, fr. OE fēgan; akin to OHGr fuogen to fit, L pangere to fasten] (before 12c) To smooth and fit together, as with two surfaces about to be lap-joined.

FDA Abbreviation for Food and Drug Administration, the U.S. agency within the Department of Health, Education and Welfare that is concerned with the safety of products marketed for consumer use, particularly those substances that might be ingested, applied to the skin, or used in therapy or prostheses. See ▶ Food and Drug Administration.

Fe Chemical symbol for the element iron (Latin: ferrum).

FE Fluorine-containing elastomers.

Feather Edging Reducing the thickness of the edge of a dry paint film, e.g., the edge of a damaged area, prior to repainting.

Feathering (1) Operation of tapering off the edges of a coat of paint by laying off with a comparatively dry brush. (2) Printing with irregular edges to the design. (3) A ragged or feather edge which shows at the edge of type or cuts. It may be caused by poor ink distribution, bad impression, excessive ink or an ink not suitable for the paper. Also called *Laying Off* (Printing ink handbook. National association of printing ink manufacturers, Inc., 5th edn. Kluwer, London, 1999; Leach RH, Pierce RJ, Hickman EP, Mackenzie MJ, Smith HG (eds) (1993) Printing ink manual, 5th edn. Blueprint, New York).

Feculose A mixture of esters prepared by treating starch with glacial acetic acid and washing the product free of uncombined acid.

Federation of Societies for Coatings Technology A not-for-profit organization comprised of affiliated members and members belonging to its constituent societies in the United States, Canada, Mexico, and England, dand located at 492 Norristown Road, Blue Bell, PA 19422–2350 (www.coatingstech.org). Members are engaged in research, engineering, technical development, quality control, supervisory production, administrative management and sales for the manufacture, use or improvement of the finished products or raw materials of the coatings and printing ink industries. Prior to 1975, it was known as the Federation of Societies for Paint Technology.

Feedback Control A system of controlling a machine or process in which the difference between a measured output variable and its target value is amplified and, through automation, causes an appropriate adjustment of an input machine variable or process condition that will move the output nearer to its target.

Feed Block In coextrusion, a massive metal block in which the streams of the several extruders are brought together to form the layers of a single stream just before it enters the die.

Feed Bushing Syn: ▶ Sprue Bushing.

Feedforward Control Process control in which early process variables are monitored and their disturbances are fed to a process model that computes adjustments o the variables needed to provide the desired process outputs. Compare ▶ Feedback Control.

Feed Hopper An inverted conical or pyramidal vessel mounted over the feed port of an extruder or injection-molding machine that contains a supply of pellets or powder being fed. Feed hoppers typically have a slender window from bottom to top along one side to permit observation of the feedstock level.

Feeding British term for livering. See ▶ Livering.

Feed Plate In injection molds, the plate contacting the injection nozzle and containing the sprue and, usually, most of the runner system. Used withy a floating cavity plate, the system provides for separation of runners and sprue from the moldings and stripping of both into separate collectors or chutes.

Feed Port An opening at the rear end of the barrel of an extruder of injection molder through which plastic powder or pellets fall into the rotating screw or, in older injection machines, in front of the withdrawn ram (raw now).

Feedscrew See ▶ Extruder Screw.

Feed Zone The first (rear) zone of an extruder screw that is fed from the hopper, usually of constant lead and greater depth than other zones and officially terminating at the beginning of the compression zone.

Feel A journeyman's term for the working qualities of a paint. See ▶ Hand.

Feldspar \\|fel(d)-|spär\ *n* [mod. of obs. Gr *Feldspath* (now *Feldspat*), fr. G *Feld* field + obs. Gr *Spath* (now *Spat*) spar] (1772) $K_2O \cdot Al_2O_3 \cdot 6SiO_2$. Natural potassium aluminum silicate which occurs in granite. The gradual decomposition of feldspar by natural agencies yields china clay and silica. Any of several anhydrous minerals containing aluminum silicates or alkali or alkaline-earth metals (Na, K, Ca, Ba) which, when ground, make lowcost, modulus-raising, nontoxic fillers for plastics (McGraw-Hill dictionary of geology and mineralogy. McGraw-Hill, New York, 2002; Kirk-Othmer encyclopedia of chemical technology: pigments-powders. Wiley, New York, 1996).

Fell (1) The end of a piece of fabric that is woven last. (2) In weaving, the last filling pick laid in the fabric at any time.

Felt *n* [ME, fr. OE; akin to OHGr *filz* felt, L *pellere* to drive, beat] (before 12c) (1) Matted fibers of wood, cotton, fur, hair, etc., individually or in combination, compacted by rolling under pressure. (2) A Nonwoven, fibrous material made up of randomly oriented fibers held together by stitching, a chemical binder, or by action of heat or moisture.

Felt and Gravel Roofing See ▶ Built-up Roofing.

Felting *n* (1686) (1) The process of exposing wool fibers alone or in combination with other fibers to mechanical and chemical action, pressure, moisture, and heat so that they tangle, shrink, and mat to form a compact material. Felting is generally carried out in a fulling mill. Also see ▶ Fulling.

Felting Down Operation of flatting down a dry film of varnish or paint by means of a pad made of felt or similar material, charged with a very fine abrasive powder and lubricated with water or other suitable liquid.

Femto- \ᵻfem(p)-tō\ [ISV, fr. Dan or N *femten* fifteen, fr. ON *fimmtān*; akin to OE *fīftēne* fifteen. (f) The SI prefix meaning × 10^{-15}.

Fenchol $C_{10}H_{17}OH$. High boiling alcohol present in terpene solvents. Bp of 201°C, sp gr, 0.964/20°C, and mp of 38°C. *Known also as Fenchyl Alcohol.* IUPAC name: 1,2,3,-Trimethyl-bicyclo[2.2.1]heptan-2-ol.

Fenchone $C_9H_{16}C=O$. Ketonic constituent of terpene solvents, bp, 193°C; sp gr, 0.945, refractive index of 1.4625; mp, 5°C.

Fenchyl Alcohol See ▶ Fenchol.
FEP Abbreviation for ▶ Fluorinated Ethylene-Propylene Resin.
Fermat's Principle The path followed by light (or other waves) passing through any collection of media from one specified point to another, is that path for which the time of travel is least.
Fermentation Alcohol See ▶ Ethyl Alcohol.
Ferric \ᵻfer-ik\ *adj* (1799) (1) Of, relating to, or containing iron. (2) Being or containing iron usually with a valence of three.
Ferric Ammonium Citrate *n* (ca. 1924) A complex salt containing varying amounts of iron and used for making blueprints.

Ferric Hydroxide *n* (1885) $Fe_2O_3 \cdot nH_2O$ A hydrate of ferric oxide that is capable of acting both as a base and as a weak acid.

Ferric Oxide *n* (1882) Fe_2O_3. Iron (III) oxide. Stable anhydrous oxide of iron. This constitutes the major ingredient in the chemically prepared synthetic reds. Indian reds, iron oxides, etc., but in the natural red oxides the proportion of ferric oxide varies considerably and is frequently associated with hydrated forms of the oxide. See ▶ Iron Oxides.

Ferriferro Cyanide See ▶ Iron Blue.
Ferrite \ᵻfer-ˌīt\ *n* (1851) (hard ferrite) A compound having the general formula $MFe_{12}O_{19}$, in which M is usually a divalent ion such as barium or strontium. These materials are strongly magnetic and can be incorporated into plastics to make bonded permanent magnets, rigid or flexible, and in many forms, including strips.
Ferrite Yellow $Fe_2O_3 \cdot xH_2O$. Pigment Yellow 42 (77492) Synthetic type iron oxide with a color range from bright yellow to dark orange. Manufactured by oxidizing ferrous hydroxide to the desired shade of hydrated ferric oxide. *Also known as Yellow Iron Oxide.* See ▶ Iron Oxides, ▶ Synthetic.

Ferrocene \ˈfer-ō-ˌsēn\ *n* [*ferro-* + *cycl-* + *-ene*] (1952) (dicyclopentadienyl iron) $(C_5H_5)_2Fe$. A coordination compound of ferrous iron and cyclopentadiene, soluble in PVC and stable to 400°C. Its uses include smoke suppression, a curing agent for silicone resins, an intermediate for high-temperature polymers, and an ultraviolet absorber (Wickson EJ (ed) (1993) Handbook of polyvinyl chloride formulating. Wiley, New York).

Ferromagnetism \-ˌmag-nə-ˌti-zəm\ *adj* (1896) A strong attraction into a magnetic field (Serway RA, Faugh JS, Bennett CV (2005) College physics. Thomas, New York).

Ferrous Oxide *n* (1873) FeO A black easily oxidizable power that is the monoxide of iron.

$$Fe^{++} \quad O^{--}$$

Ferrous Sulfate *n* (1865) $FeSO_4 \cdot 7H_2O$. *Also known as Copperas and Green Vitriol.*

Ferrule \ˈfer-əl\ *n* [alt. of ME *virole*, fr. MF, fr. L *viriola*, dim. of *viria* braclet, of Celtic origin; akin to OIr *fiar* oblique] (1611) Metal portion of the brush holding bristles, hair, feathers or other fibrous and flexible materials.

Festoon Dryer A dryer in which cloth is suspended in loops over a series of supporting horizontal poles and carried through the heated chamber in this configuration.

Festooning Method of drying employed for heavy fabrics, impregnated with varnish or oils, which involves hanging the treated material over horizontal rods or poles in large drying rooms. While the impregnant is still wet, the fabric is moved gradually to avoid excessive accumulations of impregnant in the bottom of the folds. Process by which wallpaper is hung and dried after being printed. Sticks carry the paper, in long loops, through a drying chamber (Kadolph SJJ, Langford AL (2001) Textiles. Pearson Education, New York).

Festooning Oven An oven used to dry, cure, or fuse plastic-coated fabrics with uniform heating. The substrate is carried on a series of slowly rotating shafts with long loops or "festoons" between the shafts.

FF Abbreviation for Furan-Formaldehyde Polymer. See ▶ Furan Resin.

Fiber \ˈfī-bər\ *n* [F *fibre*, fr. L *fibra*] (1540) (fibre) A single homogeneous strand of material having a length of at least 5 mm, that can be spun into a yarn or roving or made into a fabric by interlacing in a variety of methods. Fibers can be made by chopping filaments (converting). Staple fibers may be 1.2–8 cm in length with lineal density from 0.1 to 0.5 mg/m. The natural fibers used by mankind from the earliest times were first supplemented by rayon and acetate, both of which are derived from cellulose. The first commercially successful, wholly synthetic fiber was nylon, introduced in 1939. Then followed acrylic fibers in 1950, polyesters in 1951, and various other polymeric fibers in subsequent years. In 1967 the wholly synthetic, "man-made" fibers surpassed the natural fibers in volume produced (Complete textile glossary. Three Park Avenue, New York, 2000). See also ▶ Manufactured Fiber, ▶ Natural Fiber and ▶ Synthetic Fiber.

Fiber Architecture The spatial arrangement of fibers in the preform. Each architecture has a definite repeating unit.

Fiberboard *n* (1897) Building material composed of wood or other plant fibers bonded together and compressed into rigid sheets.

Fiber Content The percent by volume of fiber in a fiber-filled molding compound, or a molding or a laminate. Fiber content is sometimes started as weight percent.

Fiber Direction (Fiber orientation) In a laminate, the direction(s) in which most of the fibers' lengths lie, relative to the length axis of the part.

Fiber Distribution In a web, the orientation (random or parallel) of fibers and the uniformity of their arrangement.

Fiber Fill Manufactured fibers that have been specially engineered for use as filling material for pillows, mattress pads, comforters, sleeping bags, quilted outerwear, etc. Polyester fibers are widely used.

Fiberfill Molding (Fiberfil™) A term used for an injection-molding process employing as a molding material pellets containing short bundles of fiber surrounded by resin.

Fiberglass (Fiberglass™) See ▶ Glass-Fiber Reinforcement.

Fiber Migration See ▶ Migration, (2).

Fiber Number The linear density of a fiber expressed in units such as denier or tex. Also see ▶ Fineness.

Fiber Optics *n* (1956) A term employed for light-transmitting fibers of glass and some plastics, such as polymethyl methacrylate. Each fiber is coated with a material with a refractive index lower than that of the fiber itself, and many fibers may be gathered in a bundle that is jacketed with polyethylene or other flexible plastic. Such bundles transmit light from one end to the other even through curved. Applications are in aircraft and automobile instrument panels, telephone lines, electronics, displays, medical techniques, and packaging (Meeten GH (1986) Optical properties of polymers. Springer, New York).

Fiber Placement In general, refers to how the piles are laid into their orientation, i.e., by hand, by a textile process, by a tape layer, or by a filament winder. Tolerances and angles are specified. Microprocessor-controlled placement that gives precise control of each axis of motion permits more intricate winding patterns than are possible with conventional winding and is used to make composites that are more complex that usual filament-wound structures (Tortora PG (ed) (1997) Fairchild's dictionary of textiles. Fairchild Books, New York).

Fiber-Reactive Dyes See ▶ Dyes.

Fiber-Reinforced Plastic (FRP) Any plastic material, part or structure that contains reinforcing fibers, such as glass, carbon, synthetic, or metal fibers generally having strength and stiffness much greater than that of the matrix resin, thereby improving those properties. Because glass fibers were used so early and widely, FRP is often used to mean *glass-fiber-reinforced plastics* (Harper CA (ed) (2002) Handbook of plastics, elastomers and composites, 4th edn. McGraw-Hill, New York). See also ▶ Advanced Composites, ▶ Composite Laminates, and ▶ Reinforced Plastic.

Fiber-Resin Interface (Fiber-matrix interface) The surfaces shared by the fibers and the resin in a fiber-reinforced plastic structure. This interface, and the effects of various *sizes* and chemical treatments on the interfacial bond, are subjects of many past and ongoing studies. Because of the pretreatment of fibers with sizes, the interface has a small but finite thickness.

Fiber-Resin Ratio An expression, as a ratio of fiber to resin, of the ▶ Fiber Content.

Fiber Show (Fiber prominence) In reinforced plastics, a condition in which ends of reinforcement strands, rovings, or bundle unwetted by resin appear on or above the surface. It is believed to be caused by a deficiency in the glass, and may not appear until the part is fully cured. Remedies include measures to improve wet-out, use of resins of optimum viscosity, and reducing exotherm rates, which cause stresses within the laminate, and gel coating after the main body of the part has partly cured (Harper CA (ed) (2002) Handbook of plastics, elastomers and composites, 4th edn. McGraw-Hill, New York).

Fiber Spinning See ▶ Spinning.

Fiber Streak (Fiber whitening) A group of fibers within a translucent laminate that were incompletely wetted by resin, appearing as a whitish defect.

Fiber Stress The stress acting on the reinforcing fibers in a laminate under load (estimated) (Harper CA (ed) (2002) Handbook of plastics, elastomers and composites, 4th edn. McGraw-Hill, New York).

Fibrets Very short, fine fibrillated fibers that are highly branched and irregular resulting in very high surface area. Fibrets can be produced from a number of substances including acetate, polyester, nylon, and polyolefins. By selection of polymer type and incorporation of additives, they can be engineered to meet a range of specialized requirements (Harper CA (ed) (2002) Handbook of plastics, elastomers and composites, 4th edn. McGraw-Hill, New York).

Fibrids Short, irregular fibrous products, made by mixing a dilute polymer solution with a nonsolvent with agitation. They can also be made by flash spinning and breaking up the resulting filaments. Used in felts, in papermaking, for filtration product, etc. Also see ▶ Fibrets.

Fibril \ˈfī-brəl\ *n* [NL *fibrilla*, dim. of L *fibra*] (1664) A short threadlike element of a synthetic or natural fiber (Merriam-Webster's collegiate dictionary (2004) 11th edn. Merriam-Webster, Springfield).

Fibrillated-Film Yarn See ▶ Slit-Film Yarn.

Fibrillation The phenomenon in which a filament or fiber shows evidence of smaller-scale fibrous structure by a longitudinal raveling of the filament under rapid, excessive tensile or shearing stress. Separate fibrils can then often be seen in the main filament trunk. The whitening of polyethylene when severely strained at room temperature is a manifestation of fibrillation (Vincenti R (ed) (1994) Elsevier's textile dictionary. Elsevier, New York).

Fibrils A single crystal in the form of a fiber.

Fibrous Asbestos See Fibrous and Magnesium Silicate.

Fibrous-Glass Reinforcement See ▶ Glass-Fiber Reinforcement.

Fibrous Magnesium Silicate See Fibrous and Magnesium Silicate.

Fick's Law (Fick's first law) (1) The net diffusion rate of a gas across a membrane is proportional to the difference in partial pressure, proportional to the area of the membrane and inversely proportional to the thickness of the membrane. Fick's law is the basic law of diffusion of different molecular species into each other. (2) Combined with the diffusion rate from ▶ Graham's law, Fick's law provides the means for calculating exchange raes of gases across membranes. (3) Fick's law states that the flux of a given component will be in the direction in which the concentration of that component decreases most steeply (i.e., opposite the gradient), at a rate given by the product of the mutual diffusivity and the gradient. Fick's law, which has many equivalent forms, is the defining equation for diffusivity (Serway RA, Faugh JS, Bennett CV (2005) College physics. Thomas, New York; Perry RH, Green DW (1997) Perry's chemical engineer's handbook, 7th edn. McGraw-Hill, New York).

Field Coat The coat(s) applied at the site of erection or fabrication.

Field Diaphram The diaphram on the lamp housing which controls the size of the illuminated field of view in Köhler illumination.

Field Painting Surface preparation and painting operations of structural steel or other materials conducted at the project site.

Fiery Finish See ▶ Burned Finish.

Figure \ˈfi-gyər, *British & often US* ˈfi-gər\ *n* [ME, fr. OF, fr. L *figura*, fr. *fingere*] (13c) (Wood) The pattern produced in wood surface by irregular coloration and by annual growth rings, knots and such deviations from regular grain as interlocked and wavy grain.

Filament \ˈfi-lə-mənt\ *n* [MF, fr. ML *filamentum*, fr. LL *filare* to spin] (1594) A variety of fiber characterized by extreme length, which permits its use in yarn with little or no twist and usually without the spinning operation required for fibers (Kadolph SJJ, Langford AL (2001) Textiles. Pearson Education, New York). See also ▶ Monofilament.

Filamentary Composite A reinforced-plastic structure in which the reinforcement consists of FILAMENTS usually oriented to most efficiently withstand the stresses imposed on the structure. The filaments are not woven and in a single lamina they will all be parallel (Wallenberger FT, Weston NE (eds) (2003) Natural fibers, plastics and composites. Springer, New York).

Filament Count The number of individual filaments that make up a thread or yarn (Kadolph SJJ, Langford AL (2001) Textiles. Pearson Education, New York).

Filament Number The linear density of a filament expressed in units such as denier or tex. Also see ▶ Fineness.

Filament Winding A method of forming reinforced-plastic articles comprising winding continuous strands of resin-coated reinforcing material onto a mandrel. Reinforcements commonly used are single strands or rovings of glass, asbestos (rare today because of carcinogenicity fright), jute, sisal, cotton, and synthetic fibers. Polyester resins are most widely used, followed by epoxies, acrylics, nylon, and various others. To be effective, the reinforcing material must form a strong adhesive bond with the resin. The mandrels may be permanent structures remaining in the finished article, or of flexible or destructible material, or able to be disassembled, i.e., capable of being removed after curing. The process is performed by drawing the reinforcement from a spool or creel through a bath of resin, then winding it on the mandrel under controlled tension and in a predetermined pattern. The mandrel may be stationary, in which event the creel structure rotates about the mandrel; or it may be rotated on a lathe about one or more axes. By varying the relative amounts of resin and reinforcement, and the pattern of winding, the strength of filament-wound structures may be controlled to resist stresses in specific directions. After sufficient layers have been wound, the structure is cured at room temperature or with heat (Wallenberger FT, Weston NE (eds) (2003) Natural fibers, plastics and composites. Springer, New York; Pittance JC (ed) (1990) Engineering plastics and composites. SAM International, Materials Park).

Filament-Wound Made by ▶ Filament Winding.

Filament Yarn A yarn composed of continuous filaments assembled with or without twist. Also see ▶ Yarn.

Filiform \ˈfi-lə-ˌförm\ *adj* (1757) Slender as a thread.

Filiform Corrosion A type of corrosion that occurs under coatings on metal substrates characterized by a definite thread-like structure and directional growth (Baboian R (2002) Corrosion engineer's handbook, 3rd edn. NACE International – The Corrosion Society, Houston; Uhlig HH (2000) Corrosion and corrosion control. Wiley, New York).

Filiform Corrosion Resistance The ability of a coating to resist that type of corrosion of metal substrates characterized by a definite thread-like structure and directional growth that occurs under coatings. See ▶ Filiform Corrosion (Baboian R (2002) Corrosion engineer's handbook, 3rd edn. NACE International – The Corrosion Society, Houston).

Filing Manual filing is sometimes used to bevel, smooth, deburr, and fit the edges of plastic moldings and sheets. The process is limited to parts that cannot be tumbled easily, and to plastics with suitable hardness and heat resistance.

Fill *n, adj* Syn: ▶ Weft.

Fill-and-Wipe A decorating process for articles molded with depressed designs, wherein the general area containing the designs is coated with paint by brushing, spraying, or rolling, then surplus paint is wiped from the undepressed areas surrounding the depressions.

Filled Plastic Any plastic compound containing a significant percentage of a solid, usually not fibrous or resinous, material whose main purpose may be to dilute the resin, or to provide certain enhanced properties in the compound.

Filler \ fi-lər\ *n* (15c) (1) A pigmented composition used for filling fine cracks and indentations to obtain a smooth, even surface preparatory to painting. (2) Synonymous with extender. (3) A relatively nonadhesive substance added to an adhesive to improve its working properties, permanence, strength, and other qualities. (4) Any compounding ingredient, usually in dry, powder form, added to rubber in substantial amount to improve quality of lower cost. Fillers have various effects; some are relatively inert, like calcium carbonate and silica, and provide loading for cost reduction; others, like carbon blacks, have a definite and desirable reinforcing effect with improvement in abrasion resistance and other properties. Fillers provide are added to a plastic compound to reduce its cost per unit volume and/or to improve such mechanical properties as hardness, modulus, and impact strength. (5) A filler differs from a reinforcement in two respects. Filler particles are generally small and roughly equidimensional, and they do not markedly improve the tensile strength of a product. Reinforcements, on the other hand, are fibrous, having one dimension much longer than the others, and they do markedly improve tensile strength (Strong AB (2000) Plastics materials and processing. Prentice Hall, Columbus; Modern plastics encyclopedia. McGraw-Hill/Modern Plastics, New York, 1986; 1990, 1992, 1993 editions; Kirk-Othmer encyclopedia of chemical technology: pigments-powders. Wiley, New York, 1996).

Filler Sheet *n* A sheet of deformable or resilient material, that when placed between the assembly to be bonded and the pressure applicator, or when distributed within a stack of assemblies, aids in proving uniform application of pressure over the area to be bonded (Kirk-Othmer encyclopedia of chemical technology: pigments-powders. Wiley, New York, 1996).

Filler Coat A coat of paint, varnish, etc., used as a primer.

Filler Rod (Welding rod) A rod of plastic material used in ▶ Hot-Gas Welding, made of the same material as the plastic to be welded.

Filler Specks Visible particles of a filler, such as wood flour or asbestos, that stand out in color contrast against a background of plastic binder.

Fillet \ fi-lət\ *n* [ME *filet*, fr. MF, dim. of *fil* thread] (14c) A concavely curved transition at the angle formed by the junction of two plane surfaces, i.e., a rounded inside corner. Also, the material making up the transition. Where the surfaces are likely to endure bending toward or away from each other, the fillet distributes and reduces the stress that would otherwise be magnified at the corner.

Filling In a woven fabric, the yarn running from selvage to selvage at right angles to the warp. Each crosswise length is called a pick. In the weaving process, the filling yarn is carried by the shuttle or other type of yarn carrier.

Filling Band See ▶ Filling.

Filling Barré See Barré.

Filling Bow See ▶ Skewness.

Filling of Coated Abrasives Clogging of the abrasive coat by swarf. It can be reduced in many operations by using an open coat construction or a lubricant.

Filling Skewness See ▶ Skewness.

Filling Snarl See ▶ Snarl.

Filling Up (or Filling In) A condition in the printing of halftones where the ink fills areas between the dots, and produces a solid rather than a sharp halftone print.

Filling Up (1) Covering the nonprinting areas (on a lithographic plate) with a partially dried ink film. Effect caused by an excessive drier in the ink. (2) A condition in the printing of halftones where the ink fills areas between the dots, and produces a solid, rather sharp halftone print. They may also occur in the printing of type matter (Printing ink handbook. National association of printing ink manufacturers, Inc., 5th edn. Kluwer, London, 1999; Leach RH, Pierce RJ, Hickman EP, Mackenzie MJ, Smith HG (eds) (1993) Printing ink manual, 5th edn. Blueprint, New York).

Filling Yarn See ▶ Weft.

Fill-type Insulation Loose insulating material which is applied by hand or blown into wall spaces mechanically.

Film \ film, *Southern also* fi(ə)m\ *n* {often attributive} [ME *filme*, fr. OE *filmen*; akin to Gk *pelma* sole of the foot, OE *fell* skin] (before 12c) (1) Customarily in the plastics industry, a web of plastic that is 0.25 mm or less in thickness. Thicker webs are called *sheet*. Films are made by extrusion, casting from solution, and calendaring (Pittance JC (ed) (1990) Engineering plastics and composites. SAM International, Materials Park). (2) In convective heat transfer, the thin, supposedly stagnant layer of fluid next to a heated or cooled surface (such as a pipe wall) that contributes part (or all) of the resistance to transfer of heat from the main body of the fluid to a medium on the opposite side of the wall (or to the wall itself. A closely related concept exists in mass transfer (Perry RH, Green DW (1997) Perry's chemical engineer's handbook, 7th edn. McGraw-Hill, New York). (3) A layer of one or more coats of paint or varnish covering an object or surface. (4) Any supported or unsupported thin continuous covering or coating. (5) Unsupported, usually organic, non-fibrous, thin, flexible material of a thickness not exceeding 0.010 in. In excess of 0.010 in. thickness, such material is usually called sheet or sheeting (Wicks ZN, Jones FN, Pappas SP (1999) Organic coatings science and technology, 2nd edn. Wiley, New York).

Film Blowing (Blown-film extrusion) The process of forming thermoplastic film wherein an extruded plastic tube is continuously inflated by internal air pressure, cooled, collapsed by rolls, and subsequently wound into rolls on thick cardboard cores. The tube is usually extruded vertically upward, and air is admitted through a passage in the center of the die as the molten tube emerges from the die. An Air Ring is always employed to speed and control the initial cooling close to the die. Air is contained within the blown bubble by a pair of pinch rolls that also serve to collapse and flatten the film. Thickness of the film is controlled not only by the die-lip opening but also by varying the internal air pressure and by the rates of extrusion and take-off. Extremely thin films (<0.01 mm) and films with considerable biaxial orientation can be produced by this method (Strong AB (2000) Plastics materials and processing. Prentice Hall, Columbus).

Film Build The rheological property which coatings possess of providing thickness in applied films. (Paint/coatings dictionary (1978) Federation of societies for coatings technology. Blue Bell, Philadelphia).

Film Casting The process of making an unsupported film or sheet by casting a fluid resin, a resin solution, or a plastic compound on a temporary carrier, usually an endless belt or circular drum, followed by solidification by cooling or drying, and removal of the film from the carrier. The term *film casting* is also used for the process of extruding a molten polymer through a slot die onto a chilled roll (Strong AB (2000) Plastics materials and processing. Prentice Hall, Columbus).

Film Coefficient (1) In convective heat transfer, the rate of heat flow through a "stagnant" fluid film adjacent to a solid surface, per unit area of the film, divided by the temperature difference through the film. (2) A similarly structured definition applying to mass transfer through films at fluid interfaces (Perry RH, Green DW (1997) Perry's chemical engineer's handbook, 7th edn. McGraw-Hill, New York).

Film Die A die for the extrusion of flat or blown film. Flat-film dies are usually of the ▶ Crosshead and Coathanger designs with one lip locally adjustable so as to achieve uniform thickness across the film (see ▶ Flexible-Lip Die). Blown-film dies are cylindrical, end- or side-fed, with the concentricity/eccentricity of the core and body adjustable for circumferential uniformity of film thickness. That uniformity, which is critical in both types of films for winding even rolls, also depends on the performance of the air ring. Blown-film dies are often oscillated slowly about their axes to distribute the remaining non-uniformity evenly over the final roll width, in that way assuring that the unrolled film will lie flat in spite of slight thickness variations (Strong AB (2000) Plastics materials and processing. Prentice Hall, Columbus).

Film Extrusion Making plastic films by extruding molten plastic through a ▶ Film Die by ▶ Film Blowing or ▶ Film Casting (Strong AB (2000) Plastics materials and processing. Prentice Hall, Columbus).

Film Formation Term applied to an extruded sheet of a polymer.

Film Former A type of resin with qualities of forming a tough, continuous dry film. Example: nitrocellulose.

Film Forming Ability of a material to form a continuous dry film.

Film Integrity Continuity of a coating free of defects.

Filmogen General term for film forming materials.

Film Slitting See ▶ Slitting.

Film Thickness Thickness of any applied coating, wet or dry.

Film, Thickness Gauge Device for measuring film thickness; instruments for measuring either wet or dry films are available.

Filter Aid Inert, insoluble material, more or less finely divided, used as a filter medium or to assist in filtration by maintaining adequate porosity of the filter cake.

Filter Cake The solid mass remaining on a filter after the liquid that contained it has passed through.

Filter Cloth Any cloth used for filtering purposed. Nylon, polyester, vinyon, PBI, and glass fibers are often used in such fabrics because they are not affected by most chemicals.

Filter Fabrics See ▶ Geotextiles and ▶ Filter Cloth.

Filter, Optical Any uniform optical device which transmits radiant energy of limited wavelengths and/or intensity. Filters may be made of glass, quartz, plastic, etc., containing absorbent material. They may consist of thin metallic coatings on glass or plastic; or they may consist of carefully made screens of uniform pore size. Optical filters may be used to modify the spectral distribution of radiant energy (selective filters) or the photometric intensity of radiant energy (neutral filters) (Moller KD (2003) Optics. Springer, New York; Giambattista A, Richardson R, Richardson RC, Richardson B (2003) College physics. McGraw Hill, New York).

Filter Press Apparatus for filtering, consisting of a number of flat chambers enclosed between sheets of filter cloth (paper or metal) through which the liquids is forced by a pump, leaving the solid matter in the chambers.

Filter Spectrophotometer A spectrophotometer that uses filters of fixed, narrow band-pass transmissions of discrete wavelengths spaced across the spectrum, to measure the transmittance or reflectance of materials at these discrete wavelengths. The resulting special data arranged in order constitute an abridged spectrophotometric curve. Thus, the series of filters replaces the dispersion monochromator used in a continuous spectrophotometer (Willard HH, Merritt LL, Dean JA (1974) Instrumental methods of analysis. D. Van Nostrand, Company, New York).

Filtration \fil-ˈtrā-shən\ *n* (1605) The operation of separating suspended solids from a liquid, or gas, by forcing the mixture through a porous barrier. See ▶ Straining.

Fin *n* Overflow material protruding from surface of cured, molded articles, usually appearing at mold separation line or mold vent points. See ▶ Flash.

Findings *n* (1) Miscellaneous items attached to garments and shoes during manufacture. Included are buttons, hooks, snaps, and ornaments. (2) Miscellaneous fabrics in garments such a zipper tapes, linings, pockets, waistbands, and facings.

Fine End *n* (1) A warp yarn of smaller diameter than that normally used in the fabric. (2) A term for a defect in silk warp yarn consisting of thin places that occur when all the filaments required to make up the full ply are not present. This condition is generally caused by poor reeling.

Fine-Etching *n* A method of modifying a photomechanically prepared plate by controlled undercutting to change the size of the halftone dots.

Fine Melt Process of running copals at high temperatures as distinct from low temperature running, which is described as a slack melt.

Fineness *n* (1) A relative measure of fiber size expressed in denier or tex for manufactured fibers. For cotton, fineness is expressed as the mean fiber weight in micrograms per inch. For wool, fineness is the mean fiber width or mean fiber diameter expressed in microns (to the nearest 0.001 mm). (2) For yarn fineness, see ▶ Yarn Number. (3) For fineness of knit fabrics.

Fineness of Dispersion *n* A measure of the size and prevalence of oversize particles in the coating or ink. The Hegman Gage is a widely used tool for measuring fineness of particles for coatings and inks supplied by Paul N Gardner, Company, Inc., 316 N. E. Fifth Street, Pompano Beach, FL (www.gardco.com) (Weismantal GF (1981) Paint handbook. McGraw-Hill, New York; Printing ink handbook. National association of printing ink manufacturers, Inc., 5th edn. Kluwer, London, 1999; Leach RH, Pierce RJ, Hickman EP, Mackenzie MJ, Smith HG (eds) (1993) Printing ink manual, 5th edn. Blueprint, New York).

Fineness of Grind *n* (deprecated) The degree of dispersion of a pigment in a printing ink vehicle, usually measured on a grindometer or grind gauge (which see) (Weismantal GF (1981) Paint handbook. McGraw-Hill, New York).

Fineness of Grind Gauge *n* A device to measure the fineness of dispersion of a pigment, based on drawing a paint down over a channel of tapered depth and observing the minimum depth at which pigment particles are observed to interfere with the smooth wet surface of the paint. ASTM provides standards for fineness of grind (www.astm.org) (Tracton AA (ed) (2005) Coatings technology handbook. Taylor & Francis, New York; The Hegman Gage is a widely used tool for measuring fineness of grind for coatings and inks supplied by Paul N. Gardner, Company, Inc., 316 N. E. Fifth Street, Pompano Beach, FL (www.gardco.com). Weismantal GF (1981) Paint handbook. McGraw-Hill, New York).

Fines *n* In the classification of powdered or granular materials according to particle size, fines are in portion of the material whose particles are smaller than a stated minimum size. When the particle-size distribution is determined by ▶ Sieve Analysis, the fines are those particles passing the finest sieve and found on the pan, usually designated as "minus 000 mesh," where 000 is the mesh number of that finest sieve (Provder T, Texter J (eds) (2004) Particle sizing and chacterization. American Chemical Society, Washington, DC).

Fine Structure *n* Orientation, crystallinity, and molecular morphology of polymers, including fiber-forming polymers.

Finger Mark *n* A defect of woven fabrics that is seen as an irregular spot showing variation in picks per inch for a limited width. Causes are spreading of warp ends while the loom is in motion and pressure on the fabric between the reed and take-up drum (Kadolph SJJ, Langford AL (2001) Textiles. Pearson Education, New York; Vincenti R (ed) (1994) Elsevier's textile dictionary. Elsevier Science and Technology Books, New York).

Fingernail Test *n* Gouging a dried film with fingernail to form a subjective, qualitative estimate of the relative hardness and toughness.

Fingers \˫fiŋ-gər\ *n* [ME, fr. OD; akin to OHGR *fingar* finger] (before 12c) Classification of fossil copals, especially of the Congo type. The pieces of resin have long cylindrical shapes, similar to those of human fingers.

Finish \˫fi-nish\ *n* [ME *finisshen*, fr. MF *finiss-*, stem of *finir*, fr. L *finire*, fr. *finis*] (1) Refers to the degree of gloss or flatness of a print or any surface (Merriam-Webster's collegiate dictionary (2004) 11th edn. Merriam-Webster, Springfield). (2) (size) In reinforced plastics, a compound containing a coupling agent and (optionally) a lubricant and/or binder, used to pretreat glass fibers prior to using them as reinforcements (Lee SM (1989) Dictionary of composite materials technology. Technomic Publishing Co., Lancaster). (3) The surface texture of a molding, a machined or polished surface, or other article. When measured, it is usually stated as the root-mean-square roughness in nanometers or microinches (Ash M, Ash I (1982–83) Encyclopedia of plastics polymers, and resins, vols I–III, Chemical Publishing Co., New York). (4) The final coat in a painting system. (5) Sometimes refers to the entire coating system: the texture, color, and smoothness of a surface, and other properties affecting appearance (Weismantal GF (1981) Paint handbook. McGraw-Hill, New York).

Finish Coat *n* (1) The final layer of plaster applied over a basecoat or other substrate. (2) See ▶ Top Coat.

Finish Composition *n* (Yard) Physical and chemical analysis of the lubricant applied to yarns to reduce friction and improve processability.

Finished Fabric Fabric that is ready for the market, having passed through the necessary finishing processes (Vigo TL (1994) Textile processing, dyeing, finishing and performance. Elsevier Science, New York).

Finishing *n* (1) The removal of flash, gates, and defects from plastic articles. (2) The development of a desired texture and/or color on the surfaces of an article when such are not accomplished in compounding and forming the article. See ▶ Grind, ▶ Polishing, ▶ Sanding, and the processes listed under ▶ Decorating.

Finishing Bar *n* A noticeable streak across the entire width of a fabric, usually caused by machine stoppage during processing.

Finishing Spot *n* A discolored area on a fabric caused by foreign material such as dirt, grease, or rust.

Finish Insert (Neck insert) In blow molding bottles, a removable part of the mold that aids in forming a specific neck finish of the bottle.

Finish Turns The actual degree of twist in the final yarn product.

Finn Oil See ▶ Tall Oil.

Fire Clay *n* An earthy or stony mineral aggregate which has the essential constituent hydrous silicates of aluminum with or without free silica. It is plastic when sufficiently pulverized and wetted, rigid when subsequently dried, and of suitable refractoriness for use in commercial refractory products.

Fire Point *n* The temperature at which a material, when once ignited, continues to burn for a specified period of time. The fire point is several degrees of temperature higher than the flashpoint. It is the lowest temperature at which a liquid evolves vapors fast enough to support continuous combustion (Troitzsch J (2004) Plastics flammability handbook: principle, regulations, testing and approval. Hanser-Gardner Publications, New York; Babrauskas V (2003) Ignition handbook. Fie Science Publishers, New York; Tests for comparative flammability of liquids, UI 340 (1997) Laboratories Incorporated Underwriters, New York).

Fire Resistance *n* The property of a material or assembly to withstand fire or give protection from it. As applied to elements of buildings, it is characterized by the ability to confine a fire or to continue to perform a given structural function or both. See ▶ Flammability.

Fire-Resisting Finish *n* The preferred term is "fire-retardant coating."

Fire Resistive *n* Refers to properties of materials or designs to resist the effects of any fire to which the material or structure may be expected to be subjected. Fire resistive materials or structures are noncombustible, but noncombustible materials are not necessarily fire resistive. Fire resistive implies a higher degree of fire resistance than noncombustible.

Fire Retardant *n* Descriptive term which implies that the described product, under accepted methods of test, will significantly: (a) reduce the rate of flame spread on the surface of a material to which it has been applied; (b)

Fire-Retardant Chemical *n* A chemical or chemical preparation used to reduce flammability or to retard the spread of flame.

Fire-retardant Coating *n* A coating which will do one or more of the following: (1) reduce the flame spread over which the coating is applied, sometimes at the sacrifice of the coating. See ▶ Intumescent Coatings; (2) resist ignition when exposed to high temperature; or (3) insulate substrate to which it is applied and thereby prolong time required to reach ignition, melting or structural-weakening temperature.

First Coat First coating applied in any painting schedule; in some cases, it would be the sealing coat; in others, the priming coat.

First-Down Color *n* In a multicolor printed material this is the first color printed on the substrate usually subsequently overprinted by other colors.

First-Order Kinetics *n* A thermal transition that involves both a latent heat and a change in the heat capacity of the material.

First-Order Transition Temperature *n* The temperature at which a polymer freezes or melts.

First Quality See ▶ Yarn Quality.

Fish Eye *n* (1) Paint defect which manifests itself by the crawling of wet paint into a recognized pattern resembling small "dimples" or "fish eyes." (2) A visible fault in transparent or translucent plastics, particularly films or thin sheets, appearing as a small globular mass (gel particle) and thought to be caused either by stray resin particles of much higher molecular weight than that of the polymer as a whole, or by inclusion of foreign particles. In rubber technology, such globules are known as "cat eyes" (Koleske JV (ed) Paint and coating testing manual. American Society for Testing and Materials; Paint testing manual: physical and chemical examination of paints, varnishes, lacquers, and colors – stp 500 (1973) American Society for Testing and Materials; Hess M (1965) Paint film defects. Wiley, New York). See ▶ Pinhole.

Fish Oil *n* A natural oil extracted from fish and generally characterized by a rather large group of saturated fatty acids commonly associated with mixed triglycerides (Fish oil: a key to better coatings, Manuf Eng Mag, October 15, 1987; Gooch JW (2002) Emulsification and polymerization of alkyd resins. Kluwer/Plenum Publishers, New York). The fatty acids derived from fish oils are three principal types: saturated, monounsaturated and polyunsaturated with carbon chain lengths ranging from C_{12} to C_{24}. Both types and relative amounts of the fatty chains vary widely among different species of fish and different geographical areas. The fish oil commonly used in the coatings industry is *menhaden oil*, produced from *menhaden fish* caught along the Atlantic Coast of the United States. This oil has an iodine value of about 175. Other common types of fish oils and their iodine values are: herring – 134, anchovy – 199. In addition to glycerides of stearic and the lesser unsaturated fatty acids, fish oils, contain glycerides of clupanodonic acid, which appears to contain four double bonds. The iodine value varies over a wide range, approximately 130–1990. Fish oil is available in Asia, Europe or any where a fishing industry is active. The tendency of fish oils films to yellow considerably is due to the presence of highly unsaturated groups in the molecule (Paint/coatings dictionary (1978) Compiled by definitions committee of the federation of societies for coatings technology).

Fishtail Die Syn: ▶ Fantail Die.

Fission \ˈfi-shən, -zhən\ *n* [L *fission-, fissio*, fr. *findere* to split] (ca. 1617) A nuclear reaction from which the atoms produced are each approximately half the mass of the parent nucleus. In other words, the atom is split into two approximately equal masses. There is also the emission of extremely great quantities of energy since the sum of the masses of the two new atoms is less than the mass of the parent heavy atom. The energy released is expressed by Einstein's equation (Serway RA, Faugh JS, Bennett CV (2005) College physics. Thomas, New York).

Fissure \ˈfi-shər\ *n* [ME, fr. MF, fr. L *fissura*, fr. *fissus*] (14c) A term used in the cellular-plastics industry to denote a separation, crack, or split in a formed cellular article.

Fitch *n* Long handled small brush bound with metal, with which nearly inaccessible areas are painted.

Five Regions of Viscoelasticity *n* As an amorphous polymer is heated from an extremely low temperature it gains more dimensions of molecular motion and its mechanical behavior changes through five qualitative regions: glassy, slowed elastic, rubbery, rubbery flow, and viscous flow.

Fixation *n* The process of setting a dye after dyeing of printing, usually by steaming or other heat treatment.

Fixative *n* Solution which can be sprayed onto drawings rendered in pencil, chalk and other impermanent and easily removed materials so as to fix them and prevent smudging.

Fixed Oil See ▶ Nondrying Oil.

Flag *n* [ME *flagge* reed, rush] The end of a brush bristle which divides into two or more branches. Flagging provides the brush with the ability to hold a greater amount of liquid coatings.

Flair A subjective term applied to the change in hue of a colored material when the light source is changed. Thus, a color may appear blue in daylight but change to a purple-blue in incandescent light; ultramarine blue is an example of such a blue, so it is said to "flair" red or to have a red "flair." Opposite of color consistency; not to be confused with metamerism, which applies to a pair of colors. See ▶ Metamerism (McDonald R (1997) Colour physics for industry, 2nd edn. Society of Dyers and Colourists, West Yorkshire; Billmeyer FW, Saltzman M (1966) Principles of color technology. Wiley, New York).

Flake *n* [ME, of Scand origin; akin to Nor *flak* disk] (1) A term used to signify the dry, unplasticized, basic form of cellulosic plastics (Paint: pigment, drying oils, polymers, resins, naval stores, cellulosics esters, and ink vehicles, vol 3. American Society for Testing and Material, 2001). (2) ▶ Glass Flakes. (3) As used by Celanese, a term that refers to the granular form in which cellulose acetate and triacetate polymers exist prior to dissolving or feeding into the extrusion or molding unit (Complete textile glossary, Celanese acetate LLC. Three Park Avenue, New York, 2000).

Flake Board *n* Same as particle board.

Flake Shellac *n* Orange shellac gum in flake form.

Flake White *n* (1) Another name for white lead of fine textured prepared electrolytically. (2) Variety of white lead, frequently used by artists.

Flake Yarn *n* Yarn in which roving or short, soft staple fibers are inserted at intervals between long filament binder yarns.

Flaking *n* That phenomenon manifested in paint films by the actual detachment of pieces of the film itself either from its substrate or from paint previously applied. Flaking (scaling) is generally preceded by cracking or checking or blistering and is the result of loss of adhesion, usually due to stress-strain factors coming into play. See ▶ Peeling. Syn: ▶ Scaling.

Flaking Resistance *n* The ability of a coating to resist the actual detachment of film fragments, either from its undercoating or substrate. Flaking is generally preceded by cracking, checking or blistering and is the result of loss of adhesion. *Also known as Scaling Resistance.* See ▶ Flaking.

Flaky Web *n* A web at the card that shows thick and thin places, approximately 1–6 in.^{2} in size. This indicates that, instead of a free flow of fibers through the card, either an uneven amount has been fed into the card, or groups of fibers have hesitated in the card and then dropped back into production.

Flamboyant Finish *n* A glossy transparent coating with or without colorant, over a bright undercoat, metallic surface or metallic finish.

Flame \\ˈflām\ *n* [ME *flaume, flaumbe*, fr. MF *flamme* (fr. L *flamma*) & fr. OF, fr. *flamble*, fr. L *flammula*, dim. of *flamma* flame; akin to L *flagrare* to burn] (14c) The visible heat rays which appears when the ignition of a material is reached. Hydrogen is one of the exceptions since the heat rays are not visible.

Flame Cleaning *n* Impingement of an intensely hot flame to the surface of structural steel resulting in the removal of mill scale and the dehydration of any remaining rust, leaving the surface in a condition suitable for wire brushing followed by the immediate application of paint.

Flame (Flash) Arrestor *n* Devices utilized on vents for flammable liquid or gas tanks, storage containers, cans, gas lines or flammable liquid pipelines to prevent flashback (movement of flame) through the line or into the container, when a flammable of explosive mixture is ignited. Wire screen of 40 mesh is utilized on smaller openings. On larger openings, parallel metal plates or tubes are more effective.

Flame Hardening *n* A cheap method, obsolete today, of initially hardening flight tips of extruder screws in which the surface is rapidly chilled (tempered) after being heated with a flame (a process not easy to control). The hardness imparted is gradually lost over a few hundred hours of normal operation, requiring that the process be repeated frequently, defeating the initial saving. Also called *case hardening* and *surface hardening*.

Flame Polishing *n* A method of finishing a plastic article, particularly a just formed extrudate, in which a carefully controlled flame or stream of hot gas is directed at the surface, melting a thin skin of resin that, when quenched, has a high gloss.

Flame Pretreaters *n* Equipment used in the flame pretreatment of polyolefins, by which the polyolefin molding is subjected to prior to the application of a decorative coating (usually in ink). Flame pretreatment is performed in order that the applied coating will have acceptable and permanent adhesion. Flaming consists of exposing the surface to be decorated to a suitable oxidizing flame. This treatment brings about a change to the polymer surface that makes it wettable and permits a strong adhesive bond between the molding surface and the coating,

Flame Proofing See ▶ Fire Retardant.

Flame Resistance Tests See ▶ Flammability Tests.

Flame Resistant *adj* A term used to describe a material that burns slowly or is self-extinguishing after removal of an external source of ignition. A fabric or yarn can be flame resistance because of the innate properties of the fiber, the twist level of the yarn, the fabric construction, or the presence of flame retardants, or because of a combination of these factors. Also see ▶ Flame-Retardant and ▶ Inherent Flame Resistance.

Flame Retardant *n* (1947) A material that reduces the tendency of plastics to burn. Flame retardants are usually incorporated as additives during compounding, but sometimes applied to surfaces of finished articles. Some plasticizers, particularly the phosphate esters and chlorinated paraffins, also serve as flame retardants. *Inorganic flame retardants* include antimony trioxide, hydrated alumina, monoammonium phosphate, dicyandiamide, zinc borate, boric acid, and ammonium sulfamate. Another group, called *reactive-type flame retardants*, includes bromine-containing polyols, Chlorendic acid and anhydride, tetrabromo- and tetrachlorophthalic anhydride, tetrabromo bisphenol A, diallyl chlorendate, and unsaturated phosphonated chlorophenols. A few neat resins, such as PVC and the fluoro- and chlorofluorocarbons, are flame-retardant (Elias, H (2003) An introduction to plastics. Wiley, New York; Modern plastics encyclopedia. McGraw-Hill/Modern Plastics, New York, 1986; 1990, 1992, 1993 editions). See ▶ Flammability.

Flame Retardants *n* A chemical or surface covering material that delays ignition and reduces flame spread.

Flame Spray *n* Any process whereby a material is brought to its melting point and sprayed onto a surface to produce a coating. The process includes: (1) metallizing; (2) thermospray; and (3) plasma flame.

Flame-Spray Coating *n* A coating process utilizing powdered metals or plastics, in which the powdered materials are heated to the sintering temperature in a cone of flame enroute from a spray gun orifice to the article being coated.

Flame Spraying *n* Blowing a powder through a flame that partially melts the powder and fuses it as it hits the substrate.

Flame Spread *n* Flaming combustion along a surface; not to be confused with flame transfer by air currents.

Flame Treating A method of rendering inert thermoplastics, particularly polyolefins, receptive to inks, lacquers, paints, and adhesive by briefly bathing the surface of the article in a highly oxidizing flame. This treatment oxidizes the surface slightly, creating carbonyl and possibly peroxide groups, thereby increasing its surface energy.

Flammability \ fla-mə- bi-lə-tē\ *n* (1646) With respect to plastics, flammability is a very broad term that has been the focus of a potpourri of tests and standards generated by many organizations, predominantly in the U.S. by Underwriters Laboratory (UL) and ASTM. The behavior of various plastics when burning, and tests designed to evaluate flammability, encompass six categories: ignitability, burning rate, heat evolution, smoke production, products of combustion, and endurance of burning. Also called *Flammability and Burn Tests* (Troitzsch J (2004) Plastics flammability handbook. Hanser-Gardner Publications, New York; Shah V (1998) Handbook of plastics testing technology. Wiley, New York).

Flammability and Burn Tests These tests are conducted for the purpose of evaluating polymeric materials for combustible properties. The American Society of Testing and Materials (www.astm.org) and National Institute for Standard and Technology (www.nist.gov) provide continuously updated method of evaluating flammability and combustible properties of polymeric materials (adhesives, coatings, plastics, etc.) (Troitzsch J (2004) Plastics flammability handbook. Hanser-Gardner Publications, New York; Shah V (1998) Handbook of plastics testing technology. Wiley, New York).

Flammable \ fla-mə-bəl\ *adj* [L *flammare* to flame, set on fire, fr. *flamma*] (1813) A substance that is easily ignited, burns intensely, or has a rapid rate of flame spread. Flammable and inflammable are identical in meaning, however, the prefix "in" indicates negative in many words and can cause confusion. Flammable, therefore, is the preferred term. According to I.C.C. Regulations, liquids are flammable if their flash point is 100°F or lower (Troitzsch J (2004) Plastics flammability handbook. Hanser-Gardner Publications, New York).

Flammable Limits See ▶ Explosive Limits.

Flammable Liquid *n* Any liquid having a flp below 37.8°C (100°F), except a mixture having components with flp of 37.8°C (100°F) or higher, the volume of which make up 99% or more of the total volume of the mixture (Babrauskas V (2003) Ignition handbook. Fie Science Publishers, New York; Tests for comparative flammability of liquids, UI 340 (1997) Laboratories Incorporated Underwriters, New York; Wray HA (ed) (1991) Manual for flash point standards and their use. American Society for Testing and Materials). See ▶ Combustible Liquid.

Flange Crimping *n* Simultaneous crimping of two ends of yarn by using heated snubber pins, then combining both ends on a draw roll after they contact a rubber flange on the draw roll.

Flannel \ˈfla-nᵊl\ *n* [ME *flaunneol*] (1503) Medium weight plain- or twill-weave, slightly napped fabric, usually of wool or cotton, but may be made of other fibers.

Flapper \ˈfla-pər\ *n* (ca. 1570) The movable side of a fiber-crimping chamber that periodically opens or flaps to permit crimped fiber to be expelled from the chamber.

Flash *n* (fin) The thin, surplus web of material that is forced into the parting line between mating mold surfaces during a molding operation and which remains attached to the molded article. For methods of removing flash, see ▶ Deflashing.

Flash Aging *n* A process for rapid reduction and fixation of vat dyes obtained when the printed fabric is padded with caustic soda and sodium hydrosulfite and immediately steamed in air-free steam.

Flash Drying *n* (1) Rapid method of drying brought about by the exposure of a coating surface to an elevated temperature or forced draft for a short period of time. (2) Drying at ambient temperature, preparatory for a period of forced drying at elevated temperatures.

Flash Gate *n* A long, shallow rectangular gate in an injection mold, extending from a runner that lies parallel to an edge of a molded part along the flash or parting line of the mold.

Flash Groove *n* (spew groove) A groove in a mold force that allows the escape of excess material during a compression-molding operation.

Flashing *n* (1) A paint defect in a paint film in which patches glossier than the general finish develop, especially at joints or laps in the coating. (2) The nonuniform appearance, including spotty differences in color or gloss, usually due to improper or nonuniform sealing of a porous substrate. (3) Noncorrosive metal used around angles or junctions in roofs and exterior walls to prevent leaks.

Flash Line See ▶ Parting Line.

Flash Mold *n* A mold in which the mating surfaces are perpendicular to the claming action of the press so that, as the clamping force increases, the distance between the mating surfaces decreases, thus permitting excess molding material to escape as flash as the mold closes.

Flash Off *n* Causing the greater part of the more volatile solvents in a sprayed coat of lacquer or enamel to evaporate before proceeding with the application of another coat.

Flash-Off Time *n* Time allowed to elapse between the spray application of successive wet-on-wet coats or the time allowed for the evaporation of the bulk of the solvent before entering into a baking oven. See ▶ Flash Drying.

Flashover *n* (1892) (1) A flammability term. Flashover occurs when hot, combustible gases are generated in burning sections of a building, become mixed with sufficient oxygen upon spreading to non-burning areas, and ignite to cause total surface involvement, but without a progressive flame-spreading stage. (2) In the electrical industry, an electric discharge, around the edge or over the surface of insulation.

Flash Point *n* (1878) The lowest temperature at which a combustible liquid will give off a flammable vapor that will momentarily burn when exposed to s small flame. The flash point can be determined by the open cup or the closed cup method. The flash point determined by the open cut method is usually somewhat higher than the closed cup method (Wray HA (ed) (1991) Manual for flash point standards and their use. American Society for Testing and Materials). See ▶ Flammability.

Flash Points *n* The lowest temperature of a liquid at which it gives off sufficient vapor to form an ignitable mixture with the air near the surface of the liquid or within the container used.

Flash Ridge *n* That part of a flash mold along which the excess material escapes until the mold is fully closed.

Flash Spinning *n* See Spinning (5).

Flat *n* In carding, one of the parts forming an endless chain that partially surrounds the upper portion of the cylinder and gives the name to a revolving flat card. Flats are made of cast iron, T-shaped in section, about 1 inch wide, and as long as the width of the cylinder. One side of the flat is nearly covered with fine card clothing, and the flats are set close to the teeth of the cylinder so as to work point against point. A chain of flats contains approximately 110 flats and operates at a surface speed of about 3 in./min. See ▶ Flatting Down and ▶ Rubbing.

Flat Abrasion Tester See ▶ Stoll-Quartermaster Universal Wear Tester.

Flat Card *n* The type of card used for cotton fibers and for cotton-system processing. It is named for the flat wire brushes called flats that are assembled on an endless chain that partially surrounds the main cylinder. The staple is worked between the flats and cylinder, transferred to a doffer roll, and peeled off as a web that is condensed into a sliver. (Vincenti R (ed) (1994) Elsevier's textile dictionary. Elsevier Science and Technology Books, New York). Also see ▶ Flat.

Flat Coat *n* Coat of filler (British) also incorrectly used when meaning prime coat under enamel. The preferred term is "enamel undercoater." An intermediate coat of paint used as a base for a topcoat.

Flat Duck See ▶ Duck.

Flat Enamel *n* Pigmented coating of low specular gloss, which has the leveling characteristics of a gloss enamel. Enamel has the connotation of both gloss and flow. See ▶ Enamel.

Flat-Entry Die *n* An extrusion die in which the approach to the land has no taper, i.e., one having a 180° ▶ Entrance Angle.

Flat Film *n* Film made by extrusion from a flat die onto a polishing roll. Not to be confused with ▶ Lay-Flat Film.

Flat Finish See Flat Paint (Finish).

Flat Grain *n* Wood or veneer so sawed that the annual rings form an angle of less than 45° with the surface of the piece.

Flat Knit Fabric *n* (1) A fabric made on a flat-knitting machine, as distinguished from tubular fabrics made on a circular-knitting machine. While tricot and milanese warp-knit fabrics (non-run) are knit in flat form, the trade uses the term flat-knit fabric to refer to weft-knits fabrics made on a flat machine, rather than warp-knit fabrics. (2) A term used in the underwear trade for plain stitch fabrics made on a circular-knitting machine. These fabrics have a flat surface and are often called flat-knit fabrics to differentiate them from ribbed-knit or Swiss rib fabrics. In this case, the term refers to the texture, not the type of machine on which the fabric was knit (Kadolph SJJ, Langford AL (2001) Textiles. Pearson Education, New York; Vincenti R (ed) (1994) Elsevier's textile dictionary. Elsevier Science and Technology Books, New York).

Flat Knitting See ▶ Knitting.

Flat Knitting Machine *n* A weft-knitting machine with needles arranged in a straight line in a flat plate called the bed. The yarn travels alternately back and forth, and the fabric may be shaped or varied in width, as desired, during the knitting process. Lengthwise edges are selvages. Flat-knitting machines may be divided into two types: latch-needle machines for sweaters, scarves, and similar articles and fine spring-needle machines for full-fashioned hosiery.

Flat Lacquer *n* A lacquer having the appearance of having been rubbed after it has dried.

Flat Paint *n* (FINISH) Paint which dries to a surface which scatters the light falling on it, so as to be substantially free from gloss or sheen. *Also called Flat Finish.*

Flat Paint Brush See ▶ Flat Wall Brush.

Flat Spot *n* An imperfection on a glossy painted surface; a spot lacking gloss, usually caused by a porous spot on the undercoat.

Flat Spotting *n* A characteristic of certain tire cords. It occurs with all materials but is more noticeable with nylon cord and is associated with nylon cord by users. Nylon exerts a shrinkage force as it becomes heated in tire operation. When the tire is stopped under load, the cord in the road-contact portion of the tire is under less tension than that in other portions of the tire, and it shrinks to conform to the flat surface of the road. When cooled in this position, the cord maintains the flat spot until it again reaches its glass transition temperature in use.

Flat Stone Mill *n* Type of grinding mill in which the material to be distintegrated is fed between the grinding surfaces of two flat stones, one of which is caused to rotate. The grinding surfaces are specifically prepared with grooves or channels. These mills were used for either dry or wet grinding.

Flattening Agent See ▶ Flatting Agent.

Flatting Agent *n* Material added to paints, varnishes and other coating materials to reduce the gloss of the dried film. Any material which, when added to a paint or plastic, lowers the gloss of the final surface. See ▶ Flatting Pigment.

Flatting Down *vt* Cutting or rubbing down the surface of a paint or varnish with fine abrasives to produce a smooth, dull surface.

Flatting Oil A varnish-like composition made of heavy-bodied oil dissolved in a thinner, used to reduce paste paint to a flat paint.

Flatting or Flattening *vt* (1) Undesirable loss of gloss during drying; (2) Addition of a flatting pigment or agent to a paint or varnish; (3) Addition of a flatting oil to a paint or enamel.

Flatting Pigment *n* Any finely divided particle added to a paint formulation in order to decrease the gloss of the dried film. It may be an extruder-type (low refractive index) or a hiding pigment, and it is generally nonchromatic (neutral near-white), although not necessarily so. The reduction in gloss comes about from surface-light scattering which occurs when light strikes the pigments protruding at the surface. See ▶ Extender (Pigment) and ▶ Dry-Hiding.

Flatting (Rubbing) Varnish *n* Varnish containing a high proportion of hard resin which can be rubbed down after application, to produce a smooth foundation for a finishing coat, or serve as the finish coat itself (not to be confused with a flat varnish). *Also known as Rubbing Varnish and Polishing Varnish.*

Flat Top Card See ▶ Flat Card.

Flat Varnish *n* Varnish formulated with a flatting agent so a to dry with a dull finish (not to be confused with a flatting or rubbing varnish).

Flat Wall Brush *n* A paintbrush, usually 4–6 in. (10–15 cm) in width, with long, stiff bristles, usually made of synthetic fiber. Syn: ▶ Flat Paint Brush.

Flax \ ˈflaks\ *n* [ME, fr. OD *fleax*; akin to OHGr *flahs* flax, L *plectere* to braid] (before 12c) The plant from which the cellulosic fiber linen is obtained.

Fleece Fabric *n* A fabric with a thick, heavy surface resembling sheep's wool. It may be a pile or napped fabric of either woven or knit construction.

Fleming's Rule *n* A simple rule for relating the directions of the flux, motion, and emf in an electric machine. The forefinger, second finger and thumb, placed at right-angles to each other, represent respectively the directions of flux, emf, and motion or torque. If the right hand is used the conditions are those obtaining in a generator and if the left hand is used the conditions are those obtaining in a motor (Weast RC (ed) (1971) Handbook of chemistry and physics, 52nd edn. The Chemical Rubber Co., Boca Raton).

Fleshing See ▶ Flushing.

Flex Abrasion Tester See ▶ Stoll-Quartermaster Universal Wear Tester.

Flex-Cracking *n* Development of small cracks in flexible articles or coatings when these articles are subjected to repeated flexing or bending.

Flexibility *n* (1) Degree to which a coating after drying is able to conform to movement or deformation of its supporting surface, without cracking or flaking. (2) A term relating to the hand of fabric, referring to ease of bending and ranging from pliable (high) to stiff (low).

Flexibility Test *n* Test applied to films to ascertain if they are able to accommodate elongations without fracture. Flexibility is usually determined by bending the film, applied to a suitable piece of thin metal, around one or a series of rods or mandrels or a special conical mandrel in a specified time. Syn: ▶ Bend Test.

Flexibilizer *n* A term rarely used for an additive that makes a plastic more flexible. See ▶ Plasticizer.

Flexible Foam See ▶ Polyethylene Foam, ▶ Polyurethane Foam, and ▶ Vinyl Foam.

Flexible-Lip Die *n* In film and sheet extrusion, a die in which a deep groove, reaching almost to the inside surface of the upper die body just behind the lip, has been machined. Adjusting bolts that can either push or pull on the lip pass through the gap from the upper body. The relative flexibility of the thin steel web from which the upper lip extends facilitates die adjustment to minimize sheet-thickness variations across the sheet. See also ▶ Auto-Flex Die.

Flexible Mold *n* A mold made of rubber, elastomer, or flexible thermoplastic, used for casting thermosetting plastics or other materials such as concrete and plaster. The mold can be stretched to permit removal of the cured casting, even one with undercuts.

Flexible Wall Coverings *n* Those which are pliable, such as paper, manmade vinyls and fabrics, as opposed to "liquid" wall coatings, such as paint, or "rigid" like wood paneling or other solid wall products such as ceramic tile.

Flex Life *n* Informally, the number of bending-reversal cycles causing a part to fail in a particular service. Most specifically, the number of cycles to failure of a test specimen repeatedly bent in a prescribed manner. The ASTM test for plastics is D 671. The specimen, molded or cut from sheet, is subjected to load reversal at 30 Hz at a predetermined level of outer-fiber stress until it either fails or the test is discontinued. By setting up different stresses for successive specimens, one can develop a graph of stress at failure vs number of cycles to failure (usually plotted on semilogarithmic coordinates), i.e., the flex-life curve of fatigue curve.

Flexographic Decorating *n* Printing technique by which images are transformed from a flexible raised plate directly to the material. Flexographic printing presses can process many different materials, including polyethylene, polypropylene, oriented polypropylene, polyester, and nylon. Film thicknesses range from 0.0004 to 0.008 in.

Flexographic Ink *n* Quick drying, low viscosity ink based on volatile solvents that are used in the flexographic printing process.

Flexographic Printing *n* A rotary process employing flexible rubber or Elastomeric printing plates adhered to a roll, inked by a screen roll which in turn is coated from a feed roll immersed in ink.

Flexography \flek-ˈsä-grə-fē\ *n* [*flex*ible + -*o*- + -*graphy*] (1954) A typographic form of printing using rubber plates and relatively thin bodied resin-solvent inks. Formerly known as aniline printing.

Flexural Fatigue *n* A physical property expressed by the number of times a material can be bent on itself through a prescribed angle before it ruptures or loses its ability to recover.

Flexural Modulus *n* (flex modulus) The ratio, within the elastic limit, of the applied stress in the outermost fibers of a test specimen in three-point, static flexure, to the calculated strain in those outermost fibers, according to ASTM Test D 790 or D 790M. For a given material and similar specimen dimensions and manufacture, the modulus values obtained will usually be a little higher than those found in a tensile test such as D 638, and may differ, too, from the moduli found in the cantilever-beam test, D 747.

Flexural Rigidity *n* This measure of a material's resistance to bending is calculated by multiplying the material's weight per unit area by the cube of its bending length.

Flexural Strength *n* (flexural modulus of rupture) The maximum calculated stress in the outermost fibers of a test bar subjected to three-point loading at the moment of cracking or breaking. ASTM Test D 790 and D 790M are widely used for measuring this property. For most plastics, flexural strength is usually substantially higher than the straight tensile strength. Also called modulus of rupture, bending strength.

Flight *n* In an extruder screw, the helical ridge of metal remaining after machining the screw channel.

Flight Clearance *n* In screw extruders, half the difference between the inside diameter of the barrel and the diameter of the flight surface, usually about 0.1% of the nominal diameter in a new machine. Clearances in older machines may vary along the screw, intentionally or because of differential wear.

Flight Depth *n* (screw depth, channel depth) In screw extruders, the radial dimension, at any point along the screw, from the flight surface to the screw root. Some users of the term have taken it to mean half the difference between the internal diameter of the barrel and the screw-root diameter, but this difference is larger than the true flight depth by the amount of radial clearance between flight and barrel. In most (but not all) single-screw extruders, flight depth is much greater at the feed end, decreasing toward the delivery end of the extruder, where it is usually constant for at least several flights. See ▶ Channel-Depth Ratio and ▶ Extruder Screw.

Flint Abrasive *n* Not flint at all, but actually a natural quartz (silicon dioxide) which fractures into sharp-edged grains and is used on the common sandpaper for wood.

Flint Paper *n* An inexpensive paper using a natural quartz as the abrasive, for small sanding jobs on wood. It tends to clog quickly and cannot be used for wet sanding.

Flip-flop *n* Decorative effect with automotive metallized finishes in which there is a difference in color depth and flash when the finish is viewed at different angles. Generally used in application to automotive metallic finishes. See ▶ Travel.

Flitter *n* \ˈfil-tər\ *vt* Syn: ▶ Glitter.

Float *n* (1) The portion of a warp or filling yarn that extends over two or more adjacent filling picks or warp ends in weaving for the purpose of forming certain designs. (2) In a knit fabric, a portion of yarn that extends for some length without being knitted in. (3) A fabric defect consisting of an end lying or floating on the cloth surface instead of being woven in properly. Floats are usually caused by slubs, knot-tails, knots, or fly waste, or sometimes by ends being drawn in heddle eyes incorrectly or being twisted around heddle wires.

Floating *vt* Defect which is sometimes apparent in colored paints containing one or more mixtures of different pigments. During drying or on storage, one or more of the pigments separates or floats apart from the others and concentrates in streaks or patches on the surface of the paint, producing a variegated effect. This effect also occurs with single pigment type consisting of multiple particle sizes. See ▶ Flooding.

Floating Chase *n* A mold member, free to move vertically, that fits over a lower plug or cavity, and into which an upper plug telescopes.

Floating Coat See ▶ Brown Coat.

Floating Ends See ▶ Float.

Floating Platen *n* In compression molding, a platen located between the main head and the press table in a multi-daylight press and capable of being moved independently of them.

Floating Punch *n* A male mold member attached to the head of a press in such a manner that it is free to align itself in the female part of the mold when the mold is being closed.

Floats *n* A term used in the past for asbestos filler in the form of very fine, short fibers with associated dust.

Float Stitch See Miss-Switch.

Flocculate \ˈflä-kyə-ˌlāt\ *v* (1877) Cluster of pigment particles formed in a paint after the pigment has been wetted or dispersed. The spaces between the particles are filled with vehicle, as distinguished from agglomerates, where the spaces between the particles are filled with air. A group of two or more attached particles held together by physical force such as surface tension, adsorption, or similar forces. Flocculates are much affected by the sample preparation and technique.

Flocculation *n* The formulation of clusters of pigment particles in a fluid medium which may occur after dispersion has been effected. The condition is usually reversible and the particle clusters can be broken up by the application of relatively weak mechanical forces or by a change in the physical forces at the interface between the liquid and the solid dispersed particles. Flocculation is often visible, as a "Jack Frost" pattern in a flowout of a dispersion; microscopically, it appears as a lacework or reticulum of loosely clustered particles. It results in more rapid settling although it is usually soft, shows loss of color strength and poor dispersion. A flocculated dispersion of sufficient pigment concentration shows yield value. Surface active agents are often useful in reducing the extent of flocculation and hence the yield value. See ▶ Agglomeration.

Flocculation of the Vehicle *n* In the printing ink industry, sometimes used to mean livering.

Flock \ˈfläk\ *n* [ME *flok*, fr. OF *floc*, fr. L *floccus*] (13c) (1) Short fibers of cotton or synthetic fibers such as polyester, acrylic, or nylon. They are used as reinforcements in phenolic, allylic, and other thermosetting molding compounds, also for decorating plastics by the process of ▶ Flocking. (2) Wallcoverings imitating the surface of damask or cut velvet. Made by shaking finely chopped fibers over a pattern printed in varnish or some other sticky material.

Flock Finish *n* Finish produced by the application on a suitable adhesive coat of short fiber cotton, wood, or other fiber, giving a soft suede-like feel and appearance.

Flock Gum Special spray gun for applying flock finishes.

Flocking *vt* (flock coating) A method of finishing sometimes employed for plastics articles whereby the article is coated with a tacky, slow-drying adhesive, then is dusted with a fibrous material cut into very short lengths to give a finish resembling suede, plush, etc. Fibers for flocking are available in a wide range of materials including acrylic, nylon, polyester, polyolefins, and natural fibers. Machinery for flocking films and fabrics includes gravure printing stations for applying the adhesive in desired patterns and flock heads that distribute a precalculated layer of flock to the web, and retrieve and recirculate surplus flock.

Flood Feeding *vt* The usual way of feeding an extruder or screw-injection molder, in which the feed material flows from the feed hopper by gravity and completely fills the feed section of the screw. The actual throughput is thus controlled by screw design, die resistance and temperature conditions within the screw, in contrast to what occurs in ▶ Starve Feeding.

Flooding *vt* (1) A concentration at the surface of a paint film of one of the ingredients of the pigment portion, giving rise to a change in color at the surface. (2) A differential separation of pigments in a dispersed pigment mixture. Floating is a differential separation of pigments (a) by gravity separation in the bulk dispersion; and (b) where the flooding results in a nonuniform or mottled surface coloration. While the terms "flooding" and "floating" are often used interchangeably, they also have been defined as separate, though related, phenomena. "Flooding" refers to uniform color changes and "floating" refers to local excess of one color. (3) Process of color change in which a paint or enamel undergoes homogeneously, from the freshly applied material to the finished dried film in which one or more pigments appear to be "flooding" the surface color. See ▶ Floating. (4) An excess of ink on the printing plate caused by the ink fountain being open too much. In the case of lithographic or offset work, the use of too little water or the absence of an etching material in the water fountain.

Floor and Deck Enamel *n* An enamel designed for abrasion resistance.

Floor Paint See ▶ Deck Paint.

Floor Sealer *n* Composition of resins, with or without oils, in a solvent; designed to penetrate the wood rather than produce a surface finish.

Floor Varnishes *n* Varnishes which are formulated for the coating of flooring. Their main properties include rapid drying, tack-free surface, toughness and resistance to abrasion, washability, and receptiveness to wax and other floor polishes.

Flop Where two different painted panels appear to be a good match for color when viewed at a given angle, but appear different at all other angles. *Also known as Geometric Metamerism.*

Flop Color *n* Used to describe the color observed on paints (particularly metallic paints) when they are viewed near to the grazing angle, i.e., nearly parallel to the surface. The term is generally used in contrast to the terms "face color" or "top color," the color observed when viewing the surface oat an angle near to the perpendicular. See ▶ Face Color.

Florals *n* (1897) Any wallcovering design featuring recognizable flowers and foliage.

Florentine Lake See ▶ Crimson Lake.

Flory-Fox Theory *n* (or Fox-Flory) This theory is relates viscosity to molecular dimensions by treating the polymer molecule as a hydrodynamic sphere. In a θ-solvent in which the molecular coil is compact, these authors write the intrinsic viscosity as,

$$[\eta] = KM^{0.5}$$

where K is a constant. When the polymer is dissolved in a solvent that not a θ-solvent, the equation is replaced with

$$[\eta] = KM^{0.5}\alpha_\eta$$

where α_η is an expansions factor that takes into account the expansion of the polymer coil in going from a bad solvent to a good one. Also, the expansion in a good solvent is accompanied by an increase in viscosity (η) resulting expansion of the polymer molecule due to greater interactions between solvent and polymer.

Flory-Huggins Parameter *n* For polymers dissolved in solvents (may be monomers of the polymer), the Flory-Huggins parameter χ a measure of the interaction energy $\Delta\varepsilon$ (average energy gain per contact; $\Delta\varepsilon$ is a measure of the Gibbs energy and not of the enthalpy; and χ also contains an entropy contribution, which is often found to depend on the concentration of polymer or solute (Flory PJ (1953) Principles of polymer science. The Cornell University Press, New York).

Flory-Huggins Theory *n* (1) A thermodynamic theory of polymer solutions, first formulated independently by Flory and Huggins, in which the thermodynamic quantities of the solution are derived from a simple concept of combinational entropy of mixing and a reduced Gibbs energy parameter, the "χ parameter" (a dimensionless quantity). The Flory-Huggins theory lead to the equation,

$$\Delta F_1 = RT \ln a_1 = RT\, [\ln(1-v_2) + (1-1/m)v_2 + \chi v_2^2] \quad (a)$$

for the partial free energy of the solvent, and to the corresponding equation for the partial free energy of the polymer,

$$\Delta F_2 = RT[\ln v_2 - (m-1)v_1 + m\chi v_1^2] \quad (b)$$

When the logarithmic term in the ● equation (a) is expressed as a series in v2 and m is replaced by the degree of polymerization, P, the convenient equation results,

$$\Delta F_1 = RT[-(v_2/P) - (0.5-\chi)v_2^2 - (v_2^3/3)\ldots] \quad (c)$$

A numerical parameter, χ, employed by the Flory-Huggins equation accounts for the contribution of the non-combinational entropy of mixing and for the enthalpy of mixing (Elias, 1977). The equation assumes that the $\Delta V_{mix} = 0$, and the enthalpy of mixing does not influence the value of ΔS_{mix}. The equation represents Flory-Huggins free energy of mixing per molecule. (2) Historical development of theory and equation: Polymer solutions show enormous deviations from Raoult's law. Polymer chemists early realized that the large deviations from the ideal behavior predicted by Raoult's law resulted from the flexibility of polymer molecules. A polymer molecule continuously changes shape, and, in the course of time, takes on a tremendous number of conformations. Therefore, when a polymer is dissolved, there is, in addition to entropy of solution which results from simple interchange of position, conformational entropy. Flory and Huggins first worked out the statistical theory of polymer solutions which took conformational entropy into account. They chose a lattice picture, widely used in statistical theories of simple liquids, as a model for the polymer solution. In the lattice picture, the space occupied by the solution is divided into boxes of equal size, and polymer segments and solvent molecules are fitted into these boxes. The number of ways, Ω, that n_2 polymer molecules, which occupy n_2 sets of m contiguous boxes, can be arranged in a lattice can be computed by classical statistics. The number Ω is related to the entropy of mixing by Boltzmann's equation, $\Delta S = k \ln \Omega$, where k is Boltzmann's constant; if the entropy computed by this equation represents all of the entropy of mixing, and if ΔH^m is assumed to be positive and, following Van Laar, is set equal to $kt\chi n_1 v_2$, the free energy of mixing is given by,

$$\Delta F^m = -T\Delta S^m + \Delta H^m = kT[n_1 \ln v_1 + n_2 \ln v_2 + \chi n_1 v_2]$$

Where v_1 and v_2 are the volume fractions of solvent and polymer respectively, n_1 and n_2 are the number of molecules of solvent and polymer, and χ is a dimensionless quantity called an interaction parameter that is related to the interaction energy characteristic of a particular polymer-solvent combination (Mark JE (ed) (1996) Physical properties of polymers handbook. Springer, New York; Flory PJ (1969) Statistical mechanics of chain molecules. Interscience Publishers, New York; Huggins ML (1958) Physical chemistry of high polymers. Wiley, New York; Flory PJ (1953) Principles of polymer science. The Cornell University Press, Ithaca).

Flory-Krigbaum Theory *n* A theory of dilute polymer solutions which describes the distribution of polymer segments as clusters of segments separated by regions of pure solvent. A model for this structure is used which takes account of the excluded volume within the coils.

Flory-Rehner Equation *n* The correlation between thermodynamic parameters and the value of an equilibrium swelling of a polymer by a solvent is given by this equation:

$$\ln(1-\varphi_2) + \varphi_2 + \chi_1 \varphi_2^2 = -\frac{v_2}{V} Vs \left\{ \varphi_2^{1/3} - \frac{2\varphi_2}{f} \right\}$$

Where φ_2 = polymer volume fraction in a swollen sample,

$\frac{v_2}{V}$ = volume fraction of elasticity active chains, f = functionality, and χ = interaction parameter. The value of equilibrium swelling can be a practical criterion of solubility. The presence of even a small amount of crosslinks hinders chain separation and polymer diffusion into solution. Solvent can penetrate into the polymer and cause swelling (Wypych G (ed) (2001) Handbook of solvents. Chemtec Publishing, New York).

Flory Temperature *n* The temperature at which, for a given polymer-solvent pair, the polymer exists in its unperturbed dimensions (Elias HG (1977) Macromolecules, vol 1–2. Plenum Press, New York). Syn: ▶ Theta Temperature (Flory PJ (1960) The statisical thermodynamics of solutions. Wiley, New York).

Flow \ˈflō\ *v* [ME, fr. OE *flōwan*, akin to OHGr *flouwen* to rise, wash, L *pluere* to rain, Gk *plein* to sail, float] (before 12c) (1) Resistance to movement by a liquid material and divided rheologically into four categories: Newtonian (simple) Flow, Plastic Flow, Pseudoplastic Flow, Dilatant Flow (Gooch JW (1977) Analysis and deformulation of polymeric materials. Plenum Press, New York). (2) Movement of a coating during and after application and before the film is formed. (3) Property which a coating or ink possesses of leveling after application. Coatings may possess excessive flowing properties and may sag or run to an undesirable degree from vertical or inclined surfaces (Patton TC (1965) Paint flow and pigment dispersion. Interscience Publishers, New York, New York). See ▶ Leveling. (4) Inks of poor flow are classed as short or buttery in body, while inks of good flow are said to be long (Eirich FR Rheology – theory and applications. Academic, New York; vol 1 (1956), vol 2 (1958), vol 3 (1967), vol 4 (1967), vol 5 (1969); Green H (1949) Industrial rheology and rheological structures. Wiley, New York; Mercurio A (September 1964) Rheology of acrylic paint resins. Canadian Paint and Varnish).

Flow-Behavior Index See ▶ Power Law.

Flow Birefrigence *n* (streaming birefringence) The difference, Δ*n* between the refractive indices of a flowing polymer solution or melt in the direction of flow and a direction perpendicular to the flow. Usually measured, using one of two techniques, by directing a light beam downward through the liquid in a rotational viscometer. The amount of birefringence is related to the degree of orientation of the polymer chains, in turn related to shear stress and first normal-stress difference in the flowing medium. The measurements are useful for testing molecular theories and rheological models, also for understanding processing problems such as extrudate roughness.

Flow Coating *n* A painting process in which the article to be painted is drenched with a paint, either by pouring or by spraying with a mist in a closed or semi-closed chamber. The parts are sometimes rotated during and after drenching to avoid sags and runs. The process is used for coating metallized parts and other irregularly shaped articles that are difficult to paint by ordinary spraying methods.

Flow Curve *n* A plot of applied force against the resulting rate of flow. A more in-depth definition is a graph, usually on bilogarithmic coordinates, of shear stress vs shear rate or, sometimes, of apparent or true (corrected) viscosity vs either shear rate or shear stress. For nearly all polymer melts, log-log plots of shear stress vs shear rate of sufficient range exhibit a Newtonian region of slope = 1 at extremely low shear rates, a brief transition region of decreasing slope ("knee"), and a higher-shear region in which the slope is nearly constant or very gradually decreasing and is in the range 0.7–0.25. See also ▶ Pseudoplastic Fluid and ▶ Viscosity.

Flower of Antimony See ▶ Antimony Oxide.

Flowers of Antimony *n* Syn: ▶ Antimony Trioxide.

Flowing Varnish *n* A varnish designed to produce a smooth lustrous surface without rubbing or polishing.

Flow Line Syn: ▶ Weld Line.

Flow Mark *n* A defect in a molded article characterized by a way surface appearance, caused by improper flow of the resin into the mold, which itself may have a number of causes.

Flow Molding *n* (1) A variation of ▶ Injection Molding used for thick-walled parts. A large gate is used and pressure is maintained on the injected melt so as to fore a little more melt into the mold as each part solidifies from the outside inward, thus minimizing shrinkage and improving consistency of the parts' final dimensions. (2) A process of heating material such as a cloth-backed, plasticized PVC with a high-frequency electric field while pressing the material into a mold made of silicone rubber. In 10–15 s, the PVC is formed to the contours and surface texture of the mold, which can give it the look of hand-tooled leather or similar effects.

Flow Properties See ▶ Melt-Flow Index, ▶ Viscosity, ▶ Pseudo-Plastic Fluid, and ▶ Rheology.

Flue \ˈflü\ *n* (1582) A passageway in a chimney for conveying smoke, gases or fumes to the outside air.

Fluffing *v* (1875) A term describing the appearance of a carpet after loose fiber fragments left during manufacture have worked their way to the surface. Fluffing is not a defect; it is simply a characteristic of new carpets that disappears with vacuuming.

Fluid \ˈflü-əd\ *n* (1661) A gas or a liquid, or, in the supercritical region, a hybrid.

Fluidity *n* \flu-ˈi-də-tē\ *n* (1603) (1) The ease with which a liquid flows under stress. (2) Specifically, the reciprocal of viscosity. The SI unit of fluidity is $(Pa·s)^{-1}$ or $1/(Pa·s)$, replacing the deprecated cgs unit, the *rhe*. 1 rhe = 10 $(Pa·s)^{-1}$ (3) The reciprocal of viscosity. The cgs unit is the rhe, the reciprocal of the poise. Dimensions $[m^{-1}lt]$.

Fluidization *vt* (ca. 1855) A gas-solid or liquid-solid contacting process in which a stream of fluid is passed upwards through a bed of small solid particles, causing them to lift, expand, and behave as a boiling liquid. The process is widely used in the chemical industry for performing reactions in which the solid is either a reactant or a catalyst. In the plastics industry, the main application is in ▶ Fluidized-Bed Coating.

Fluidized Bed *n* An expanded bed of solid particles, fluidized to create a system with similar properties of a liquid. Fluidized beds are used successfully in a multitude of processes both catalytic and noncatalytic. Among the catalytic uses are hydrocarbon cracking and reforming, oxidation of naphthalene to phthalic anhydride, and ammoxidation of propylene to acrylonitrile. A few of the noncatalytic uses are coating of sulfide ores, coking of petroleum residues, drying, and classification.

Fluidized-Bed Coating *n* The process of applying plastics coatings to objects of other, higher-melting materials, often metals, wherein a powdered resin is placed in a container provided with a porous or perforated bottom through which a gas is directed upward to keep the resin particles in a state of agitated levitation. The part to be coated is preheated above the resin's softening temperature and lowered into the fluidized bed until a deposit of the desired thickness has formed, then the part is withdrawn and allowed to cool.

Fluorescein Dye See ▶ Acid Dyes.

Fluorescence \flü-ˈre-sᵊn(t)s\ *n* [*fluor*spar + opal*escence*] (1852) Fluorescence is a luminescence that is mostly found as an optical phenomenon in cold bodies, in which a molecule asborbs a high-energy-photons (low wavelength such as ultraviolet) and emits photons at as a lower-energy photon (longer-wavelength) and usually visible light. The energy difference between the absorbed and emitted photons (hv) manifests itself as molecular vibrations or heat in the absorbing matter Usually the absorbed photon is ultraviolet and the emitted light (luminescence) is in the visible range so that humans can observe, but this depends on the absorbance curve and Stokes shift of the particular fluorosphore. Fluorescence is named after the mineral fluorite (calcium fluoride) that exhibits this phenomenon The expression below shows the fluorescence phenomenon,

$$S_1 \rightarrow S_2 + hv$$

where, h = Planck's constant, and v = frequency of the fluorescing light. The Kasha-Vavilov rule describes the quantum yield of luminescence that is independent of the of the wavelength of exciting radiation, and the Jablonski diagram decribes most the relaxation mechanism for excited state molecules (Serway RA, Faugh JS, Bennett CV (2005) College physics. Thomas, New York; www.Wikipedia.org).

Fluorescent Brightening Agent See ▶ Brightening Agents.

Fluorescent Dyes *n* luminescence which is mostly found as an According to the Paint/Coatings Dictionary, dyestuffs which exhibit the phenomenon of fluorescence in the visible region of the spectrum. Dyes are characterized as fluorescent if: they absorb invisible radiation, such as ultraviolet and emit visible light; they also absorb primarily visible light and emit visible light; or they absorb visible light and emit invisible radiation, generally infrared. Some dyes fluoresce more strongly than others, i.e., some are only slight fluorescent and some are strongly fluorescent. The former may go undetected as fluorescent dyes. In order to measure the color materials containing fluorescent dyes, the material must be illuminated with light of defined spectral quality, and the light reflected or transmitted by the materials must be passed through the monochromating device after reflectance or transmission, using so-called "reversed optics." Strongly fluorescent pigments are frequently referred to as high visibility pigments; their combined emission and reflectance is geneally much higher at certain wavelengths than the reflectance of the perfect diffuser (Paint/coatings dictionary (1978) Compiled by definitions committee of the federation of societies for coatings technology; Morgans WM (1977) Pigments for paints and inks. Selection and Industrial Training Administration Ltd., London; Lubs HA (ed) (1965) The chemistry of synthetic dyes and pigments (ACS monograph). Hafner Publishing Co., New York).

Fluorescent Inks *n* Inks which exhibit fluorescence, resulting in a very brilliant effect.

Fluorescent Paint *n* See ▶ Luminous Paint and ▶ High Visibility Paints.

Fluorescent Pigment *n* (1) Inorganic, usually coarse, crystalline, materials which emit light when activated by ultraviolet radiation, e.g., zinc sulfide, combined zinc and cadmium sulfides. A type known as *daylight fluorescent pigments* (*dayglo pigments*) responds to radiation in both the ultraviolet and visible ranges, causing the effect of glowing in normal daylight. These pigments are comprised of fluorescent dyes incorporated in a clear-resin matrix, ground to powder form. Urea and melamine resins have been used as matrices, also a modified sulfonamide resin. (2) Organic dyes or pigments used to manufacture luminescent materials. These include spirit-soluble dyes, oil-soluble dyes and metallic organic compounds, e.g., rhodamine, eosine, rhodamine tungstate and zinc salt of 8-hydroxyquinoline. By absorbing unwanted wavelengths of light and converting them into light of desired wavelength, these pigments have the appearance of possessing an actual glow of their own. (3) Pigments which exhibit the phenomenon of fluorescence in the visible region of the spectrum. Few commercial mineral pigments are fluorescent; most of the fluorescent pigments are made from fluorescent dyes incorporated into an insoluble matrix and ground to a small particle size suitable for use as a pigment. See ▶ Fluorescent Dyes.

Fluorinated Ethylene-Propylene Resin *n* (FEP, PFEP) This member of the fluorocarbon family is a copolymer of tetrafluoroethylene and hexafluoropropylene, possessing most of the desirable properties of PTFE, yet truly meltable and, therefore, processable in conventional extrusion and injection-molding equipment. It is available in pellet form for those operations and as dispersions for spraying and dipping.

Fluorine-Containing Polymers *n* Polymers in which one or more hydrogen atoms have been replaced by fluorine. Often used in high-temperature wire and cable insulation, due to their resistance to chemicals and oxidation and broad useful temperature range.

Fluorite Objective *n* This objective is corrected both for spherical aberration and chromatic aberration at two wavelengths.

Fluorocarbon \ ˈflúr-ō-ˌkär-bən\ *n* (1937) Teflon tetrafluoroethylene (TFE) and Teflon fluorinated ethylene propylene (FEP). Both have remarkable chemical resistance.

Fluorocarbon Blowing Agent *n* A family of inert, noncorrosive liquid compounds containing carbon, chlorine and fluorine, originally developed as refrigerants. They are compatible with all resins and leave no residues in molds. For years they were widely used in structural-foam extrusion, in which they were incorporated with the polymer by direct injection through the barrel of the first of two tandem extruders. Fluorinated hydrocarbons are numbered by a three-digit system developed by DuPont for use with its trade name Freon®. The first digit is the number of carbon atoms in the molecule −1, omitted when it is 0 (the methane group), leaving just two digits for them. The second digit equals the number of hydrogens in the molecule +1; the third digit is the number of fluorine atoms. The remaining atoms in these saturated compounds are chlorine. Thus the once-common blowing agents were Freon 11 (trichlorofluoromethane), Freon 12 (dichlorodifluoro-methane), Freon 113 (trichloro-trifluoroethane), and Freon 114 (dichlorotetra-fluoroethane). Because there is strong evidence that these compounds when released to the atmosphere, migrate to the upper levels and catalyze the destruction of UV-blocking ozone by chain reactions, they are being phased out of use of international agreement and are being replaced, in the manufacture of plastics foams, by hydrocarbons, such as neopentane, and by hydrohalocarbons.

Fluorocarbon Elastomer *n* (fluoroelastomer) Any fluorocarbon polymer of low T_g and no crystallinity, therefore rubbery. As rubbers, these materials are resistant to high temperature and most chemicals and solvents. Most of the commercial materials are copolymers. In the U.S., DuPont's Viton® materials are most familiar.

Fluorocarbon Resin *n* Any of a family of thermoplastics chemically similar to the polyolefins, with all the hydrogen atoms replaced by fluorine. They are made by addition polymerization from olefinic monomers composed only of fluorine and carbon. The main members of the family are ▶ Polytetrafluoroethylene, ▶ Fluorinated Ethylene-Pro-Pylene Resin, and ▶ Polyhexafluoropropylene.

Fluorocarbons *n* Group of compounds containing fluorine atoms, fluoroplastics, or solvents.

Fluoroethylene Syn: ▶ Vinyl Fluoride.

Fluorohydrocarbon Resin *n* Any resin polymerized from an olefinic monomer composed of carbon, fluorine, and hydrogen only. Included are ▶ Polyvinylidene Fluoride, ▶ Polyvinyl Fluoride, ▶ Poly-Trifluorostyrene and copolymers of halogenated and fluorinated ethylenes.

Fluoroplastic *n* (Fluoropolymer) A plastic based on polymers made from monomers containing one or more atoms of fluorine, or copolymers of such monomers with other monomers, the fluorine-containing

monomer(s) being in the greatest amount by mass (ASTM D 883). This is a broad family including:
Polytetrafluoroethylene
Fluorinated ethylene-propylene resin
Polychlorotrifluoroethylene
Polyvinylidene Fluoride
Fluorocarbon Resin
Fluorohydrocarbon Resin
Chlorofluorocarbon Resin
Chlorofluorohydrocarbon Resin
Ethylene-tetra-fluoroethylene Copolymer
Ethylene-chloro-tri-fluoroethylene Copolymer
Perfluoroalkoxy Resin

Fluoroplastics *n* Plastics which contain the monomer of fluorine. See ▶ Fluorocarbons.

Fluoropolymers *n* Polymers whose repeating units contain fluorine. Such polymers often have outstanding thermal, thermo-oxidative and chemical resistance, both to chemical attack and to swelling by solvents.

Fluorosilicones *n* Polymers with chains of alternating silicon and oxygen atoms and trifluoropropyl pendant groups. Most are rubbers.

Flushed Color *n* A color base in paste form prepared by flushing.

Flushed (Flushing) Colors *n* Pigments which have been flushed. See ▶ Flushed Pigment and ▶ Flushing.

Flushed Pigment *n* A pigment obtained by the direct transfer of pigment particles that have been precipitated, from an aqueous phase to a nonaqueous stage (no intermediate drying or pulverizing stage being involved). In the process, the press cake (aqueous or water-wet pigment phase) is mixed and agitated with a nonaqueous vehicle (oil, solvent, and/or resin phase) in a heavy duty mixer. The pigment particles preferentially transfer (flush) to the nonaquerous phase, and the bulk of the essentially clear water is poured off. Any residual water is then vacuumed off at a temperature of about 50°C (122°F).

Flushing *v* A method of transferring pigments from dispersions in water to dispersions in oil by displacement of the water by oil. The resulting dispersions are known as flushed colors. Syn: ▶ Fleshing. See ▶ Flushed Pigment.

Fluted Core An integrally woven reinforcing material consisting of ribs between two skins, thus providing unitized sandwich construction.

Fluted Mixing Section In extrusion, a screw section in which several short barrier flights are placed so that the plastic must flow through the high-shear clearance between the flight tip and the barrel. Fluted sections may have the flights parallel to the screw axis (Maddock-Union Carbide) or at an angle to the axis (Egan).

Fluting *n* (1611) Parallel concave grooves used as furniture decoration. Vertical and spiral flutings are common.

Flux \ˈfləks\ *n* [ME, fr. MF & ML; MF, fr. ML *fluxus*, fr. L, flow, fr. *fluere* to flow] (14c) (1) In chemistry and metallurgy, a substance, e.g., borax or fluorspar, used to promote fusion of metals or minerals. (2) In plastics compounding, the term flux is sometimes used for an additive that improves flow properties, e.g., coumarone-indene resin in the milling of vinyl compounds. (3) In heat and mass transfer, the rate of transfer per unit of cross-sectional area perpendicular to the direction of transfer. (4) ▶ Luminous Flux.

Flux *v* (15c) To melt, fuse, or make liquid. In the early years of the vinyl-plastisol art this term was often used for "fuse" before the latter term came into general use.

Fluxing Temperature Syn: ▶ Fusion Temperature.

Flux, Luminous Radiant flux weighted for the relative luminous (visual) efficiency.

Flux, Radiant See ▶ Radiant Flux.

Fly *n* The short, waste fibers that are released into the air in textile processing operations such as picking, carding, spinning, and weaving.

Flyer *n* (1) A device used to insert twist into subbing, roving, or yarn, and to serve as a guide for winding it onto a bobbin. The flyer is shaped like an inverted U that fits on the top of the spindle and revolves with it. One arm of the U is solid and the other is hollow. The yarn enters through the top of the hollow arm, travels downward, and emerges at the bottom where it is wound around a presser finger onto the take-up package. (2) See ▶ Loom Fly.

Flyer Spinning *n* A method of spinning by means of a driven flyer. It is used primarily for spinning worsted and coarser yarns. Also see ▶ Flyer, (1).

Flyer Spinning Frame See ▶ Spinning Frame.

Flyer Waste *n* During the roving operation, flyer waste refers to fibers that free themselves by centrifugal force from the regular bulk of roving and accumulate on the flyers and adjacent machinery.

Fly Frame See ▶ Roving Frame.

Flying *n* A condition wherein a fine mist or spray of ink is thrown off rapidly moving ink rollers. Syn: ▶ Misting and ▶ spraying.

FMC Color Difference Equations *n* Letters "FMC" are the initials of Friele, MacAdam, and Chickering. There are two different equations, referred to as FMC-1 and FMC-2: FMC-1 is the same as FMC-2, except that the terms K_1 and K_2 are not included: FMC-2 is:

$$\Delta E = [(\Delta C)^2 + (\Delta L)^2]^{1/2}$$

where:

$$\Delta L = K_2 \Delta L_2$$

$$\Delta C_1 = [(\Delta C_{rg}/a)^2 + (\Delta C_{yb}/b)^2]^{1/2}$$

$$\Delta C_{rg} = (Q\Delta P - P\Delta Q)/(P^2 + Q^2)$$

$$\Delta C_{yb} = S\Delta L_1/(P^2 + Q^2)^{1/2} - \Delta S$$

$$\Delta L_1 = (P\Delta P + Q\Delta Q)/(P^2 + Q^2)^{1/2}$$

$$\Delta L_2 = 0.279 \Delta L_1/a$$

$$K_1 = 5.5669 \cdot 10^{-1} + 4.9434 \cdot 10^{-2}Y - 8.2575 \cdot 10^{-4}Y^2 + 7.9172 \cdot 10^{-6}Y^3 - 3.0087 \cdot 10^{-8}Y^4$$

$$K_2 = 1.7548 \cdot 10^{-1} + 2.7556 \cdot 10^{-2}Y - 5.7262 \cdot 10^{-4}Y^2 + 6.3893 \cdot 10^{-6}Y^3 - 2.6731 \cdot 10^{-8}Y^4$$

$$a^2 = 1.73 \cdot 10^{-5}(P^2 + Q^2)/[1 + 2.73 P^2 Q^2/(P^4 + Q^4)]$$

$$b^2 = 3.098 \cdot 10^{-4}(S^2 + 2.015 \cdot 10^{-1}Y^2)$$

$$P = 0.724X + 0.382Y - 0.098Z$$

$$Q = -0.48X + 1.37Y + 0.1276Z$$

$$S = 0.686Z$$

X, Y, Z are tristimulus values for the standard
X_1, Y_1, Z_1 are tristimulus values for the sample
ΔP, ΔQ, ΔS are above formulate for P, Q, S substituting
$\Delta X = X_1 - X$ for X, etc.

FMQ See ▶ Methylfluorosilicones.

Foam \ ˈfōm\ *n* [ME *fome*, fr. OE *fām*; akin to OHGr *feim* foam, L *spuma* foam, *pumex* pumice] (before 12c) Dispersion of gas in a liquid or solid. The gas bubbles may be any size. The term covers a wide range of useful products such as insulating foam, cushions, etc. It also describes the undesirable froth in polymer melts, dyebaths, etc. See ▶ Cellular Plastic.

Foam-Backed *n* A term describing a fabric laminated to or coated with a layer of rubber or plastic foam.

Foam Casting *n* (foam molding) A process withy many variations, depending on the polymers used. In general, a fluid resin or prepolymer containing catalyst is foamed before or during molding by mechanical frothing, or by gas dissolved in the mixture or vapor from a low-boiling liquid. See also ▶ Reaction Injection Molding.

Foamed Plastic Plastics with an apparent density which are significantly decreased by the presence of numerous cells disposed throughout its mass usually by an expanding gas foam blowing agent. Syn: ▶ Cellular Plastic.

Foam Extrusion See ▶ Extruded Foam.

Foam Fabrication The process of cutting large slabs, logs, or "buns" of foamed plastics into sections of desired dimensions. The raw slab is conveyed through an array of saws, knives, or hot wires that first remove the uneven top, bottom and sides, then slice the remaining rectangular block into boards, finally cross-cutting the boards to standard lengths.

Foaming A dispersion of gas in a liquid or solid. Foams can be made by mechanically incorporating air, as in the food industry for whipped cream, egg whites, and ice cream.

Foaming Agents (1) Materials that increase the stability of a suspension of gas bubbles in a liquid medium. (2) Blowing agents in rubber or plastics. See ▶ Blowing Agents.

Foam-in-Place Refers to deposition of foams at the site of the work, e.g., between two containing walls for insulation, as opposed to bringing the work to the foaming machine.

Fog \ ˈfóg\ *n* [prob. Scand origin, akin to Dn *fog* spray, shower] (1544) A colloidal dispersion of a liquid in a gas. See ▶ Bloom and ▶ Blush.

Fogged Coat Noncontinuous spray coat. See ▶ Mist Coat.

Fogged Metal Metal, the luster of which has been sharply reduced by a film of corrosion products.

Fogging See ▶ Bloom and ▶ Blushing.

Foil *n* [ME, leaf, fr. MF *foille* (fr. L *folia*, plural of *folium*) & *foil*, fr. L *folium*] (14c) (1) Refers to very thin membranes (less than 6 mils) of metal such as aluminum. Above 6 mils, the thin metal is called a sheet. (2) In wallcoverings, a very thin sheet of flexible metal on a paper or fabric back. Can be printed with transparent or opaque color, and mottled to resemble marble, tortoise shell, etc.

Foil Decorating See ▶ In-Mold Decorating.

Folded Selvage *n* A curled selvage.

Folded Yarn See ▶ Plied Yarn.

Folding Machine *n* A machine that folds sheet plastics such as cellulose acetate into shapes such as identification-card envelopes, sheets for ring binders, visible indexes, and the like. An electrically heated blade softens the plastic and folds it into a tight, 180° crease.

Fold Testing *n* Mechanical test to determine the ability of a polymer to retain its strength after being folded back and forth.

Food and Drug Administration (FDA) *n* The FDA is the governmental body that is responsible for the approval of all food additives. All inks, coatings, and other packaging materials coming in direct contact with food or drugs must be shown to be non-migrating, or must be made only from raw materials which are known to be harmless and which are listed in the Code of Federal Regulations, Title 21, in the paragraph covering the intended end use. See ▶ FDA.

Footcandle *n* (1906) A deprecated unit of surface-lighting intensity, equal to 10.76391 lux. The lux (lx) is defined as 1 lumen per square meter (lm/m^2). See also ▶ Luminous Flux.

Footing *n* Concrete base on which a foundation sits.

Foot Lambert *n* Foot lambert is the unit of photometric brightness (luminance) equal to $1/\pi$ candle per square foot.

Foots *n* Sediment which settles from an oil on standing; chiefly albuminous matter. Originally used to describe those solid impurities which precipitate from raw linseed oil during storage and then settle to the bottom or "foot" of a storage tank. Later used to denote material insoluble in a mixture of equal parts of acetone and linseed or other oil under test, and insoluble in calcium chloride solution under the specific conditions of test method. Also, the term is used to denote that material which is precipitated from the oil by phosphoric acid and which is insoluble in acetone under the specific conditions of test method. See also ▶ Tailings.

Force \ **¹**fôrs, **¹**fôrs\ *n* [ME, fr. MF, fr. (assumed) VL *fortia*, fr. L *fortis* strong] (14c) (1) Either half of a compression mold (top force or bottom force), but usually the half that forms the concave surfaces of the molded part. (2) The male half of the mold, which enters the cavity and exerts pressure on the resin, causing it to flow. (3) A basic familiar quantity, familiar to everyone, customarily defined as that which changes a body's state of rest or motion, according to Newton's second law of motion:

$$F = \frac{d}{dt}(mv)/g_c = ma/g_c$$

The second form applies when mass *m* is constant, as in all industrial applications. Here F = the force, v = the velocity, t = time, a = the body's acceleration = dv/dt, and $1/g_c$ is the proportionality constant relating the dimensions of the three primary quantities. Historically the units of the primary quantities, in whatever system, have been devised so that g_c is exactly 1 and it is often omitted from the equation. In the SI system, 1 N is defined as the force that will accelerate a 1-kg mass 1 (m/s) per second, making g_c exactly 1.0000 N·s^2/(kg·m). In the English system, however, where "pound" is used to mean either the pound-mass or the pound-force (lb or lb$_f$), the value of g_c = 32.174 lb$_f$·s^2/(lb·ft)! *Weight* is the downward force exerted on bodies in the earth's gravitational field and results from the as yet unexplained mutual attraction of masses as set forth in Newton's law of universal gravitation. Because the earth is nearly spherical, the distances between bodies on and close to its surface and the earth's center are almost constant and so, therefore is the earth's attractive force on a unit mass at its surface. For convenience, the force is replaced by g, the standard "acceleration due to gravity," 9.806650 m/s^2 (32.174 ft/s^2). Weight, then, = $m·g/g_c$. For precise work, corrections are made for the variation in g with latitude and altitude, but g_c is always constant, even in outer space. Force has other aspects that are unconnected with rate of change of momentum. For example, the forces developed in a tightened bolt or the force exerted by a hydraulic ram on molding compound in a closed mold. One should keep in mind that force is *not identical* to mass times acceleration.

Force Between Two Charges n, Coulomb's Law If two charges q and q' are at a distance r in a vacuum, the force between them is,

$$F = \frac{qq'}{r^2}$$

The force will be given in dynes if q and q' are in electrostatic units and r in centimeters.

Force Between Two Magnetic Poles *n* If two poles of strength m and m' are separated by a distance r in a medium whose permeability is μ (unity for a vacuum), the force between them is,

$$F = \frac{mm'}{\mu r^2}$$

Force will be given in dynes if r is in centimeters and m and m' are in cgs units of pole strength. The strength of a magnetic field at a point distance r from an isolated pole of strength m is

$$H = \frac{m}{\mu r^2}$$

The field will be given in gauss if m and r are in cgs units.

Forced Drying *n* Drying coatings at a temperature between room temperature and 65.6°C (150°F) as opposed to air drying or baking.

Forced Drying Temperature *n* Temperature between room temperature and 65.6°C (150°F). See ▶ Baking.

Force Feeder See ▶ Crammer-Feeder.

Force Plate *n* The plate that carries the plunger or force plug of a mold and the guide pins or bushings. Since the force plate is usually drilled for steam or water lines, it is sometimes called the *steam plate*.

Force Plug *n* (plunger, piston) The portion of a mold that enters the cavity block and exerts pressure on the molding compound, designated as the *top force* or *bottom force* by position in the assembly.

Ford Cup *n* Efflux cup used for measuring the viscosity of paint or varnishes. See ▶ Efflux Viscometer and ▶ Viscometer.

Ford Viscosity Cups *n* A series of three cylindrical cups with conical bottoms, differing only in the diameters of the orifices at the apexes of the cones, each cup having a capacity of about 100 ml. From the time or efflux, the sample volume and the orifice diameter, the kinematic viscosity of a liquid may be estimated. ASTM Test D 1200 (sec. 06.01) describes the Ford cup procedures to be used with paints and varnishes.

Foreign Matter *n* Anything visibly unrelated to the true nature of the substance under examination.

Foreign Waste *n* Thread waste or lint that is twisted in the yarn or woven in the fabric. If such foreign matter is of a different fiber, it may dye differently and thus show plainly.

Forgeability *n* The ability of a solid material to flow without rupture under sudden intense compression. Some plastics have been forged into useful articles.

Forging See ▶ Cold Forming and ▶ Solid-Phase Forming.

Form *n* (sometimes Forme) \ˈfórm, ˈfórm\ *n* (15c) (1) Type and other matter locked in a chase ready for printing. (2) In crystallography, a group of similar faces, e.g., cube, prism, dipyramid, etc. Also, a given polymorphic form of a substance, e.g., Form II of calcium carbonate is aragonite.

Formability *n* The relative ease with which a plastic sheet or rod may be given another permanent shape. See ▶ Thermoformability.

Formal Charge *n* A somewhat arbitrary but useful way of indicating the approximate electrical characteristic or charge of an atom.

Formaldehyde \fór-ˈmal-də-ˌhīd, fər-\ *n* [ISV *form-* + *aldehyde*] (1872) (formic aldehyde, methanal, oxymethylene) HCHO. A colorless gas with a pungent, suffocating odor, obtained most commonly by the oxidation of methanol or low-boiling petroleum gases such as methane, ethane, etc. The gas is difficult to handle, so it is sold commercially in the form of aqueous solutions (formalin), solvent solutions, as its oligomer, paraformaldehyde, and as the cyclic trimer, 1,3,5-trioxane (α-trioxym-ethylene). High-molecular-weight, commercial polymers of formaldehyde are called *polyoxymethylene* or ▶ Acetal Resin. Formaldehyde is also used in the production of other resins such as ▶ Phenolic Resin (phenol-formaldehyde) and ▶ Amino Resin (urea formaldehyde). See ▶ Formalin.

$$\mathrm{=O}$$

Formalin \ˈfór-mə-lən, -ˌlēn\ *n* [*Formalin*, a trademark] (1893) (formol) HCHO. Formaldehyde gas commercially available as a 37% wt solution in water with a small amount of methanol for inhibiting polymerization. See ▶ Formaldehyde.

Formed Fabric See ▶ Nonwoven Fabric.

Form Grinding *n* A method of forming circularly symmetrical parts from plastic rod or tubing, employing a grinding wheel shaped to the inverse of the desired contour, a smaller hardened-steel regulating wheel that presses the plastic rod or tube against the grinding wheel, and work-rest blade that supports the work between the two wheels. Water is usually supplied as the coolant, from which the scrap powder can be recovered and reused.

Formica® \fór-ˈmī-kə, fər-\ A trademark of the Formica Corporation for high-pressure, decorative laminates of melamine-formaldehyde, phenolic and other thermosetting resins with paper, linen, canvas, glass cloth, etc., often misused by the public in a generic manner.

Formic Acid *n* (Methanoic acid) HCOOH. The first of the aliphatic acids, with a pungent odor and probably existing mostly in the dimmer form, formic acid is produced by the reaction of carbon dioxide and dry sodium hydroxide followed by treatment with sulfuric acid. It is a solvent for phenol-formaldehyde resins, some polyesters, polyurethanes and nylons.

Forming A general term encompassing processes in which the shapes of plastics pieces such as sheets, rods, or tubes are changed to a desired configuration, usually with the acid of heat. The term is not usually applied to operations such as molding, casting, or extrusion in which shapes or articles are made from molding materials and liquids. See also ▶ Fabricate and ▶ Thermoforming.

Forming Box See Vacuum Sizing.

Forming Cake *n* In filament winding, the collection (*package*) of glass-fiber strands on a mandrel during the operation.

Form Lacquer *n* Thin lacquer or varnish used to coat concrete forms to prevent concrete from adhering to the forms.

Formol Syn: ▶ Formaldehyde.

Form Roller *n* The form roller is that roller in the ink distribution system of a printing press which is in direct contact with the printing plate and transfers the ink to it.

Formula \fór-myə-lə\ *n* [L, dim. of *forma* form] (1618) (Chemical) A combination of symbols with their subscripts representing the constituents of a substance and their proportions by weight.

Formula Unit The group of atoms indicated by the formula of a substance.

Formula Weight *n* (ca. 1920) (molecular weight) Of a chemical compound, the number obtained by summing the atomic weights of all the compound's atoms. The term is not used with polymers, which are always mixtures of compounds of different molecular weights.

Formula Weight The sum of the masses of the atoms indicated in a formula, expressed in atomic mass units; the mass of one formula unit.

Fossil Gum Resin *n* Fossil gums are the so-called hard gums or copals with were exuded from living plants in the form of liquids or semisolids, which have lain in the ground for centuries and thus hardened or fossilized with time. Some of the fossil gum resins are: Amber, Zanzibar, Kauri, Manila, Pontianak and Congo. See ▶ Resin.

Fossil Resin *n* A natural resin obtained from fossilized remnants of plant or animal life. An example is AMBER, a fossilized resin derived from an extinct species of pine.

Foster Charts *n* Graphs for determining color differences in modified MacAdam units from tristimulus and ratio tristimulus measurements made on a tristimulus colorimeter. They are based on the Simon-Goodwin Charts and give results which are a close approximation to those which would be obtained with the S-G charts.

Foulard \fu-lärd\ *n* [F] (1830) A lightweight, lustrous 2/2 twill that is usually printed with small figures on a solid background, foulard is frequently used in men's ties. Foulards are made of silk, filament polyester, acetate, etc.

Fouling *n* Sessile marine organisms on the hulls of ships.

Foundation Lower parts of walls on which the structure is built. Foundation walls of masonry or concrete are mainly below ground level.

Foundry Resin *n* A thermosetting resin used as a binder for sand in metals founding. The types most commonly used as water-soluble phenol-formaldehyde resins that become insoluble when cured, and cold-setting furfuryl alcohol resins that cure in the presence of an acid catalyst.

Fountain \faún-t$^\partial$n\ *n* [ME, fr. MF *fontaine*, fr. LL *fontana*, fr. L, feminine of *fontanus* of a spring, fr. *font-,-fons*] (14c) Part of a printing press which contains the ink to be fed to the distributing system. In lithography, it is also the part which feeds the water or fountain solution to the dampening rollers.

Fountain Roller *n* The roller that revolves in the ink fountain. In lithography it is also the roller that revolves in the dampening solution.

Fountain Solution *n* In lithography, generally a mixture of water, acid, buffer and a gum to prevent the non-printing areas of the plate from receiving ink. To some fountain solutions, alcohols are added. *Also known as Dampening Solution.*

Four-Dimensional Braid *n* A type of braided reinforcement used to achieve specially directed strength and resistance to interlaminar shear, often involving mixed fibers of different materials.

Fourier Number *n* (N_F) A dimensionless group important in analysis of unsteady heat transfer in solids, such as sheets being heated or cooled in thermoforming, or cooling of extrudates and moldings, $N_F = \alpha \cdot t / x^2$, where α = the material's thermal diffusivity, t = the heating or cooling time, and x is a thickness or half-thickness in the direction of heat flow.

Fourier's Law of Heat Conduction *n* The fundamental equation for steady heat flow through solids. It is

$$q = -kA \frac{dT}{dx}$$

where q = the rate of heat flow, k = the thermal conductivity of the material at temperature T, and A = the area through which heat flow is occurring, normal to coordinate x within the material in the direction of heat flow. this is the defining equation for ▶ Thermal Conductivity. The quotient q/A is known as the *heat flux* or *thermal flux*. If the thermal conductivity is constant over the range of temperatures involved or is linearly dependent on temperature, an integrated, simpler version of the law, with an average conductivity, can be used to find the heat flow through a layer of thickness Δx:

$$q = k_{avg} A (T_2 - T_1)$$

where T_2 and T_1 are the temperatures on the hot and cold sides of the layer.

Fourier-Transform Infrared Spectroscopy *n* (FTIR) Infrared spectroscopy (IR) is most commonly used for the identification of unknown pure organic compounds. In FTIR, infrared radiation of a broad range of wavelengths is passed through an interferometer and a pathlength difference is introduced into one part of the light beam. This IR beam is then passed through the sample, which absorbs light energies corresponding to various bond-vibration and –rotation frequencies. The beam is then focused on a detector, and a computer calculates the absorption of the IR frequencies by the sample, identifying compounds present and their concentration in the sample. An example of a FTIR spectroscope is the Spectrum 100 FTIR Spectrophotometer by PerkinElmer, Inc., image courtesy of PerkinElmer, Inc.

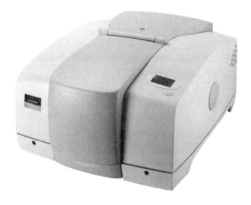

Foxy Pigment crystals are sometimes arranged in different order by method of application, causing different shades of color when viewed from different angles; this condition is often referred to as "foxy."

F.P.S. System n The foot-pound-second system of units. The British system of physical units derived from the three fundamental units of length, mass, and time, i.e., the foot, pound, mass, and the second.

Fraction \ˈfrak-shən\ n [ME *fraccioun*, fr. LL *fraction-*, *fractio* act of breaking, fr. L *frangere* to break] (14c) Solvent of definite boiling range obtained by fractional distillation.

Fractional Distillation n Separation of a liquid into its different components by collecting fractions or distillates of restricted boiling range.

Fractionation n A method of determining the molecular-weight distribution of polymers based on the fact that polymers of higher molecular weight are less soluble than those of lower molecular weight. Two basic methods in use are (1) precipitation fractionation, in which phases are separated from a solution of the polymer by incremental addition of nonsolvents, stepwise lowering of the solution temperature, or volatilization of the solvent; and (2) extraction fractionation, in which fractions of increasing molecular weight are preferentially extracted from a layer of polymer that has been deposited on a substrate. In either method, a series of fractions is obtained which must be recovered and characterized with respect to molecular weight. A third method that has had some success is ultracentrifugation. See ▶ Ultracentrifuge. Fractionation is also used to prepare polymer specimens having narrow molecular-weight distributions.

Fracture \ˈfrak-chər, -shər\ n [ME, fr. L *fractura*, fr. *fractus*] (15c) The separation of a body, usually characterized as either brittle or ductile. In brittle fracture, the crack propagates rapidly with little accompanying plastic deformation. In ductile fracture, the crack propagates slowly, usually following a zigzag path along planes on which a maximum resolved shear stress occur, and their is substantial plastic deformation. Slower loading and higher temperature favor ductile behavior.

Fracture Mechanics The study, both theoretical and experimental, of the behavior of cracks in stressed bodies. A basic principle is that fracture is driven by the energy released by the growth of the crack, which begins at a small imperfection such as is found in all bodies.

Fracture Toughness n (K_c) The critical value of the ▶ Stress-Intensity Factor in a material beyond which a crack will start to grow.

Frame n (1) A general term for many machines used in yarn manufacturing such as the drawing frame, roving frame, and spinning frame. (2) See ▶ Tenter Frame.

Framework Knitting See ▶ Knitting, ▶ Weft Knitting.

Frankel-Acrivos Equation n An equation, derived wholly from theoretical considerations, giving the relative viscosity of suspensions of monodisperse spheres in Newtonian liquids. It is

$$\eta_r = \frac{\eta}{\eta_o} = \frac{9}{8}\left[\frac{(f/f_m)^{1/3}}{1-(f/f_m)^{1/3}}\right]$$

where η_o = the viscosity of the pure liquid, η = the viscosity of the liquid containing a volume fraction f of spheres whose density is close to that of the liquid. f_m represents the maximum attainable volume fraction, which depends on the assumed geometry of packing, about 0.7–0.74. The equation well represents the best available data to loadings approaching $f/f_m = 1$. Compare ▶ Eilers Equation and ▶ Mooney Equation.

Frankincense \ˈfraŋ-kən-ˌsen(t)s\ *n* [ME *fraunk encense*, fr. OF *franc encens*, fr. *franc* (perhaps in sense "of high quality") + *encens* incense] (14c) A fragrant gum resin from trees of a genus (*Boswellia* of the family Burseracae) of Somalia and coastal Arabia that is a an important incense resin and was used in ancient times in religious rites and embalming. See ▶ Gum Thus.

Fraunhofer's Lines *n* When sunlight is examined through a spectroscope it is found that the spectrum is traversed by an enormous number of dark lines parallel to the length of the slit. These dark lines are known as Fraunhofer's lines. Kirchoff conceived the idea that the sun is surrounded by layers of vapors which act as filters of the white light arising from incandescent solids within and which abstract those rays which correspond in their periods of vibration to those of the components of the vapors. Thus reversed or dark lines are obtained due to the absorption by the vapor envelop, in place of the bright lines found in the emission spectrum.

Fraying The slipping or raveling of yarns from unfinished edges of cloth.

Free Energy *n* Also Gibb's Free Energy or change in free energy of a reaction;

$$-\Delta G = T\Delta S - \Delta H$$

where T is temperature, S is entropy, and H is enthalpy; if −DG the reaction can proceed, but if +DG then the reverse reaction can proceed; also referred to as the driving force of a reaction; and the decrease in free energy of a constant temperature, constant-pressure process is the measure of the tendency of the process to proceed spontaneously.

Free Energy of Formation *n*, ΔG_f The free energy change for a reaction in which one mole of a compound is formed from its uncombined elements. For the chemical reaction,

$$1/2 N_2 + 2H_2 + 1/2 Cl_2 \longrightarrow NH_4Cl$$

$$\Delta G = \Delta H - T\Delta S$$
$$\Delta G = (-314.4 \times 10^3 \text{ J mol}^{-1})$$
$$\quad -(298.2)(-373.8 \text{ J K}^{-1}\text{mol}^{-1})$$
$$\Delta G = 2.030 \times 10^5 \text{ J mol}^{-1}$$

−ΔG indicates that the reaction occurs spontaneously.

Free Energy *n*, F (Gibbs Free Energy, G) A thermodynamic quantity which measures the energy in a system which is available for doing work of expansion. The free energy of a system is defined as the difference between its enthalpy, H, and its temperature-entropy product: At constant temperature and pressure, $\Delta G = H - T\Delta S$. If $\Delta G < 0$, then a spontaneous reaction occurs, $\Delta G > 0$, nonspontaneous or a reversible reaction, $\Delta G = 0$, the system is at equilibrium. An example of $\Delta G < 0$ is the melting of ice above 0°C, and $\Delta G > 0$ for the freezing of water above 0°C. Consider a material undergoing a "change of state" at its equilibrium temperature. Gold metal undergoes a change of state (fusion) at its equilibrium temperature (melting temperature, T_m) and

$$\Delta S = \frac{\Delta H}{T_m}$$

T_m can be calculated as,

$$T_m = \frac{12.36 \times 10^3 \text{ J mol}^{-1}}{9.250 \text{ K}^{-1}} = 1336 \text{ K}$$

and 1336 K − 273 = 1063°C

Free Expansion *n* Expansion of a substance, usually a gas, against no opposing pressure.

Free-Falling-Dart Test *n* A method of measuring the impact resistance of thermoplastic films by dropping a dart with a hemispherical head onto a film specimen held in a clamping frame. As described in ASTM D 1709, the dart is dropped from a fixed height onto each of ten specimens and the percent failure is noted. Another increment of weight is added and ten more specimens are tested. The process is repeated until 50% of the specimens fail. The weight of the dart at this point, times the drop height, is a measure of the film's impact strength. In ASTM D 4272, another falling-dart test for films, the dart is instrumented and the energy consumed in penetrating the film is computed. ASTM D 3029 describes a similar test for rigid sheets. See also ▶ Falling-Dart Impact Test.

Free Fatty Acid *n* Unreacted fatty acid present in a coating vehicle.

Free Forming *n* A variant of ▶ Sheet Thermoforming in which a bubble is blown into the clamped, heat-softened sheet, either by applying a vacuum to the side that will be convex or pressure to the underside. The method has been used most with cast-acrylic sheet for applications where the best possible optical properties are foremost, such as airplane canopies.

Free Phenol *n* The uncombined phenol existing in a phenolic resin after curing, the amount of which is indicative of the degree of cure. The presence of such free phenol can be detected by the Gibbs Indophenol Test.

Free Radical *n* (1900) An atom or group of atoms having at least one unpaired electron. Most free radicals are short-lived intermediates with high reactivity and high energy, difficult to isolate. They are important agents in many polymerization processes and have been detected on corona-discharge-treated films by electron-spin resonance (ESR).

Free Radical Chain n Polymerization of an atom or molecule which has at least one electron which is not paired with another electron.

Free-Radical Polymerization n A reaction initiated by a ▶ Free Radical derived from a polymerization catalyst. Polymerization proceeds by the chain-reaction addition of monomer molecules to the free-radical ends of growing chain molecules. Major polymerization methods such as bulk, suspension, emulsion, and solution polymerization involve free radicals. The free-radical mechanism is also useful in copolymerization, in which alternating monomeric units are promoted by the presence of free radicals.

Free Silica n SiO_2. Silica generally present in small amounts in natural deposits of clay-like minerals and diatomaceous earth and usually considered to be a contaminant.

Free-Thaw Resistance n Extent to which water-base paints, utilizing synthetic lattices or synthetic resin emulsions as vehicles: (1) retain their original properties, free from detrimental changes in consistency, and (2) resist coagulation, or the formation of lumps and specks, when subjected to freezing and subsequent thawing.

Free Volume n In a liquid or amorphous solid, the specific volume minus the volume occupied by the molecules themselves. Free volume increases with rising temperature, causing viscosity to diminish. The f_{WFL} is calculated from the specific volume of the amorphous polymer (v^o_{am}) and the specific volume of the liquid polymer (v^o_t) in a polymer-monomer mixture during polymerization (Elias, 1977).

$$f_{WFL} = \frac{v^o_{am} - v^o_t}{v^o_{am}}$$

The vacant sites of the solid polymer (amorphous) are called "free volume"; no long range order in extensive regions occurs in the amorphous state (e.g., amorphous material is not x-ray crystalline) although indications are that these regions possess a certain order; and theoretically a certain number of vacant sites must be present in the solid polymer. Free volume is mainly responsible for the compressibility of liquids and solids. *Also called Free Volumes f: Vacant Free Volume f_{vac}, Williams-Landell-Ferry Free Volume f_{WFL}, Fluctuation Free Volume f_{fluc}, and Free Volume for Thermal Expansion f_{exp}.* The effect of system variables on solubility is discussed in Handbook of Solvents, 2001. Miller (1968) noted that "the concept of free volume is easy to grasp, but, quantitatively, its definition runs into snares. Is free volume the specific volume of the liquid (solution) minus the volume of the molecules computed from Van Der Waals radii, or minus the volume swept out by the segments as they rotate, or is it some other volume?" The free volume is generally, but necessarily, about 2.5% for all polymers.

Free-Wheeling n In reference to rolls, spinning without the application of either driving or braking force.

Freeze Grinding See ▶ Cryogenic Grinding.

Freeze Line Syn: ▶ Frost Line.

Freezing Point n (1747) Temperature at which a liquid material and solid are in equilibrium with one another, i.e., at a lower temperature the liquid will solidify. Amorphous materials, such as paints or paint vehicles, normally solidify over a temperature range referred to as freezing range.

Freezing-Point Depression n $\Delta T = -K_{FP}m$, where T is freezing point depression, K_{FP} is 1.86 for water and m is molality (moles solute/1,000 g solvent), The lowering of the freezing point of a solvent brought about by the presence of a solute.

Frenchback n A fabric with a corded twill backing of different weave than the face. The backing, which is frequently of inferior yarn, gives added weight, warmth, and stability to the cloth.

French Blue See ▶ Ultramarine Blue.

French Chalk n $3MgO \cdot 4SiO_2 \cdot H_2O$. It is distinguished by its soapy or greasy feel. Chemically, it is substantially a hydrated magnesium silicate. It is supplied commercially as a white or gray powder with sp gr of 2.4–2.6. Known also as *Talc, Soapstone, or Steatite.* (Electromagnetic radiation) to produce an exceedingly high gloss to wood surfaces. The use also of a little linseed oil on the polishing cloth is involved.

French Process Zinc Oxide n Zinc oxide pigment made from zinc metal. See ▶ Zinc Oxide.

French Ultramarine n Synthetic ultramarine blue. See ▶ Ultramarine Blue.

French Varnish n A hard-rubbed high-gloss finish, achieved by multiple applications of such varnish.

French Veronese Green See ▶ Hydrated Chromium Oxide.

Freon® \ˈfrē-ˌän\ See ▶ Fluorocarbon Blowing Agent.

Frequency \ˈfrē-kwən(t)-sē\ n (1600) Number of vibrations per second, equal to the velocity of light (or any electromagnetic radiation) divided by the wavelength:

$$v = \frac{c}{\lambda}$$

where v is the frequency, c is the velocity (2.99793×10^8 m/s (for light and for all electromagnetic radiation)), and λ is the wavelength which is specific for each type of radiation.

Frequency Factor *n* The quantity preceding the exponential term in the Arrhenius equation.

Frequency of Vibrating Strings *n* The fundamental frequency of a stretched string is given by

$$n = \frac{1}{2l}\sqrt{\frac{T}{m}}$$

where l is the length, T, the tension and m the mass per unit length. For a string or wire of circular section of length l, tension T, density d, and radius r, the frequency of the fundamental is

$$n = \frac{1}{2rl}\sqrt{\frac{T}{\pi d}}$$

The frequency in vibrations per second will be given if T is in dynes, r and l in centimeters and d in grams per cubic centimeters.

Fresco \fres-(▮)kō\ *n* [It, fr. *fresco* fresh, of Gmc origin; akin to OHGr *frisc* fresh] (1598) Method of painting in which pigments mixed only with water are applied to a freshly laid surface of lime plaster. The painted surface is rendered durable by the pigment absorbing calcium hydroxide from the wet plaster, which is converted into insoluble calcium carbonate as the surface dries. The term is sometimes used incorrectly to denote any form of mural painting.

Fresco Color *n* Water paints used for wall decorations.

Fresnel Reflection *n* Phenomenon of reflection at the surface or interface where media or materials of different refractive indices join. Fresnel laws state that the angle of reflection is equal and opposite to the incident angle, and that the magnitude of the reflection depends on the angle of the incident light, on the relative refractive indices, and on the polarization of the incident light. When the surfaces are smooth and planar, specular (mirror like) reflection occurs, which is subjectively referred to as specular gloss. When the surfaces are rough, the reflectance at the air-material interface is diffuse, and the surface is observed to be matte-like.

Fret *vt* [ME, back-formation fr. *fret*, *fretted* adorned, interwoven, fr. MF *freté*, fr. OF, pp of *freter*, *ferter* to tie, bind, prob. fr. (assumed) VL *firmitare*, fr. L *firmus* firm] (14c) A geometric band or border motif, consisting of interlacing or interlocking lines. Also known as a Key or Meander Pattern.

Friction \▮frik-shən\ *n* [ME, fr. MF or L; MF fr. L *friction-*, *frictio*, fr. *fricare* to rub; akin to L to crumble] (1704) (1) The force resisting the sliding or rolling of one body relative to another with which it is in contact. The *coefficient of friction*, μ, is the quotient of the force required to cause or maintain motion divided by the normal force N exerted by the bodies upon each other, i.e., $\mu = F/N$. With most materials, the force required to initiate motion is somewhat greater than that required to maintain it, so a distinction is drawn between *static* and *dynamic* coefficients of friction. Coefficients of *rolling* friction are generally much lower than coefficients of sliding friction. Dynamic sliding friction is the most important one in plastics processing, but all three have roles in product design and operation. (2) In flowing fluids, particularly liquids, interlaminar friction is believed to be the basis of viscosity.

Frictional Coefficient *n* Resistance to sliding or rolling of surfaces of solid bodies in contact with each other.

Frictional Heating *n* (1) Heat evolved when two surfaces rub together. For a frictional force F and relative velocity V, the rate of heat evolution is F·V (SI: J/s). (2) Viscous dissipation within liquids undergoing shear flow. At any point in a laminar-flowing liquid, the rate of viscous dissipation, per unit volume, is equal to the product of the shear rate times the shear stress, which is also equal to the product of the local viscosity and the square of the shear rate. Both these types of functional heating are important mechanisms of plastication in extruders and injection molders.

Friction Calendering *n* A process in which an elastomeric compound is forced into the interstices of woven or cord fabrics while passing through the rolls of a calender. The heated compound is fed into the top opening of three adjacent rolls, so that it will cling to the middle roll. The fabric to be impregnated is fed into the lower opening between the rolls. The distance between the rolls is regulated so as to squeeze the fabric without crushing it, and the rolls are operated at slightly different speeds so that the compound is wiped by friction into the meshes of the fabric.

Friction Coefficient *n* The coefficient of friction between two surfaces is the ratio of the force required to move one over the other to the total force pressing the two together. If F is the force required to move one surface over another and W, the force pressing the surfaces together, the coefficient of friction,

$$k = \frac{F}{W}$$

Friction False-Twist Texturing See ▶ Texturing.

Friction Spinning *n* A spinning system in which the yarn receives its twist by being rolled along the longitudinal axis in the nip between two revolving surfaces. The surfaces may rotate at the same or different speeds in the same or opposite directions depending on the particular machine design. Potential advantages include high production capacity, low stress on the fiber in processing, and the capacity to produce very fine counts

Friction Top Can *n* Can in which the lid is held in place by friction.

Friction Welding *n* (angular welding) A term encompassing ▶ Spin Welding and the newer process of applying rapid angular oscillations to heat the plastic parts to be joined. This variation of the spin-welding process is used for parts that are not symmetrical about an axis of rotation. The equipment must be programmed to stop when the parts of properly positioned for joining.

Friedel-Catalysts *n* Strongly acidic metal halides such as aluminum chloride, aluminum bromide, boron trifluoride, ferric chloride, and zinc chloride, used in the polymerization of unsaturated hydrocarbons, e.g., olefins. (Friedel-Crafts reactions using such catalysis are named for Charles Friedel and James Crafts, who first used them in 1877.) These acidic halides are also known as *Lewis acids*.

Friedel-Crafts Catalyst *n* Lewis Acid catalysts such as aluminum chloride or ferric chloride.

Friel-MacAdam-Chickering Color Difference Equations See ▶ Fmc Color Difference Equations.

Friezé \ˈfrēz *or* frē-ˈzā\ *n* [ME *frise*, fr. MF, fr. MDu *vriese*] (15c) (1) Any decorative horizontal band, as along the upper part of a wall in a room. (2) A type of wallpaper popular in the early 1900s. Generally, a pictorial border which ran above door height, or, in dining rooms, above the plate rail. (3) A term applied when the pile of a velvet, plush, velour, or other pile fabric is uncut. A friezé fabric is sometimes patterned by shearing the loops at different lengths. Friezé fabrics are widely used for upholstery. (4) A cut-pile carpet made of highly twisted yarns normally plied and heat-set. A kinked or curled yarn effect is achieved. Excellent durability results from the hard-twist pile yarns.

Fringed Micelle *n* A model for a partially crystalline polymer widely accepted as explaining many of the properties of such a polymer. Consists of crystallites (the "micelles") embedded in an amorphous matrix.

Frisket \ˈfris-kət\ *n* [F *frisquette*, fr. MF] (ca. 1898) Device, usually of gummed paper placed on the working surface, to block out subsequent painting.

Frosting *n* (1) Formation of a translucent finely wrinkled surface on a film of oil, ink, or paint during drying, particularly when exposed to gas fumes, etc. See ▶ Color Abrasion. (2) Salt like deposit on the surface of a topcoat. (3) A clouding of the surface of some rubber and synthetic rubber goods, appearing within a few hours or days after vulcanization. The frosted appearance is different from bloom or blush and cannot be readily removed by washing with a solvent. It may disappear if the article is heated moderately but will generally reappear on cooling. It is thought to be caused by ozone in the air which produces a maze of minute cracks. Some antioxidants have definite anti-frosting effects. See also ▶ Chalking, ▶ Bloom, and ▶ Haze.

Frosting Marks *n* A defect of woven fabric consisting of surface highlights that give a frosted appearance. Frost marks are caused by improper sizing or insufficient warp tension as a result of uneven bending of some warp ends over the picks.

Frost Line *n* (freeze line) In the extrusion of blown film, a ring-shaped transition zone of frosty appearance located at the level at which the film reaches its final diameter and is changing from melt to solid.

Frothing *n* A technique for applying urethane foam in which blowing agents or small air bubbles are introduced under pressure into the liquid mixture of foam ingredients.

Frozen-in Strain *n* (residual strain) Strain that remains in an article after it has been shaped and cooled to its final form, due to a non-equilibrium configuration of the polymer molecules. Such strains occur when cooling its carried below a certain temperature before stresses of a molding or forming operation have been allowed to relax, Frozen-in strains/stresses can cause warping of large flat members and can lead to crazing at low applied stress levels. Aside from measures that can be taken during processing, post-annealing, sometimes done with the aid of shape-holding jigs, is used to complete the relaxation of stresses.

FRP *n* Abbreviation for ▶ Fiber-Reinforced Plastic.

FSCT *n* Abbreviation for ▶ Federation of Societies For Coatings Technology. (Paint Technology prior to 1975).

Fuchsin \ˈfyük-sən, -ˌsēn\ *n* [F *fuchsine*, prob. fr. NL *Fuchsia*; fr. its color] (1865) Synthetic rosaniline dyestuff, a mixture of rosaniline and pararosaniline hydrochlorides. Dark green powder of greenish crystals with a bronze luster; faint odor. Soluble in water and alcohol. Used in textile and leather industries; as a red dye; and as a pharmaceutical (Vigo TL (1994) Textile processing, dyeing, finishing and performance. Elsevier Science, New York). Also known as *Basic Fuchsin Magenta*.

Fugitive Colors *n* (1) Coloring matter that exhibits color change or color transfer while wet or during subsequent drying. (2) Inks made from pigments or dyes which are not permanent, and change or lose color rapidly when exposed to light, heat, moisture, or other conditions. (3) Colorant, pigment or dyestuff, which changes color (fades) rapidly following exposure to light.

Fugitometer See ▶ Fadeometer.

Full-Bodied See ▶ Body.

Full Coat *n* (1) Maximum film thickness of a particular coating which can be properly applied. (2) Application of a coating at a specified film thickness designed to achieve a desired effect.

Fuller's Earth *n* $3MgO \cdot 1.5Al_2O_3 \cdot 8SiO_2 \cdot 9H_2O$. Hydrated magnesium aluminum silicate (attapulgite) used for decolorizing solutions and oils, and as a substitute for absorbent charcoal and as a dusting powder. Density, 2.36 g/cm^3 (19.7 lb/gal). Refractive index, 1.50–1.55 (Ash M, Ash I (1998) Handbook of fillers, extenders and dilutents. Synapse Informtion Resources, New York). See ▶ Hydrated Magnesium Aluminum Silicate.

Full-Fashioned *n* A term applied to fabrics produced on a flat-knitting machine, such as hosiery, sweater, and underwear, that have been shaped by adding or reducing stitches.

Full-Flighted Screw *n* An extruder screw in which the flights extend over the entire length of the screw.

Full Gloss See ▶ Gloss.

Fulling *n* A finishing process used in the manufacture of woolen and worsted fabrics. The cloth is subjected to moisture, heat friction, chemicals, and pressure which cause it to mat and shrink appreciably in both the warp and filling directions, resulting in a denser, more compact fabric (Vigo TL (1994) Textile processing, dyeing, finishing and performance. Elsevier Science, New York).

Fumaric Resin A synthetic, hard resin formed by the reaction of fumaric acid and rosin (Ash M, Ash I (1982–83) Encyclopedia of plastics polymers, and resins, vols I–III. Chemical Publishing Co., New York).

Fumed Silica *n* (pyrogenic silica) An exceptionally pure form of silicon dioxide made by reacting silicon tetrachloride in an oxyhydrogen flame. Individual particles of fumed silica, ranging in size from 7 to 40 nm, tend to link together by a combination of fusion and secondary bonding to form chain-like aggregates with high surface areas that retard the flow of liquids in which they are dispersed. Thus fumed silica is a useful thickening agent, imparting thixotropy to liquid resins that are normally Newtonian, e.g., certain polyesters. Fumed silica is also used in dry molding powders to make them free-flowing and easier to disperse with colorants. Improved electrical properties, prevention of blocking, and reduction of plasticizer migration are other benefits attributed to fumed silica in vinyl compounds (Solomon DH, Hawthorne DG (1991) Chemistry of pigments and fillers. Krieger Publishing Co., New York; Ash M, Ash I (1998) Handbook of fillers, extenders and dilutents. Synapse Informtion Resources, New York; Kirk-Othmer encyclopedia of chemical technology: pigments-powders. Wiley, New York, 1996; Parfitt GD (1969) Dispersion of powders in liquids. Elsevier Publishing Co., New York).

Fume Fading See ▶ Gas Fading.

Fumes *n* A gas-like emanation containing minute solid particles arising from the heating of a solid body such as lead, in distinction to a gas or vapor. This physical change is often accompanied by a chemical reaction, such an oxidation. Fumes flocculate and sometimes coalesce. Odorous gases and vapors should not be called fumes. As distinguished from dusts, fumes are finely divided solids produced by other methods of subdividing, such as chemical processing, combustion, explosion, or distillation. Some solids, when heated to a liquid produce a vapor which, while arising from the molten mass, immediately condenses to a solid without returning to its liquid state. Fumes are much finer than dusts, containing particles from 0.1 to 1.0 μm in size (Tests for comparative flammability of liquids, UI 340 (1994) Laboratories Incorporated Underwriters, New York).

Functional Group *n* A group of atoms in a molecule which cause the molecule to undergo a set of characteristic reactions (Morrison RT, Boyd RN (1992) Organic chemistry, 6th edn. Prentice Hall, Englewood Cliffs).

Functionality The ability of a molecule or group to form covalent bonds with another molecule or group in a chemical reaction. Compounds may be mono-, di-, tri-, or polyfunctional, depending on the number of functional groups capable of participating in a reaction. (Morrison RT, Boyd RN (1992) Organic chemistry, 6th edn. Prentice Hall, Englewood Cliffs).

Fundamental Units See ▶ Mass and ▶ Length.

Fungi \ˈfəŋ-gəs\ *n pl*, *-sing*. Fungus. See ▶ Fungus

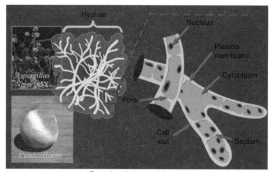

Fungi cellular structure
Black, J.G., Microbiology, 5th Ed., John Wiley & Sons, New Work

Fungicidal Paint *n* Some pigments are used as biocides. For example, zinc oxide I sused as a fungicide. Paint which discourages the growth of fungi on its dry applied film is a fungiciial paint or coating. The fungicidal properties are normally conferred by special additives, although certain pigments such zinc oxide, commonly used in paints, may themselves contribute to the fungicidal properties of the paint (Wicks ZN, Jones FN, Pappas SP (1999) Organic coatings science and technology. 2nd edn. Wiley, New York).

Fungicidal Wash Wash containing fungicides used before painting and designed to kill existing spores or germinations or to prevent their inception. Many of these substances are toxic (to human beings) used when in concentrated form and therefore need careful handling. Use of the term "antiseptic wash" is deprecated.

Fungicide \ˈfən-jə-ˌsīd\ *n*, - *pl* fungicides, [ISV] (1889) An agent incorporated in a plastic compound to control fungus growth, usually by killing the organisms. Most plastics with a few exceptions, notably some of the cellulosics, are inherently resistant to fungus attack. However, many plasticizers are highly susceptible to attack. Examples of fungicides used in plastics are copper-8-quinolinate and tributyltin oxide. Agents that retard fungal growth without killing the organisms are called *fungistats* (Black JG (2002) Microbiology, 5th edn. Wiley, New York). See also ▶ Biocide.

Funginertness (Fungus resistance) Not susceptible to the formation of fungus growth (James F (ed) (1993) Whittington's dictionary of plastics., Carley, Technomic Publishing Co.).

Fungistat *n* An agent incorporated in plastics compounds to control fungus growth without killing the fungi. See also ▶ Biocide.

Fungistatic \ˌfən-jə-ˈsta-tik\ *adj* (1922) Preventing the growth of a fungus by the presence of some chemical or physical agent but not fungicidal.

Fungus \ˈfəŋ-gəs\ *n sing.*, *-n pl* fungi [L] (1527) Multicellular plant or plants of the phylum, *thallophytae* that contain no chlorophyll and reproduce by sexual or asexual spore. Fungi cells are eucarotic, each cell contains a nucleus. Fungi cause discoloration and growth in or on a coatings surface, i.e., mildew, mold (Black JG (2002) Microbiology, 5th edn. Wiley, New York).

Fungus Resistance See ▶ Mildew (Fungus) Resistance.

Furan *n* Furfuran; oxole; tetrole; divinylene oxide, C_4H_4O, Molecular Weight: 68.07. Occurs in oils obtained by the distillation of rosin contg pine wood. Prepared by decarboxylation of 2-furancarboxylic acid (Quin LD (2010) Fundamentals of Heterocyclic Chemistry, Wiley, New York; Williams A (1973) Synthesis and Applications, Noyes Data Corporation, New York; Wilson (1941) Org syn coll, vol I, 2nd end, p 274). Furan has been prepd directly from furfural over hot soda-lime or by dropping furfural on a fused mixt of sodium and potassium hydroxides (Hurd et al (1932) J Am Chem Soc 54:2532). Other physical properties: Liquid. $d_4^{19.4}$ 0.9371. bp_{760} 31.36°; bp_{758} 32°. nD^{20} 1.4216. Flash pt, closed cup: $-32°F$ ($-35°C$). Absorption spectrum: Purvis (1910) J Chem Soc 97:1648, 1655. Insoluble in water. Freely soluble in alcohol and ether. Stable to alkalies; resinifies on evaporation or when in contact with mineral acids. Boiling point: bp_{760} 31.36°; bp_{758} 32° (Odian GC (2004) Principles of polymerization. Wiley, New York; Merck Index (2001), 13th edn. Merck and Company, Whitehouse Station).

Furan Resin *n* (furfuryl resin) A dark-colored, thermosetting resin obtained primarily by the condensation polymerization of furfuryl alcohol in the presence of a strong acid, sometimes in combination with formaldehyde or furfural (2-furaldehyde). The term also includes resins made by condensing phenol with furfuryl alcohol or furfural, and furfuryl-ketone polymers. The resins are available as liquids in a wide range of viscosities that cure to highly crosslinked, brittle substances. They are used for impregnating cured plaster structures, as binders for foundry-sand cores, for binding high explosives, and as wood adhesives. The cured resins exhibit good resistance to chemicals and solvents. Improved resin/catalyst systems have made the older furan systems obsolete and enabled the use of fire-retardant furans in hand layup, sprayup, and filament-winding, competing with polyesters (Odian GC (2004) Principles of polymerization. Wiley, New York;

Harper CA (ed) Handbook of plastics, elastomers and composites, 4th edn. McGraw-Hill, New York).

Furan Prepreg *n* Latent catalysts announced in 1972 made it possible to produce furan-resin prepregs of acceptable shelf life, which avoided the difficulties experienced with the wet-layup process. These composites possess good heat and chemical resistance, excellent surface hardness and fire resistance (Harper CA (ed) (2002) Handbook of plastics, elastomers and composites, 4th edn. McGraw-Hill, New York).

Furfural \ˈfər-f(y)ə-ˌral\ *n* [L *furfur* bran + ISV ^{3}-*al*] (1879) (2-furaldehyde, ant oil, furfuraldehyde) $C_4H_3O \cdot HO$. A liquid aldehyde with a bp of 161°C, mp of −36.5°C; sp gr of 1.159/22°C, obtained by distilling acid-digested corn cobs or the hulls of oats, rice, or cottonseed, and having the structure shown below. Furfural is colorless liquid when first distilled, but darkens on exposure to air. It is used as a solvent, and in the production of furans and tetrahydrofurans. See also ▶ Furan Resin.

Furnace Black *n* A form of carbon black obtained by decomposing natural gas and/or petroleum oil under controlled conditions in a furnace and precipitating the pigment in special chambers. See ▶ Carbon Black.

Fuse Heating a coating component of dispersed resin(s) to a temperature at which it becomes homogeneous.

Fused Acetate *n* (1) A hard particle of acetate material of almost any shape or size other than recognizable fiber. Sometimes fused acetate particles resemble rock-like, hardened drops of acetate dope; in other cases fused acetate consists of particles covered with fiber clusters and completely hardened in the center. (2) Acetate yarns in which the individual filaments are coalesced (Tortora PG (ed) (1997) Fairchild's dictionary of textiles. Fairchild Books, New York).

Fused Congo *n* Congo copal which has been subjected to a fusing or running process, in order to induce oil solubility.

Fused Driers *n* Driers which are prepared by the direct interaction of the main constituents, stimulated by heat, in the absence of added water. The term is used to distinguish them from driers which are made by precipitation from aqueous solutions of metallic salts and alkali soaps (Weismantal GF (1981) Paint handbook. McGraw-Hill, New York).

Fused Ribbon *n* Acetate fabrics in wide widths may be cut into narrow ones by the application of heat. A hot knife blade caused the edges to sear and bead, thereby doing away with selvages on the edges of the goods.

Fusel Oil \ˈfyü-zəl\ *n* An arid oily liquid occurring in insufficiently distilled alcoholic liquors, consisting mainly of amyl alcohol, and used especially as a source of alcohols and as a solvent. Alcohols such as normal, iso, and tertiary butyl alcohols, propyl and isopropyl alcohols, and well as amyl and isoamyl alcohols may be present (Merriam-Webster's collegiate dictionary (2004) 11th edn. Merriam-Webster, Springfield).

Fusing *n* (1) Melting. (2) Uniting, as by melting together.

Fusion \ˈfyü-zhən\ *n* [L *fusion-*, *fusio*, fr. *fundere*] (1555) With respect to vinyl plastisols and organosols, fusion is the state attained by heating when the resin particles have completely dissolved in the plasticizers and solvents present, so that upon cooling a homogeneous solid solution results (Wypych G (ed) (2003) Plasticizer's data base. Noyes Publication, New York; Wickson EJ (ed) (1993) Handbook of polyvinyl chloride formulating. Wiley, New York).

Fusion *n* (atomic) A nuclear reaction involving the combination of smaller atomic nuclei or particles into larger ones with the release of energy from mass transformation. This is also called a thermo-nuclear reaction by reason of the extremely high temperature required to initiate it.

Fusion Bonding *n* Any of several methods that create a thin layer of melt on the plastic surfaces to be joined, which are then pressed together. Fusion bonding is limited to identical or melt-compatible polymers (Strong AB (2000) Plastics materials and processing. Prentice Hall, Columbus).

Fusion Point Temperature of fusion or melting.

Fusion Temperature *n* In vinyl dispersions, the temperature at which ▶ Fusion occurs. The *optimum fusion temperature* is that at which thermal degradation has not occurred and maximum physical properties are obtained in the final product (Wickson EJ (ed) (1993) Handbook of polyvinyl chloride formulating. Wiley, New York).

Fustics \ˈfəs-tik\ *n* [ME *fustyk* smoke tree, fr. MF *fustoc*, fr. Arabic *fustuq*, fr. Gk *pistakē* pistachio tree] (15c) Yellow coloring matters of natural origin. Old fustic is obtained from the species, *Morus tinctorial*, and consists of moric and morintannic acids. Young fustic is obtained from *Rhus cotinus*, and consists of fustic.

Fuzz *n* An accumulation of short, broken filaments collected from passing glass strands, yarns, or rovings over a contact point. The fuzz may be collected, weighted and used as an inverse measure of abrasion resistance.

Fuzz Ball See ▶ Balling Up.

Fuzziness *n* (1) A term describing a woven fabric defect characterized by a hairy appearance due to broken fibers or filaments. Principle causes are underslashed warp; rough drop wires, heddles, or reed; fabric slippage on take-up drum; rough shuttles; cut glass, dents, or reeds in warped; and damage in slashing. (2) A term describing a fabric intentionally made with a hairy surface; such fabrics are usually produced from spun yarns.

G

g \jē\ *n* (1) SI abbreviation for ▶ Gram, (2) Symbol for the acceleration due to gravity at the earth's surface, and, in particular, for the standard value, 9.806,650 m/s². It is sometimes used, erroneously, for the proportionality constant (g_c) in Newton's law of momentum change. See ▶ Force (3).

G *n* (1) SI abbreviation for ▶ Giga-. (2) Symbol for shear modulus. In dynamic testing, G′ symbolizes the "real" or in-phase part of G' the "imaginary" or out-of-phase component.

Gabardine \ga-bər-dēn\ *n* [MF *gaverdine*] (1520) A firm, durable, warp-faced cloth, showing a decided twill line, usually a 45°C or 63°C right-hand twill.

Gable \gā-bəl\ *n* [ME, fr. MF, of Gr origin; akin to ON *gafl* gable] (14c) Triangle formed by the edge of a ridged roof and the similarly shaped wall enclosed by the horizontal and raking cornices; called a pediment in classic architecture. (Harris CM (2005) Dictionary of architecture and construction. McGraw-Hill, New York).

Gableboard ▶ See Bargeboard.

Gadoleic Acid *n* $CH_3(CH_2)_9CH = CH(CH_2)_7COOH$. Fatty acid from marine oils.

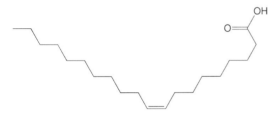

Gage \gāj\ Alternate spelling of ▶ Gauge.

Gahn's Ultramarine See ▶ Cobalt Blue.

Gaiting *n* The spacing of the needles in the dial and cylinder in relation to each other on rib (double-knit) and interlock knitting machines. In rib gaiting, the dial needles are midway between the cylinder needles. For interlock gaiting the dial and cylinder needles are in direct alignment.

Gal *n* [*Galileo* Galilei] (1914) 1 gal = cm/s/s. Therefore, where the value of gravity is 980 this is the same as 980 gal. The milligal is now quite commonly used since it is approximate one part in a thousand of the normal gravity of the earth.

Galatea \ga-lə-tē-ə\ *n* [L, fr. Gk *Galeteia*] A sturdy, serviceable, warp-effect, five-shaft, left-hand twill-weave fabric, frequently cotton or a cotton blend, used for children's play clothes. (Kadolph SJJ, Langford AL (2001) Textiles. Pearson Education, New York).

Galipot Another name for Bordeaux turpentine.

Galley \ga-lē\ *n* [ME *galeie*, fr. OF *galie*, *galee*, ultimately fr. MGk *galea*] (13c) A shallow metal tray used for holding type.

Galley Proof A proof taken of type standing in a galley, before it is made up into pages. (Printing ink manual, 5th edn. Leach RH, Pierce RJ, Hickman EP, Mackenzie MJ, Smith HG (eds). Blueprint, New York, 1993).

Galliolino *n* Old name for Naples yellow, which is substantially lead antimoniate.

Gallon, Imperial *n* English gallon, equivalent to 277.77 in.³ or 3.54 L. Equals 1.2 US gallons. US gallon = 0.833 Imperial gallons.

Gallon, US *n* (1) Volume equal to 231 in.³ For paint, varnish, lacquer and related products this is measured at 25°C (77°F). (2) Volume of 8.33 lb of water. (3) Equivalent to 3.785 L.

Galvanic Anode \gal-va-nik a-nōd\ Source of current for cathodic protection provided by a metal less noble than the one to be protected, e.g., magnesium, zinc or aluminum as used for cathodic protection of steel. (Corrosion engineer's handbook, 3rd edn. Baboian R (ed). NACE International – The Corrosion Society, Houston, TX, 2002). See ▶ Cathodic Protection.

Galvanic Cell *n* (1) Cell consisting of two or more dissimilar metals or alloys in contact with the same body of an electrolytic solution such as seawater. Upon electrically connecting the dissimilar metals, a current flows as the result of accelerated corrosion of the more active of the dissimilar metals or alloys. (2) Electrolytic cell capable of producing electrical energy by electrochemical action. (Corrosion engineer's handbook, 3rd edn. Baboian R (ed) NACE International – The Corrosion Society, Houston, TX, 2002).

Galvanic Corrosion *n* Increased corrosion above normal corrosion of a metal that is associated with the flow of current to a less active metal in the same solution and in contact with the more active metal. (Corrosion engineer's handbook, 3rd edn. Baboian R (ed) NACE International – The Corrosion Society, Houston, TX, 2002).

Galvanic Protection See ▶ Cathodic Protection.

Galvanized Steel Zinc plated steel. See ▶ Galvanizing.

Galvanize \gal-və-nīz\ *v* (1802) To coat iron or steel with zinc by immersing in molten zinc to produce a coating of zinc–iron alloy.

Jan W. Gooch, *Encyclopedic Dictionary of Polymers*, DOI 10.1007/978-1-4419-6247-8,
© Springer Science+Business Media LLC 2011

Galvanizing *n* Application of a coating of zinc to steel by a variety of methods: barrel-galvanized dipped, electrogalvanized, electro-zinc plated, flake-galvanized, hot-dipped, hot-dip galvanized, mechanically plated, peer-coated galvanized, tumbler-galvanized, and wean-galvanized.

Gama (γ) Ray *n* (1903) High-energy electromagnetic radiation emitted from a nucleus. (Giambattista A, Richardson R, Richardson RC, Richardson B (2003) College physics. McGraw Hill Science/Engineering/Math, New York).

Gamboge \gam-▮bōj, -▮büzh\ *n* [NL *gambogium*, alteration of *cambugium*, fr. or akin to Portuguese *Camboja* Cambodia] (1712) $C_{30}H_{35}O_6$. Resinous exudation of intense yellow color obtained from the species, *Guttifera*, which grown in India, Ceylon, Siam and the East Indies. The product is a mixture consisting chiefly of a water insoluble resin (up to 80%), with a water soluble gum (up to 27%). The resin portion is chiefly gambogic acid. *Also known as Cambogia.*

Gamma -*adj* (1896) A prefix usually abbreviated as the Greek letter γ- and usually ignored in alphabetizing compound names, signifying that the so-labeled substitution is on the third carbon away from the main functional group of the molecule, and generally synonymous with the label "4-". They may be emitted from radioactive substances. They are quanta of electromagnetic wave energy similar to but of much higher energy that ordinary x-rays. The energy of a quantum is equal to hv ergs, where h is Planck's constant (6.6254×10^{-27} erg/s) and v is the frequency of the radiation. Gamma rays are highly penetrating, an appreciable fraction being able to traverse several centimeters of lead. (Giambattista A, Richardson R, Richardson RC, Richardson B (2003) College physics. McGraw Hill Science/Engineering/Math, New York).

Gamma Cellulose *n* One of the three forms of cellulose. With beta cellulose it is called hemicellulose. (Paint: Pigment, drying oils, polymers, resins, naval stores, cellulosics esters, and ink vehicles, vol 3, American Society for Testing and Material, 2001) Also see ▶ Alpha Cellulose and ▶ Beta Cellulose.

Gamma Distribution *n* A random variable X with density

$$f(x) = \frac{1}{\Gamma(\alpha)\beta^\alpha} x^{\alpha-1} e^{-x/\beta}$$

is called a Gamma random variable with parameter α and β, and its distribution function is called a gamma distribution with parameter α and β.

Gamma Protein *n* Protein derived from soya bean meal. The commercial product contains approximately 54% of actual protein. It disperses in water in the presence of a small amount of alkali and is used as an extender for casein in water paints and distempers.

Gamma Ray *n* (1903) A quantum of electromagnetic energy emitted from some radioactive materials as they decay. Gamma rays are similar to, but much shorter wavelengths (typically ≈ 0.03 nm) and much higher energy than ordinary X-rays. They are highly penetrating, capable of passing through several centimeters of lead. (Serway RA, Faugh JS, Bennett CV (2005) College physics. Thomas, New York).

Gamma Transition Syn: ▶ Glass Transition.

Gamut, Color *n* Total variety of colors that can be produced by any prescribed method, such as the mixtures of any particular sets of colorants. Gamuts are frequently described by plotting them on a chromaticity diagram.

Gap \▮gap\ *n* [ME, fr. ON, chasm, hole; akin to ON *gapa* to gape] (14c) In filament winding, an unintentional space between two windings that should lie next to each other.

Garcia Nutans Oil *n* Vegetable drying oil obtained from the species, *Euphorbiaceae*, which grows in Mexico and Venezuela. It contains over 90% of eleostearic acid and gels in less than 8 min at 282°C. It possesses the following constants: sp gr of 0.942/15°C, refractive index of 1.525, iodine value of 177.9, and saponification value of 189.

Garden Rocket Oil *n* Oil with fair drying properties. It has an iodine value of 155, a sp gr of 0.928, and a saponification value of 192.

Gardner Color Standards *n* Liquid yellow to brown color standards for describing the color of drying oils, ASTM Method D-1544.

Gardner-Holt Bubble Viscometer See ▶ Air-Bubble Viscometer.

Gardner-Holt Viscosity Tubes *n* Series of selected glass tubes of constant, standard diameter, which is filled with liquids of various viscosities, except for a small air space at the top. When these tubes are inverted, the air bubble travels through the liquid, and the rate of travel is a measure of the viscosity of the liquid. Empty tubes of similar standard dimensions are filled with liquids of which the viscosity is required, and the rates of travel of the bubbles compared with those of the tubes containing the liquids of known viscosities. Viscosities are compared under controlled temperature conditions. (Paul N. Gardner, Company, Inc., 316 N. E. Fifth Street, Pompano Beach, FL, www.gardco.com).

Gardner Impact Test *n* (falling-weight test) A test for the impact resistance of rigid plastic sheets or parts. ASTM

D 3029 describes two methods, F and G. In F, a weight with a hemispherical nose falls through a tube to strike the specimen, and G (for which the Gardner Impact Tester is approved), in which the falling weight hits a round-nose striker resting on the specimen. Either the weight or the height of drop may be varied, the former being recommended because of the sensitivity of most plastics to the velocity of impact (determined by height). Several tests on new specimens may be made at each impact-energy level (height × weight), and the fraction of breaks at each level is noted. This procedure efficiently provides information on both the mean energy to break and the standard deviation. Where the mean is of greatest interest, the ▶ Up-and-Down Method provides a good estimate with less testing. (Paul N. Gardner, Company, Inc., 316 N. E. Fifth Street, Pompano Beach, FL, www.gardco.com).

Gardner Mobilometer *n* Viscometer used for determining the viscosity of all types of varnishes, oils, and pigmented compositions. The Mobilometer consists of a cylinder into which is poured the paint product of which the viscosity is required. Into this is fitted a closely fitting disc attached to the end of a guide rod. Provision is made at the other end of the rod for weighting it, if required, when dealing with viscous products, by including a shallow tray into which weights may be placed. The time is determined for the disc and rod to travel a definite distance into the paint product, and is a measure of its viscosity.

Garnet Abrasive \ˈgär-nət-\ Almandite, a type of garnet mineral occurring in New York State, is used in coated abrasives. Hardness and toughness are increased by heat treating. It fractures along the cleavage planes of the crude crystals, and the resulting grains have very sharp edges.

Garnet Lac *n* Form of shellac prepared from sticklac by a special refining process. The term "garnet" is merely an indication of the deep red color of the material when dissolved in alcohol.

Garnet Paper *n* (ca. 1902) The paper generally favored for hand sanding on wood. The abrasive is the same mineral as the semiprecious garnet gem, almandite.

Garnetting *n* A process for reducing various textile waste materials to fiber by passing them through a machine called a garnett, which is similar to a card.

Gas \ˈgas\ *n* [NL, alter. of L *chaos* space, chaos] (1779) A state of matter in which the material has a very low density and viscosity, can expand and contract greatly in response to changes in temperature and pressure, easily diffuses into other gases, and readily and uniformly distributes itself throughout any container. A gas can be changed to the liquid or solid state ink only by the combined effect of increased pressure and decreased temperature (below the critical temperature).

Gas Black *n* Another term for carbon black, so called from its manufacture from petroleum gas. See ▶ Carbon Black.

Gas Blue *n* Variety of Prussian blue, so called from its origin in certain by-products of coal gas manufacture.

Gas Checking *n* Film phenomenon associated with compositions containing vegetable drying oils, or derivatives therefrom, and demonstrated by the formation of crowsfoot-like wrinkles, or wrinkles which have the appearance of crystalline patterns. The phenomenon occurs when susceptible films are allowed by dry in an atmosphere containing the products of combustion of coal gas or natural gas; hence the term, "gas checking." Syn: ▶ Crystallizing, ▶ Crowsfooting, ▶ Webbing, and ▶ Frosting.

Gas Chromatography *n* (GC) (1952) A method of chemical analysis in which the specimen is vaporized and introduced into a stream of carrier gas (usually helium), the stream then passed through a column packed with adsorbent particles that separate the stream into its constituent molecules. These fractions pass through the column – first adsorbed from, then desorbed into the ongoing stream of carrier gas – at characteristic rates. Their concentrations in the effluent are measured by a detector such as a thermal-conductivity cell. The recorded concentrations are displayed on a strip chart with time as the abscissa. From the positions and areas of the peaks the identities and relative concentrations of the constituents in the sample may be determined. An example of a gas chromatograph combined with a mass spectrometer is the Clarus 500 GC Mass Spectrometer.

GC MS no hose

(Willard HH, Merritt LL, Dean JA (1974) Instrumental methods of analysis. D. Van Nostrand, New York) See also ▶ Chromatography.

Gas Crazing *n* The wrinkling of a tung oil film under certain drying conditions. See ▶ Gas Checking.

Gas Fading *n* A change of shade of dyed fabric caused by chemical reaction between certain disperse dyes and acid gases from fuel combustion, particularly oxides of nitrogen.

Gas-Injection Molding *n* A specialized technique for molding low-density structures in which a mixture containing cork particles, or glass or phenolic microspheres, glass fibers, and a thermosetting resin is injected into a mold by fluidizing it in a gas stream.

Gas-Liquid Chromatography *n* (1952) A variation of gas chromatography in which the chromatographic column is packed with a finely divided solid impregnated with a nonvolatile organic liquid. The sample to be analyzed is injected into the inlet of the column where it is quickly and completely vaporized. The carrier-gas stream carries it into the packed section, where the vapors contact the impregnated solids. They are absorbed by the nonvolatile liquid phase, and then later desorbed into the carrier gas. The vapor of each compound spends a characteristic fraction of time in the condensed phase and the remainder in the mobile gas phase. Each chemical species will tend to migrate at its own rate and will be separated from other species in the time of emergence from the column. The detector senses their concentrations in the effluent and its signal strength is recorded and displayed versus time on a strip chart. (Willard HH, Merritt LL, Dean JA (1974) Instrumental methods of analysis. D. Van Nostrand, New York).

Gas Permeability *n* The ease with which a gas or vapor passes through a membrane, e.g., a plastic sheet or film. See ▶ Permeability.

Gas-Phase Polymerization *n* A polymerization process developed by Union Carbide for high-density polyethylene, particularly for a grade for making paper-like films. Purified ethylene and a highly active chromium-containing catalyst in dry-powder form are fed continuously into a fluidized-bed reactor. The resin forms as a powder, thereby avoiding the gel, discoloration, and contamination problems often associated with conventional polymerization processes.

Gasproof *n* Coating composition which does not gas check when exposed to an atmosphere containing the products of combustion of coal gas. See ▶ Gas Checking.

Gassing *v* (1852) See Singeing.

Gas Thermometer *n* Where P_o, P_s, and P_z represent the total pressure with the bulb at 0°C, at the boiling-point of water and at the unknown temperature respectively, t_s the temperature of steam and t_z the unknown temperature,

$$t_x = t_s \frac{P_x - P_o}{P - P_{os}}$$

(approximately). The total pressure on the gas in the bulb is the algebraic sum of barometric pressure at the time and that measured by the manometer.

Gas-Transmission Rate *n* (GTR) The quantity of a given gas passing through a unit area of the parallel surfaces of a plastic film in unit time under the conditions of the test. These conditions, including temperature and partial pressure of the gas on both sides of the film, must be stated. The SI unit of GTR is mol/(m^2 s) but others, some of them involving mixed metric and English units, are still in common use. See ▶ ASTM (www.astm.org) for standard method of gas-transmission rate. See also ▶ Permeance and ▶ Permeability.

Gas Welding See ▶ Hot-Gas Welding.

Gate \ ˈgāt\ *n* [ME, fr. OE *geat*; akin to ON *gat* opening] (before 12c) In injection and transfer molding, the channel through which the molten resin flows from the runner into the cavity. It may be of the same cross section as the runner, but most often is restricted to 3 mm or even much less. A gate whose diameter is less than 0.5 mm is known as a *pinpoint gate*. A *submarine gate* is shaped to conduct the melt below the parting line of the mold and into the cavity at a point just below its edge. Other types of gates are the *fan gate* (a diverging, thin gate), and the *tab gate* (one that extends the runner into the molded part). The term gate is also used for the portion of the plastic molding formed by the gate orifice. (Strong AB (2000) Plastics materials and processing. Prentice Hall, Columbus, OH).

Gate Blush *n* (gate splay) A blemish or disturbance in the gate area of an injection-molded article. It occurs when the melt fractures as it emerges from the gate due to sudden release of large elastic stresses.

Gauffrage *n* Printing by pressure alone, without the use of a pigment, producing an embossed effect on the paper.

Gauge \ ˈgāj\ *n* [ME *gauge*, fr. ONF] (15c) (gage) (1) Any instrument that measures and indicates such quantities as thickness, pressure, temperature, or liquid level. (2) The thickness of a plastic sheet or film, usually given in mils or mm. (3) Any of the standard wire and sheet-metal scales in which the gauge numbers are inversely related to wire diameter or sheet thickness. See ▶ Strain Gauge and ▶ Wire Gauge. (4) The number of wales per inch in a knit fabric. (5) On spinning or twisting frames, the distance from the center of one spindle to the center of the next spindle in the same row.

Gauge Band *n* A term used in the packaging-film industry for a thickness irregularity found in rolls of film. A thick area at some locality over the width of a flat film will produce a raised ring in a finished roll. Similarly, a thin area will cause a depressed ring. Such films, when unwound, tend not to lie perfectly flat. With shrinkable films, thin areas are troublesome because it is more likely that they will cause tearing or burn-through during the shrink cycle.

Gauge Length *n* On a dogbone-shaped, tensile-test specimen before stress is applied, the distance between two marks on the narrow part ("waist") of the specimen, perpendicular to the direction of pull that will be used to measure elongation. If the specimen is of uniform thickness and width, gauge length may be taken as the nip-to-nip distance between the clamping fixtures.

Gauge Wire *n* Used with an extra filling yarn during weaving, this type of standing wire controls the height of fabric pile.

Gauss \gaüs\ *n* [Karl French *Gauss*] (1882) The cgs of magnetic induction (flux density). It is equal to 1 maxwell/cm^2. It has such a value that if a conductor 1 cm long moves through a magnetic field at a velocity of 1 cm, in an induction mutually perpendicular, the induced emf is one abvolt. (Giambattista A, Richardson R, Richardson RC, Richardson B (2003) College physics. McGraw Hill Science/Engineering/Math, New York).

Gaussian Distribution *n* (1905) *Also called Normal Curve.*

Gauze \góz\ *n* [MF *gaze*] (1561) A thin, sheer-woven fabric in which each filling yarn in encircled by two warp yarns twisted around each other, gauze is similar to cheesecloth. It may by made of silk, cotton, wool, or manufactured fibers. Cotton gauze is primarily for surgical dressings.

Gavan \gä-vən\ Persian name for the plant which yields gum tragacanth.

Gay-Lussac's Law \gā-lə-sak-\ [Joseph-Louis *Gay-Lussac*, 1778–1850, F chemist and physicist] Syn: Charles' Law. (Goldberg DE (2003) Fundamentals of chemistry. McGraw-Hill Science/Engineering/Math, New York).

Gay-Lussac's Law of Combining Volumes *n* If gases interact and form a gaseous product, the volumes of the reacting gases and the volumes of the gaseous products are to each other in simple proportions, which can be expressed by small whole numbers. (Goldberg DE (2003) Fundamentals of chemistry. McGraw-Hill Science/Engineering/Math, New York).

Gear Box *n* (gear reducer) Syn: for ▶ Speed Reducer.

gc The proportionality constant in Newton's law of momentum change, g_c is often needed to convert viscosities between force units and mass units: $\mu_f = g_c \mu_m$. In SI, $g_c = 1.000,000$ N s^2/(kg m), so a viscosity of, say 100 Pa s is also equal to 100 kg/(m s). See ▶ Force.

Gear Crimping See ▶ Texturing.

Gear Pump A pump consisting of a sturdy housing within which two intermeshing toothed wheels rotate, and inlet and outlet ports. The gears typically have widths about equal to their diameters and may be spur, helical, or, rarely because of their cost, herringbone bears. The inlet port is attached at the waist of the 8-shaped casing where the gears are moving apart; the outlet is at the opposite side. Liquid trapped between the tight-fitting teeth and the casing is borne around to the discharge side where it is displaced and expelled by the meshing of the teeth. Small gear pumps have long been used in producing staple fiber, requiring extremely high melt pressures (up to 100 MPa). Since about the mid-1970s, larger ones have come into increasing use for general extrusion operations where die resistances or filtration requirements generate high pressures, or where close dimensional tolerances on extrudates are desired. Gear pumps have been successfully retrofitted to extruders, providing net higher output, better product quality, and quick return on the additional investment.

Geiger Counter \gī-gər-\ *n* [Hans *Geige*r † 1945 Gr physicist] (1924) Detector for radioactivity depending upon ionized particles that affect its mechanism. As it indicates, it both detects and make a count of them possible. (Goldberg DE (2003) Fundamentals of chemistry. McGraw-Hill Science/Engineering/Math, New York).

Gel \jel\ *n* [*gelatin*] (1899) A semisolid system consisting of a network (three-dimensional network polymer) of solid aggregates within which a liquid is trapped. The initial, jelly-like solid phase that develops during the formation of a resin from a liquid. With respect to vinyl plastisols, gel is a state between liquid and solid that occurs in the initial stages of heating, or upon prolonged storage. NOTE — All three types of gel have very low strength and do not flow like liquids. They are soft, flexible, and may rupture under their own weight unless supported externally (ASTM D 883). See also ▶ Gelation. A defect in plastic film such as polyethylene or PVC characterized by tiny, hard, glassy particles that appear in an otherwise clear film. These gel particles are believed to be bits of resin of much higher-than-average molecular weight, perhaps crosslinked (Bryson HC (1956) Paint faults and remedies. Scientific Surveys, London; Hess M (1965) Paint film defects. Wiley, New York; Paint/coatings dictionary, Compiled by Definitions Committee of the Federation of Societies for Coatings Technology, 1978). Designed gels (three-dimensional network systems swell with solvents and change size predictably, and these properties can be of commercial value such as ion absorbers from aqueous, Physical properties of polymeric gels, Addad JPC (ed). Wiley, New York, 1996).

Gelatin \ ˈjel-lə-tᵊn\ *n* [F *gélatine* edible jelly, gelatin, fr. It *gelatina*, fr. *gelato*, pp of *gelare* to freeze, fr. L] (1800) It is a protein product derived through partial hydrolysis of the collagen extracted from skin, bones, cartilage, ligaments, etc. The natural molecular bonds between individual collagen strands are broken down into a form that rearranges more easily. Gelatin melts when heated and solidifies when cooled again. Together with water it forms a semi-solid colloidal gel. Gelatin (also gelatine) is a translucent brittle solid substance, colorless or slightly yellow, nearly tasteless and odorless, which is created by prolonged boiling of animal skin and connective tissue. It has many uses in food, medicine, and manufacturing. Substances that contain or resemble gelatin are called gelatinous. Gelatin is protein type material extracted from animal skins, sinews, tendons, and from bones. Gelatin may be regarded as a pure glue. It dissolves in hot water to give a viscous solution with pronounced gelling characteristics. Inedible gelatin has useful applications in the production of sizes for various purposes, and in emulsions, (Morrison RT, Boyd RN (1992) Organic chemistry, 6th edn. Prentice Hall, Englewood, Cliffs, NJ).

Gelatinous \jə-ˈlat-nəs, -ˈla-tᵊn-əs\ *adj* (1766) Having the consistency of a very soft elastic solid, the nature and appearance of gelatin.

Gelation \ji-ˈlā-shən\ *n* [L *gelation-*, *galatio*, fr. *gelare*] (1854) The formation of a ▶ Gel. With regard to vinyl plastisols and organosols, gelation is the change of state from the liquid suspension of the solid condition that occurs in the course of heating and/or aging, when the plasticizer has been mostly absorbed by the resin, resulting in a dry but weak and crumbly mass. Within normal proportions of resin and plasticizer, this state is attained when the resin particles have soaked up so much plasticizer that they touch each other. As heating progresses, the swollen particles begin to fuse together, resulting in some cohesive strength. Gelation is considered to continue until useful levels of mechanical properties are attained, such as have developed at the ▶ Clear Point. Much confusion has existed regarding the meaning of gelation because in the early years of the art, especially in Europe, *gelation* was used in place of the term fusion, now employed in the US. (Handbook of polyvinyl chloride formulating. Wickson EJ (ed). Wiley, New York, 1993).

Gelation Time *n* With reference to thermosetting resins, the interval of time between the mixing in of a catalyst and the formation of a gel.

Gel Cellophane *n* Regenerated cellulose which has never been allowed to dry out, but has always been stored swollen with water or other solvent. Thin films are frequently used as the semi-permeable membrane in membrane Osmometry.

Gel Coat *n* (1) In reinforced plastics, a thin outer layer of resin, sometimes containing pigment, applied to give the structure its surface gloss and finish. It also serves as a barrier to liquids and ultraviolet radiation. The gel coat is the first to be applied to the mold in the layup process (after any mold-release agent), becoming permanently bonded to the succeeding layers of reinforcement and resins. In ▶ Sprayup, the gel coat is applied last, sometimes after partly curing the mains structure. (2) High-build, chemical-resistant, thixotropic polyester coating. (Starr TF (1993) Data book of thermoset resins for composites. Elsevier Science and Technology Books, New York).

Gel Dyeing See ▶ Dyeing.

Gel Effect *n* The increase in molecular weight as conversion progresses. See ▶ Autoacceleration and ▶ Gel.

Gel Filtration *n* Method of polymer fractionation, both preparative and analytical, in which a sample of the polymer in dilute solution is injected into a stream of flowing solvent at the top of a packed column. The elution of the polymer is governed by the size of the polymer molecule. The polymer solute is frequently a biopolymer.

Gel, Gel Point, Gelation *n* The point at which a thermosetting system attains an infinite value of its average molecular weight; infinite viscosity and non-flowing. (Allcock HR, Mark J, Lampe F (2003) Contemporary polymer chemistry. Prentice Hall, New York; Alger MSM (1989) Polymer science dictionary. Elsevier Science, Barking, Essex, England; Elias HG (1977) Macromolecules vol. 1–2, Plenum, New York; Lenz RW (1967) Organic chemistry of synthetic high polymers. Interscience, New York) See ▶ Trommsdorf Effect.

Gelling *n* Any process whereby paint or varnish thickens to jelly-like consistency. Also see ▶ Livering.

Gelling Agent See ▶ Thickening Agent.

Gel Particle See ▶ Gel (4).

Gel-Permeation Chromatography *n* (1966) This method is used to provide the molecular weigh and molecular weight distribution of a polymer by a fractionation technique. The polymer must be dissolved in a suitable solvent. It can also be defined as chromatography in which macromolecules such as polymers in a solution are separated by size on a column packed with a gel such as polystyrene. An example of a GPC chromatogram of a typical polymer is shown.

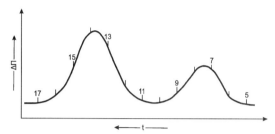

Hypothetical GPC chromatogram of a typical polymer

Original name for, but now fading synonym of Size-Exclusion Chromatography. An HPLC chromatograph equipped with a gel permeatioin column can measure the molecular weight of polymers. The highest molecular weight fraction exits the column first. (Yau WW, Kirkland JJ, Bly DD (2001) Modern size-exclusion chromatography. Wiley, New York, January; Collins EA, Bares J, Billmeyer FW Jr. (1973) Experiments in polymer science. Wiley-Interscience, New York).

Gel Point The stage at which a liquid begins to exhibit elastic properties and increased viscosity. This stage may be observed from the inflection point on a viscosity – time plot. See ▶ Gel (2) (ASTM D 883).

Gel Spinning See ▶ Spinning (2).

Gel Time n **(of a Drying Oil)** Time required for the oil to form a solid gel under specified conditions of temperature.

Geminal adj Description of substitution on adjacent atoms (1,2-chloro-) as opposed to same atom or ▶ Vicinal (e.g., 1,1-chloro-) positions.

General Purpose Polystyrene n General purpose polystyrene is an amorphous thermoplastic prepared by homopolymerization of styrene. It has good tensile and flexural strengths, high light transmission and adequate resistance to water, detergents and inorganic chemicals. It is attached by hydrocarbons and has a relatively low impact resistance. Processed by injection molding and foam extrusion. Used to manufacture containers, health care items such as pipettes, kitchen and bathroom housewares, stereo and camera parts and foam sheets for food packaging. *Also called crystal polystyrene.*

Genuine Ultramarine n Natural ultramarine blue. See ▶ Ultramarine Blue.

Geodesic \ˌjē-ə-ˈde-sik, -ˈdē-, -zik\ n (1821) Pertaining to the shortest distance, lying on a surface, between two points on that surface, e.g., a straight line on a plane or great circle on a sphere.

Geodesic Isotensoid A filamentary structure in which there exists a constant stress in any given filament at all points in its path. (Iaac MD, Ishal O (2005) Engineering mechanics of composite materials. Oxford University Press, UK).

Geodesic-Isotensoid Contour n In filament-wound, reinforced-plastic pressure vessels, a dome contour in which the filaments are placed on geodesic paths so that the filaments will experience uniform tensions throughout their lengths under pressure loading. (Iaac MD, Ishal O. (2005) Engineering mechanics of composite materials. Oxford University Press, UK).

Geodesic Ovaloid n A contour for end domes, the fibers forming geodesic lines on the surface of revolution. The forces exerted by the filaments are proportioned to meet hoop and meridional stresses at any point. (Iaac MD, Ishal O. (2005) Engineering mechanics of composite materials. Oxford University Press, UK).

Geometric Metamerism \-mə-ˈta-mə-ˌri-zəm\ n Phenomenon exhibited by a pair of colors which appear to be a color match at one angle of illumination and viewing but which are no longer a match when the angle of illumination or viewing is changed. (Colour physics for industry, 2nd edn. McDonald, Roderick, Society of Dyers and Colourists, West Yorkshire, England, 1997) See ▶ Goniometachromatism, ▶ Flop, and ▶ Metamerism.

Geotextiles n Manufactured fiber products made into fabrics of various constructions for use in a wide variety of civil engineering applications.

GEP n Glass-fiber-reinforced epoxy resins.

Geranium Lake \jə-ˈrā-nē-əm-\ n Type of colorant produced by precipitation of certain acid dyestuffs with barium chloride followed by lead nitrate.

Germicidal-Lamp Test n A quick screening method for evaluating the relative resistance of vinyl compounds to discoloration and degradation upon exposure to light and weather. The specimens are placed approximately three inches below a germicidal lamp with a principal UV wavelength of 253.7 nm. The materials are rated according to the degree of discoloration and plasticizer spewing after a specified interval of time, e.g., 24 or 48 h.

Gesso \ˈje-(ˌ)sō\ n [It, literally, gypsum, fr. L *gypsum*] (1596) Traditional gesso is a material made from chalk and gelatin or casein glue, and painted on wooden panels as a surface for tempera work. Polymer gesso provides a flexible ground that can be applied to canvas and other materials.

GFK *n* Glass-fiber-reinforced plastics.

Ghatti Gum *n* Gum obtained from India and Sri Lanka, from *Anogeissus latifolia*, from which it exudes. *Known also as Ghati Gum and Indian Gum.*

Ghosting *n* (ca. 1957) (1) A condition found in wall paints of low sheen and moderately strong colors where there is a slight but definitely perceptible bleaching in irregular patterns. (2) Presence of a faint image of a printed design appearing in areas which are not intended to receive that portion of the image. (Paint/coatings dictionary. Federation of Societies for Coatings Technology, Philadelphia, Blue Bell, PA, 1978).

Gibbs' Free Energy See ▶ Free Energy.

Gibbsite See ▶ Aluminum Oxide, ▶ Hydrated.

Gibbs' Phase Rule *n* $F = C + 2 - P$, *E*, the number of degrees of freedom of a system, is the number of variable factors (temperature, pressure and concentration) of the components, which must be arbitrarily fixed in order that the condition of the system may be perfectly defined. *C*, the number of the components of the system, is chosen equal to the smallest number of independently variable constituents by means of which the composition of each phase participating in the state of equilibrium can be expressed in the form of a chemical equation; the components must be chosen from among the constituents which are present when the system is in a state of true equilibrium, and which take part in that equilibrium; as components are chosen the smallest number of such constituents necessary to express the composition of each phase participating in the equilibrium, zero and negative quantities of components being permissible; in any system the number of components is definite, but may alter with changes in conditions of experiment; a qualitative but not quantitative freedom of selection of components is allowed, the choice being influenced by suitability and simplicity of application. *P*, the number of phases of the system, are the homogeneous, mechanically separate and physically distinct portions of a heterogeneous system; the number of phases capable of existence varies greatly in different systems; there can never be more than one gas or vapor phase since all gases are miscible in all proportions, a heterogeneous mixture of solid substances forms as many phases as there are substances present. (Watson P (1997) Physical chemistry. Wiley, New York).

Giga - \ˈji-gə, ˈgi-\ {*combining form*} [ISV, fr. Gk *gigas* giant] (G) SI prefix meaning $\times 10^9$. (CRC handbook of chemistry and physics. Lide DR (ed) CRC, Boca Raton, FL, 2004 Version).

Gigging See ▶ Napping.

Gilbert \ˈgil-bərt\ *n* [William *Gilbert*] (1893) The cgs emu of magnetomotive force. 1 gilbert = $10/4\pi$ A-turns.

Gilding \ˈgild\ *vt* [ME, fr. OE *gyldan*; akin to OE *gold* gold] (14c) Art of covering substances, such as wood, with layers of gold leaf.

Gilling See ▶ Pin Drafting.

Gilsonite \ˈgil-sə-ˌnīt {*trademark*} A black asphaltic resinous material found in Utah and Colorado used in black printing ink and as an ingredient in the manufacture of cold molding compounds, paints, etc. It is hard, bituminous, and finds application in black japans, Brunswick blacks, stains, baking enamels and electrical insulating varnishes of many types. (Hibbard MJ (2002) Mineralogy. McGraw-Hill, New York; Handbook of fillers, extenders and dilutents. Ash M, Ash I (eds) Synapse Informtion Resources, New York, 1998) *Also called Vintahite.*

g-Index *n* A measure of polydispersity of molecular-weight distribution in polymers given by

$$g = (M_z/M_w - 1)^{0.5}$$

where M_z and M_w are the Viscosity-Average and ▶ Weight-Average Molecular Weights. (Slade PE (2001) Polymer molecular weights, vol 4. Marcel Dekker, New York).

Gingelli Oil *n* Another name for sesame oil.

Gingham \ˈgiŋ-əm\ *n* [modification of Malay *genggang* striped cloth] (1615) A woven fabric characterized by a block or check effect produced by weaving in dyed yarns at fixed intervals in both the warp and the filling.

Glacé \gla-ˈsā\ *adj* [F, fr. pp of *glacer* to freeze, ice, glaze, fr. L *glaciare*, fr. *glacies*] (1847) A lustrous, glossy effect imparted to fabrics by finishing.

Glance Pitch *n* Another name for Manjak.

Glass \ˈglas, ˈglás\ *n* {*often attributive*} [ME *glas*, fr. OE *glœs*; akin to OE *geolu* yellow] (before 12c) Any of a wide variety of rigid, amorphous solid materials, but in particular, any of the transparent vitreous glasses formed by cooling molten mixtures of silica, sodium oxide, other alkali-metal and alkaline-earth oxides, lead oxide, alumina, and, for special purposes, other oxides. (Many organic polymers experience "glassy states" at low temperatures but they are seldom referred to as "glasses"). Glass compositions used in fibrous form as plastics reinforcements include vitrified quartz (pure silica), ▶ C Glass, ▶ E Glass, and ▶ S Glass. The glasses as a family are characterized by high moduli, low tensile strength (except as fibers), very low ultimate tensile elongation, brittleness, and good electrical properties and chemical resistance. (Iaac MD, Ishal O (2005) Engineering mechanics of composite materials. Oxford University Press, UK; Lee SM (1989) Dictionary of composite materials technology. Technomic, Lancaster, PA).

Glass Bead *n* Glass beads range in size from 5 to 5,000 μmm, but normally are about 30 μmm in diameter. They improve the flexural modulus, abrasion resistance, compressive strength, mold flow, and corrosion resistance of plastics; reduce mold shrinkage and cycle time. The beads are made from various kinds of glass including A type and borosilicate and can be surface modified with silane coupling agents to improve adhesion to the polymer matrix. Used in housewares, machine parts, bearings, molds, and auto parts.

Glass Fiber *n* (1882) A manufactured fiber in which the fiber-forming substance is glass (FTC definition). In the continuous filament process, glass marbles are melted in an electric furnace and the liquid flows in fine streams through small orifices at the bottom of the melting chamber. The resultant filaments are caught and drawn by a high-speed drawwinding mechanism. In the staple fiber process, the streams of molten glass are attenuated into fibers by jets of high-pressure steam or air. These fibers are gathered on a revolving drum and then wound on tubes to form staple fiber sliver or bands that can be drafted, twisted, and plied. Glass fiber is incombustible and will tolerate heat up to 1,000°F without material damage. Potential strength is not realized in woven fabrics or even in yarns, because the fiber is brittle and fracture points may develop, but nevertheless, very high tensile strength is obtained in woven fabrics, and is retained at elevated temperatures. The fiber originally was difficult to color but methods have been developed to accomplish this. Moisture absorption is low. Electrical and insulation resistance is high. Glass fiber is used for heat and electrical insulation, filter cloth in the chemical and dye industries, reinforcing belts in tires, novelty fabrics, tablecloths, and fireproof draperies. Because of its brittleness, it is not used in wearing apparel or in household fabrics that have to withstand frequent flexing. (Lee SM (1989) Dictionary of composite materials technology. Technomic, Lancaster, PA).

Glass-Fiber Reinforcement *n* Any of a family of reinforcing materials for plastics based on single, drawn filaments of glass ranging in diameter from 3 to 20 μmm. Single filaments are produced by mechanically drawing molten-glass streams, then the filaments are usually coated with a ▶ Size or ▶ Coupling Agent which protects them from abrasion (to which the uncoated filaments are very vulnerable) and improves bonding with resins. The filaments are then gathered into bundles called *strands* or larger bundles called *rovings*. The strands may be used in continuous form for filament winding or pultrusion; or chopped into short lengths for incorporation into molding compounds or during sprayup; or formed into fabrics and mats of various types for use in hand layup, matched-die molding, and other laminating processes. Glass-fiber reinforcements are classified according to their special properties. E glass is electrical-grade glass, the most common general-purpose type. C glass is the chemically resistant grade, and S glass denotes high tensile strength. Glass fibers coated with nickel by electron-beam deposition are used making molding compounds for electrically conductive articles. Another form of glass reinforcement is ▶ Glass Flakes. (Lee SM (1989) Dictionary of composite materials technology. Technomic, Lancaster, PA).

Glass Filler *n* Glass fillers are a widely used family of fillers in the form of beads, hollow spheres, flakes, or milled particles. They increase dimensional stability, chemical resistance, moisture resistance, and thermal stability of plastics.

Glass Finish (size) See ▶ Coupling Agent.

Glass Flakes *n* A reinforcing filler produced by blowing molten Type E glass into very thin tubes, then smashing the tubes into small fragments. The flakes pack closely in thermosetting-resin systems, producing strong products with good moisture resistance. (Lee SM (1989) Dictionary of composite materials technology. Technomic, Lancaster, PA).

Glassine \gla-ˈsēn\ *n* (1916) Thin transparent paper treated with urea-formaldehyde resin, used for packaging.

Glass Mat *n* (fibrous-glass mat) A thin web of Nonwoven glass fibers that may or may not contain a small percentage of resin binder.

Glass Microspheres See ▶ Microspheres.

Glass-Reinforced Plastics (GRP) See ▶ Reinforced Plastic.

Glass-Rubber Transition Syn: ▶ Glass Transition.

Glass Spheres *n* Solid glass spheres of diameters ranging from 5 to 5,000 μmm are used as fillers and/or reinforcements in both thermosetting and thermoplastic compounds. The size used most frequently passes a 325-mesh sieve, with an average sphere diameter of 30 μmm. The spheres are available coated with various silane coupling agents to improve bonding between polymer and glass. The spheres improve a physical properties and reduce costs of materials and end products. Some small glass spheres are considered to be

▶ Microspheres. (Lee SM (1989) Dictionary of composite materials technology. Technomic, Lancaster, PA).

Glass Stress In a filament-would part, typically a pressure vessel, the stress calculated from the load and the cross-sectional area of the reinforcement only.

Glass Transition *n* (gamma transition, second-order transition, glass–rubber transition) A reversible change that occurs in an amorphous polymer or in amorphous regions of a partly crystalline polymer when it is heated from a very low temperature into a certain range, peculiar to each polymer, characterized by a rather sudden change from a hard, glassy, or brittle condition to a flexible or elastomeric condition. Physical properties such as coefficient of thermal expansion, specific heat, and density, usually undergo changes in their temperature derivatives at the same time. During the transition, the molecular chains, normally coiled, tangled, and motionless at the lower temperatures, become free to rotate and slip past each other. This temperature varies widely among polymers. (Groenewoud WM (2001) Characterization of polymers by thermal analysis. Elsevier Science and Technology Books, New York; Physical properties of polymers handbook. Mark JE (ed). Springer, New York, 1996).

Glass-Transition Temperature *n* (T_g) The approximate midpoint of the temperature range over which the ▶ Glass Transition occurs. Tg is not obvious (like a melting point), and is detected by changes, with rising temperature, in secondary properties such as the rate of change with temperature of specific volume or electrical or mechanical probe. (Groenewoud WM (2001) Characterization of polymers by thermal analysis. Elsevier Science and Technology Books, New York).

Glass Wood *n* (1879) Glass fibers in a mass resembling wool and used for thermal insulation and air filters.

Glassy State *n* A solid state anhedral in shape and with conchoidal fracture: if free from strain, isotropic. Moreover, the observed T_g can vary significantly with the specific property chosen for observation and on experimental details such as the rate of heating or electrical frequency. A reported T_g should therefore be viewed as an estimate. The most reliable estimates are normally obtained from the loss peak in dynamic-mechanical tests or from dilatometric data. They change to the crystalline state, i.e., devitrify, in time or by the action of temperature, shock, scratching, etc. Glassy materials do not melt at a specific temperature but soften gradually and continuously on heating. (Groenewoud WM (2001) Characterization of polymers by thermal analysis. Elsevier Science and Technology Books, New York; Physical properties of polymers handbook. Mark JE (ed). Springer, New York, 1996).

Glaze \ˈglāz\ *v* [ME *glasen*, fr. *glas* glass] (14c) Very thin coating of a paint product, usually a semitransparent tinted coating, applied on a previously painted surface to produce a decorative effect.

Glazing \ˈglā-ziŋ\ *v* [ME *glasen*, fr. *glas* glass] (14c) (1) A finishing process that produces a smooth, highly polished, or lustrous surface on a fabric such as chintz. The fabric is treated with starch, glue, paraffin, or shellac, then friction calendered. Synthetic resins are used for a more permanent finish. (2) A shiny fabric appearance produced unintentionally, e.g., by pressing at excessive temperatures. (Kadolph SJJ, Langford AL (2001) Textiles. Pearson Education, New York).

Glazing Compound *n* A dough-like material consisting of pigment and vehicle, used for sealing window glass in frames. It differs from putty in that it retains its plasticity for an extended period. (Weismantal GF (1981) Paint handbook. McGraw-Hill, New York).

GLC *n* Abbreviation of ▶ Gas-Liquid Chromatography; a technique by which the separation and identification of volatile components of a mixture is possible.

Glitter \ˈgli-tər\ *vt* [ME *gliteren*, fr. ON *glitra*; akin to OE *geolu* yellow] (14c) (flitter, spangles) A family of decorative pigments comprising light-reflective flakes of sizes large enough so that each flake produces an individually seen sparkle or reflection. They are incorporated into plastics during compounding.

Global Radiation *n* The wavelength distribution of sunlight under a given environment (e.g., under windowpane glass).

Gloss \ˈgläs, ˈglós\ *n* [prob. of Scand origin; akin to Icelandic *glossa* to glow, akin to OE *geolu* yellow] (1538) Subjective term used to describe the relative amount and nature of mirror-like (specular) reflection. Different types of gloss are frequently arbitrarily differentiated, such as sheen, distinctness-of-image gloss, etc. Trade practice recognizes the following stages, in increasing order of gloss: *Flat* (or *matte*) – practically free from sheen, even when viewed from oblique angles (usually less than 15 on 85° m); *Eggshell* – usually 20–35 on 60° m; *Semigloss* – usually 35–70 on 60° m; *Full gloss* – smooth and almost mirrorlike surface when viewed from all angles, usually above 70 on 60° m. Gloss of plastics is measured with a glossmeter described in ASTM Test D 523. The same meter is used in measuring the resistance of shiny plastic surfaces to abrasion (D 673). See also ▶ Satin Finish.

Gloss, Distinctness-of-Image *n* The sharpness with which outlines are reflected by the surface of an object.

Glossing Up *n* Undesirable burnishing. See ▶ Burnish and ▶ Burnish Resistance.

Gloss Ink *n* An ink that dries with a minimum of penetration into the stock and yields a high luster.

Glossmeter *n* An instrument for measuring the degree of gloss in relative terms. Such instruments measure the light reflected at a selected specular angle, such as 20° (from the perpendicular), 45°, 60°, 75°, or 85°. Results obtained are very dependent on the instrument design, calibration, technique used, types of samples, etc. In essence, a glossmeter is an abridged goniophotometer. (Paint and coating testing manual. Koleske JV (ed). American Society for Testing and Materials, 1995; Paul N. Gardner, 316 N. E. Fifth Street, Pompano Beach, FL, www.gardco.com).

Gloss Oil *n* Solution of limed rosin or lined rosin acids in a volatile solvent, used chiefly in surface coatings (when made from tall oil, the source is usually indicated).

Gloss Retention *n* Degree to which the original sheen of a coating is retained. (Paint and coating testing manual. Koleske JV (ed). American Society for Testing and Materials, 1995).

Gloss White *n* A white mineral pigment used as an ink extender, made by co-precipitation of alumina hydrate and blanc fixe.

Glow Discharge See ▶ Corona Discharge.

Glowing Combustion *n* Burning of a solid material without flame but with emission of light from the zone of burning.

Glue \ glü\ *n* [ME *glu*, fr. MF, fr. LL *glut-*, *glus*; akin to L *gluten* glue] (14c) Originally, a hard gelatin obtained from hides, tendons, cartilage, bones, etc., of animals. Also, an adhesive prepared from this substance by heating with water. Through general use the term is now synonymous with the term "adhesive." (See also ▶ Adhesive, ▶ Mucilage, ▶ Paste, and ▶ Sizing). *(v)* See ▶ Bond.

Glue Line (bond line) *n* The layer of adhesive which attaches two adherents.

Glycerides \ gli-sə- rīd\ *n* (ca. 1864) The esters derived from glycerol. Glycerol is a trihydroxy alcohol and as such is capable of reacting with three monobasic acids. Such fully satisfied esters are described as the triglycerides, and the drying oils are typical members of this class. If only one hydroxy group is reacted with an acid, the resultant product is a monoglyceride, and if two are reacted a diglyceride is obtained. (Polymeric materials encyclopedia. Salamone JC (ed). CRC, Boca Raton, FL, 1996).

Glycerin \ glis-rən, gli-sə-\ *n* [F *glycerine*, fr. Gk *glykeros* sweet; akin to Gk *glykys*] (1838) See ▶ Glycerol.

Glycerol \ gli-sə- rόl, - rōl\ *n* [*glycer*in + -*ol*] (1884) (1,2,3-propanetriol, glycerin, glycerine, glycyl alcohol) HO-CH$_2$CHOHCH$_2$OH. The term *glycerol* applies to the pure product; *glycerin* (or *glycerine*) applies to commercial products containing at least 95% glycerol. Glycerol is a colorless, viscous liquid with a sweet taste, long produced as a by-product of soap manufacture (animal fats and vegetable oils are triglycerides of long-chain "fatty" acids). More recently, glycerol has been synthesized from propylene or sucrose (sugar). It uses in the plastics industry include the manufacture of alkyd resins (esters of glycerol and phthalic anhydride), the plasticizing of cellophane, and the production of urethane polymers. Several of its esters are plasticizers for various resins. It has a bp of 290°C, mp of 17°C, and sp gr of 1.260/15°C. (Polymeric materials encyclopedia. Salamone JC (ed). CRC, Boca Raton, FL, 1996) *Also known as Glycerin.*

Glycerol Diacetate *n* Plasticizer; solvent for cellulose derivatives, alkyds, and shellac. It has a bp of 260°C, sp gr of 1.176/15°C, flp of 146°C (295°F), refractive index of 1.44. See ▶ Diacetin.

Glycerol Ether Acetate *n* C$_3$H$_5$(OCH$_2$CH$_2$OOCCH$_3$)$_3$. A plasticizer for cellulosics and polyvinyl acetate.

Glycerol Monoacetate *n* CH$_3$COOCH$_2$CHOHCH$_2$OH. A plasticizer for cellulose acetate, cellulose nitrate, and vinyl resins.

Glycerol Monolactate Diacetate *n* CH$_3$CHOHCOO C$_3$H$_5$(OOCCH$_3$)$_2$. A plasticizer for cellulose acetate, imparting resistance to gasoline.

Glycerol Monolaurate *n* (glyceryl monolaurate, lauryl glycerin) C$_{11}$-H$_{23}$COOCH$_2$CHOHCH$_2$OH. A plasticizer for cellulosic and vinyl resins and polystyrene.

Glycerol Monooleate *n* C$_{17}$H$_{33}$COOCH$_2$CHOHCH$_2$OH. A yellow oil, approved by the FDA as a plasticizer for food processing.

Glycerol Monoricinoleate *n* C$_6$H13CHOHC$_{10}$H$_{18}$COO CH$_2$CHOHCH$_2$-OH. A plasticizer for cellulose nitrate, ethyl cellulose, and polyvinyl butyral.

Glycerol-Phthalic Anhydride Resin *n* An alkyd resin made by modifying glycerol phthalate with an equal portion of oil, fatty acid, and natural or synthetic resin. It is used in varnishes, lacquers, and enamels.

Glycerol Triacetate \-(▎)trī- ▎a-sə- ▎tāt\ (triacetin) A plasticizer for cellulosic resins, polymethyl methacrylate, and polyvinyl acetate. It has been approved by the FDA for food-contact use.

Glycerol Tribenzoate *n* (tribenzoin) A colorless, crystalline solid, a solid plasticizer for PVC.

Glycerol Tributyrate *n* (tributyrin) A plasticizer for cellulose esters.

Glycerol Tripropionate *n* A plasticizer for cellulosic resins and polyvinyl acetate.

Glyceryl Tri-(Acetoxyricinoleate) *n* $C_3H_5(OOCC_{10}H_{18}CHOHC_6H_{13}O\text{-}COCH_3)_3$. A plasticizer for cellulose nitrate, ethyl cellulose, and PVC.

Glyceryl Tri-(12-Acetoxystearate) *n* $C_3H_5(OOCC_{17}H_{34}OCOCH_3)_3$. A plasticizer for cellulosic and vinyl resins.

Glycidol *n* (2,3-epoxy-1-propanol). A stabilizer for vinyl resins.

γ-Glycidoxypropyltrimethoxysilane *n*

A coupling agent for fibrous glass used in reinforced thermosetting and thermoplastic resins.

Glycidyl- *n* The terminal epoxy group.

Glycidyl-Ester Resin *n* Any of a family of epoxides resins derived from the condensation of epichlorohydrin with polycarboxylic acids, first available in commercial quantities in 1968. The preferred during agents for these resins are anhydrides, polycarboxylic acids, aromatic amines, and phenolics. They are resistant to electrical tracking and weather, have high strength and high modulus, yet are tough, have low viscosity, long pot life, and high reactivity at moderately elevated temperatures. Limitations are poorer properties above 100°C, higher shrinkage, poor alkali resistance, and inability to cure at room temperature. (Polymeric materials encyclopedia. Salamone JC (ed). CRC, Boca Raton, FL, 1996).

Glycidyl-Ether Resin See ▶ Epoxy Resin.

Glycol \ˈglī-ˌkól, -ˌkōl\ *n* [ISV *glyc-* + *-ol*] (1858) (1) An organic compound having hydroxyl groups on adjacent carbon atoms, for example, propylene glycol, $CH_3CHOHCH_2OH$. The name is also used for nonadjacent dihydric alcohols such as the 1,3-isomer of the above compound, $HOCH_2CH_2CH_2OH$, known as *trimethylene glycol*. (2) Specifically, ethylene glycol.

Glycol Phthalate Resin *n* A type of thermoplastic polyester used mainly for fibers and oriented films. See ▶ Saturated.

Glyptal \ˈglip-t°l\ *n* (*trademark*) An alkyd resin, the reaction product of glycerol and phthalic anhydride (hence: gly p t al) An early product was the basis of a household and laboratory cement, trade named Glyptal. It has been used as a European generic term for alkyd resins and as an alternative to the term "alkyd."

Godet Roll \gō-ˈdet-\ *n* Roll used for transporting and controlling the movement of bundles of fibers and yarns in the processing of these materials.

Godet Stand *n* A device used in pairs downstream from the extruder when making monofilament and split-film fibers. One stand is placed before, and another following, a conditioning oven, each consisting of a vertical motor housing driving two or more horizontal drums. The filaments are wrapped repeatedly around the drums to prevent slipping. The post-oven drums are driven at speeds several times those of the first pre-oven set, causing the filaments to be stretched, oriented, and to become much stronger after cooling.

Gold Bronzes *n* Metallic powders of gold color produced from aluminum or copper and its alloys.

Golden Chrome See ▶ Chrome Yellow, and ▶ Light.

Gold Ink *n* A printing ink whose principal pigment consists of bronze powder.

Gold Lacquer *n* Vehicle of such color that it will give the effect of gold when applied over aluminum leaf.

Gold Leaf *n* (ca, 1741) Very thin leaves of metallic gold used for decorative purposes.

Gold Leafing *v* In antiquing, the art of applying a very thin sheet of gold over a "sized" area. It became fashionable in France in the 1700s.

Gold Number *n* A measure of the protective ability of hydrophilic colloids. Specifically, the number of milligrams of colloids which, when added to a standard gold solution, is just insufficient to prevent a change in color from red to blue.

Gold Size *n* Oleoresinous varnish which dries rapidly to a tacky condition, but which hardens slowly. It is used chiefly as an adhesive for fixing gold leaf to a surface.

Goniochromatic *n* Adjective used to describe a colored material which exhibits goniochromatism.

Goniochromatism *n* The phenomenon where the color of a material changes as the angle of illumination and/or viewing is changed. (Colour physics for industry, 2nd edn. McDonald, Roderick, Society of Dyers and Colourists, West Yorkshire, England, 1997).

Goniometachromatic *n* Adjective used to describe a pair of samples exhibiting goniometachromatism.

Goniometachromatism *n* The phenomenon where the colors of *two* goniochromatic materials match exactly at one set of illumination-viewing angles but no longer appear to match if either the illumination or viewing angles, or both, are changed. Such pairs of samples are also called *Geometric Metamers*.

Goniophotometer *n* An instrument used to measure the amount of light reflected as a function of the viewing angle when the incident angle is kept constant. Generally, the incident angle may be changed.

Goniophotometry *n* A procedure for evaluating the manner in which materials geometrically redistribute light. (See www.astm.org for relevant standards).

Goniospectrophotometer *n* An instrument used to measure a spectrophotometric curve at various angles of incidence and reflectance. The angles of incidence and reflectance may be changed. On goniochromatic samples, the results obtained are dependent on the angles of incidence and reflectance, and on samples exhibiting specular reflectance (glossy materials), the results are also dependent on the total subtended angle of viewing relative to the specular angle. (Willard HH, Merritt LL, Dean JA (1974) Instrumental methods of analysis. D. Van Nostrand, New York).

Gordon-Taylor Constant *n* This constant has been applied to polymer blends for the purpose of predicting the resultant glass transition temperature (T_g) mixing multiple polymers with different glass transition temperatures. A typical expression using the constant is

$$T_g = (w_1 T_{g1} + kw_2 T_{g2})/(w_1 + kw_2)$$

where the Gordon-Taylor constant $k = 5.2$, T_g is the resultant glass transition temperature, T_{g1} and T_{g2} are glass transition temperature of the two different polymers and w_1 and w_2 are respective weight fraction of the individual polymers. The above relationship is sometimes referred to as the Fox, Gordon-Taylor and Couchman equations. (Gordon M, and Taylor JS (1952), Ideal co-polymers and the second order transitions of synthetic rubbers. I. Non-crystalline polymers, J Appl Chem 2:493.)

Gouache \ˈgwäsh\ *n* [F, fr. It *guazzo*, literally, puddle, prob. fr. L *aquatio* watering place, fr. *aquari* to fetch water, fr. *aqua* water] (1882) Method of painting with opaque colors which have been ground in water and tempered with a preparation of gum. Also, a term used to describe a method of water color painting in which white is employed as a pigment by contrast with the method of attaining whites by allowing the paper to show through. (Merriam-Webster's collegiate dictionary, 11th edn. Merriam-Webster, Springfield, MA, 2004).

Gout \ˈgaút\ *n* [ME *goute*, fr. OF, gout, drop, fr. L *gutta* drop] (13c) Foreign matter that is accidentally woven into a fabric. It is usually fly or waste that drops into the loom during weaving or that catches in yarns during spinning.

GP *n* (1) Abbreviation for General-Purpose, sometimes used to denote types of resins and molding compounds suitable for a wide range of applications. (2) Abbreviation for Gutta-Percha.

GPC *n* Abbreviation for ▶ Gel-Permeation Chromatography. See ▶ Size-Exclusion Chromatography.

GR-1 *n* Former symbol for ▶ Butyl Rubber.

Grab Strength Test *n* A method for measuring the breaking strength of a fabric sample by mounting the sample in the tensile tester so that only a part of the width of the specimen is gripped in the clamps.

Gradient Elution *n* Elution in chromatography using a solvent whose composition is varied during a run.

Gradient-Tube Density Method *n* (density-gradient method) A convenient method, described in ASTM D 1505, for routinely measuring densities of small resin samples, e.g., a single pellet of molding compound. A graduated, vertical glass tube (the gradient tube) is carefully filled with graded solutions of two liquids, the densest at the bottom, the next densest above that, etc., to the least dense at the top. A specimen particle is introduced in the tube and falls to a position of equilibrium that indicates its density by comparison with the positions of calibrated hollow-glass floats. Over weeks or months, the gradient is gradually diminished by diffusion and the column must be reconstituted. See www.astm.org for a description of this method.

Graffito \grə-▎fē-(▎)tō, gra-, grä-\ *n* [It, incised inscription, fr. *graffiare* to scratch, prob. fr. *grafio* stylus, fr. L *graphium*] (1851) Last plaster coat of a fresco ground which is colored before application. (Merriam-Webster's collegiate dictionary, 11th edn. Merriam-Webster, Springfield, MA, 2004).

Graft Copolymer *n* A copolymer having branches of varying length made up of different monomer units on a common "backbone" chain. (Odian GC (2004) Principles of polymerization. Wiley, New York).

Grafting Ratio *n* In a graft copolymer, the weight of grafted side chains divided by the weight of the original polymer. (Odian GC (2004) Principles of polymerization. Wiley, New York).

Graft Polymer or Copolymer *n* A high polymer, the molecules of which consist of two or more polymeric parts of different composition, chemically united. Such a graft copolymer may be produced, for example, by polymerization of a given kind of monomer with subsequent polymerization of another kind of monomer onto the product of the first polymerization. The union of two different polymers by chemical reaction between their molecular end groups or by reaction producing crosslinks between the different molecules would also produce a graft copolymer. (Odian GC (2004) Principles of polymerization. Wiley, New York).

Grahamite *n* Naturally occurring asphaltum which varies considerably in composition. Some types are completely soluble in carbon disulfide, whereas others may contain as much as 50% of mineral matter.

Graham's Law *n* [Thomas *Graham*, 1805–1869, Scottish chemist] (1833) The relative rates of diffusion of gases under the same conditions are inversely proportional to the square roots of the densities of those gases. (Whitten KW, Davis RE, Davis E, Peck LM, Stanley GG (2003) General chemistry. Brookes/Cole, New York).

Grain *n* The direction, size, arrangement, appearance or quality of the fibers in fibrous material such as paper, wood or veneer. To have a specific meaning the term must be qualified. Also, the direction of molecular orientation in a non-fibrous material. See types of gain, i.e., ▶ Coarse Grain, ▶ Cross Grain, ▶ Diagonal Grain, and ▶ Interlocked Grain

Grain Alcohol *n* (1889) Syn: ▶ Ethyl Alcohol.

Graining *v* (1530) (1) A process for simulating a grain such as that of wood or marble on a painted surface by applying a stain, either translucent or pigmented, then working it into suitable patterns with tools such as graining combs, brushes, and rags. (Weismantal GF (1981) Paint handbook. McGraw-Hill, New York; Paint/coatings dictionary. Federation of Societies for Coatings Technology, Philadelphia, Blue Bell, PA, 1978).

Grain Raising *n* Swelling and standing up of the wood grain caused by absorbed water or solvent. Roughness of wood caused by the application of stains or other materials. Fibers of the wood raise due to swelling by water or certain solvents.

Gram \▎gram\ *n* [F *gramme*, fr. LL *gramma*, a small weight, fr. Gk *grammat-*, *gramma* letter, wirting, a small weight fr. *graphein* to write] (1810) Subunit of mass in the metric system of weights and measures, equal to 1/1,000 of a standard kilogram (453.6 g = one lb). (CRC handbook of chemistry and physics. Lide DR (ed). CRC, Boca Raton, FL, 2004 Version; Merriam-Webster's collegiate dictionary, 11th edn. Merriam-Webster, Springfield, MA, 2004).

Gram-Atom *n* (1927) (gram-atomic weight) The mass of an element in grams numerically equal to the elements atomic weight.

Gram Break Factor See ▶ Break Factor.

Gram Equivalent *n* (ca. 1897) The weight in grams of a substance displacing or reacting with 1.00797 g of hydrogen or combining with 15.9994 g of oxygen.

Gram Molecular Weight or Gram Molecule *n* (ca. 1902) A mass in grams of a substance numerically equal to its molecular weight. Gram mole.

Gram Mole, Gram Formula Weight, Gram Equivalent *n* Mass in grams numerically equal to the molecular weight, formula weight or chemical equivalent, respectively.

Granular (or Powder) \▎gran-yə-lər\ *adj* (1794) Particles having equidimensional non-spherical shapes.

Granular Polymerization *n* A polymerization process in which the monomer, or mixture of monomers, is dispersed by mechanical agitation in a liquid phase, usually water, in which the monomer droplets are polymerized while they are dispersed by continuous agitation. Also called *Suspension Polymerization*. (Odian GC (2004) Principles of polymerization. Wiley, New York).

Granular Structure *n* Nonuniform appearance of finished plastic material due to retention of, or incomplete fusion of, particles of compound either within the mass or on the surface; or to the presence of coarse filler particles.

Granulate *v* (1666) To reduce plastic sheet, chunks, or scrap to particles about 2–5 mm in size. (2, *n*) A molding compound in the form of small spheres or pellets.

Granulator *n* (scrap grinder) A machine for cutting waste material such as sprues, runners, excess parison material, trim scarp from extrusion and thermoforming, and rejected parts into particles that can be reprocessed. The most common type of granulator is comprised of several thick knives bolted to a heavy cylindrical core, parallel to the core's axis, which is also its axis of rapid rotation. The rotating knives graze stationary knives mounted in the machine's housing, given an action that combines impact with shearing. A screen placed in the discharge opening controls the final particle size.

Granule \ˈgran-(ˌ)yü(ə)l\ *n* [LL *granulum*, dim. of L *granum* grain] (1652) A small particle produced in various size and shapes by hot or cold cutting, of extruded strands or scrap, or by certain polymerization methods. (Perry's chemical engineer's handbook. 7th edn. Perry RH, Green DW (eds). McGraw-Hill, New York, 1997) Also see ▶ Pellets.

Grapeseed Oil *n* Semidrying vegetable oil. Of the fatty acids present, up to 55% may be linoleic acid. Approximate constants for oil from various sources are: sp gr of 0.925/15°C, refractive index of 1.4899/15°C, saponification value of 190, iodine value of 125–157.

Graphic Arts Coatings *n* Coatings which are marketed solely for application to indoor or outdoor signs and include lettering enamels, poster colors and bulletin colors.

Graphite \ˈgra-ˌfit\ *n* [Gr *Graphit*, fr. Gk *graphein* to write] (1796) (black lead, plumbago) A crystalline form of carbon with atoms arranged hexagonally, characterized by a soft, greasy feel. It occurs naturally, but is produced by heating petroleum coke or other organic materials under controlled atmospheric conditions. In powder form, graphite is used as a lubricating filler for nylon and fluorocarbon resins. Too, it is added to compounds to make them electrically conductive. Pyrolytic graphite fibers, made by decomposing organic filaments at high temperatures in controlled atmospheres, are used as reinforcements for high-performance applications. The best graphite fibers are among the strongest and stiffest of all fibrous reinforcements, with strengths to 2.5 GPa and moduli to 500 GPa (360 kpsi and 72 Mpsi). (Pierson HO (1994) Handbook of carbon, graphite, diamond and fullerenes. Noyes Data Corporation/Noyes Corporation, New York).

Graphite Fiber *n* Although the terms carbon and graphite are used interchangeably to describe these fibers, graphite fibers are more accurately defined as fibers that are 99+% carbonized (crystalline structure as opposed to non crystalline carbon fiber) while the term carbon is used for any fiber carbonized to 93–95% or more. (Pierson HO (1994) Handbook of carbon, graphite, diamond and fullerenes. Noyes Data Corporation/Noyes Corporation, New York) See ▶ Carbon Fiber.

Graphite Paint *n* A painting compound consisting of powdered graphite and oil; used to coat metallic structures to inhibit corrosion. (Pierson HO (1994) Handbook of carbon, graphite, diamond and fullerenes. Noyes Data Corporation/Noyes Corporation, New York).

Graphitic Mica See ▶ Aluminum Potassium Silicate.

Graphitization \ˌgra-fə-tə-ˈzā-shən\ *n* (1899) Corrosion of gray iron in which metallic iron is converted into corrosion products, leaving a residue of intact graphite mixed with iron corrosion products and other insoluble constituents of cast iron and having approximately the original dimensions of the casting. Also known as *Graphitic Corrosion*.

Graphitize *vt* (1899) The pyrolysis of an organic material such as pitch or Polyacrylonitrile (PAN) in an inert atmosphere at temperatures between 2,500°C and 3,000°C to produce graphite. Pyrolysis of PAN fibers is a principal source of graphite fibers.

Grass \ˈgras\ *n* {*often attributive*} [ME *gras*, fr. OE *græs*; akin to OHGr *gras* grass, OE *grōwan* to grow] (before 12c) Sessile algae on the side of a ship.

Grass Cloth *n* (1857) Originally a handmade product imported from Japan, made by gluing woven native grasses onto paper backing. Printed wallpapers or dimensional wall coverings, especially vinyl, which simulates the same.

Grasshopper *n* A stiff bunch of parallel strands in a fibrous mat.

Grass-Tree Gum See ▶ Accroides.

Gravel \ˈgra-vəl\ *n* [ME, fr. OF *gravele*, dim. of *grave*, *greve*, pebbly ground, beach] (13c) Rounded or semirounded particles of rock that will pass a 3-in. (76.2 mm) sieve and be retained on a No. 4 (4.75 mm) US standard sieve.

Gravel Roofing See ▶ Built-Up Roofing.

Gravimetric Feeder \ˌgra-və-ˈme-trik-\ (weigh feeder) A device for feeding extruders that continuously meters, by weighing, the rate at which feedstock enters the feed port of the extruder.

Gravitation \ˌgra-və-ˈtā-shən\ n (ca. 1645) The universal attraction existing between all material bodies. The force of attraction between two masses m and m', separated by a distance r, k being the constant of gravitation,

$$F = k \frac{mm'}{r^2}$$

(If m and m' are given in grams and r in centimeters, F will be in dynes if $k = 6.670 \times 10^{-8}$). (Giambattista A, Richardson R, Richardson RC, Richardson B (2003) College physics. McGraw Hill Science/Engineering/Math, New York).

Gravure \grə-ˈvyür, grā-\ n [F, fr. *graver* to grave, of Germanic origin; akin to OHGr *graban* to dig, engrave] (1893) One of the methods of printing using the intaglio process, i.e., the ink is placed in cells below the plate surface. Gravure inks are quick drying, low viscosity inks based on volatile solvents. (Printing ink manual, 5th edn. Leach RH, Pierce RJ, Hickman EP, Mackenzie MJ, Smith HG (eds). Blueprint, New York, 1993).

Gravure Coating n (engraved-roll coating) A roller-coating process in which the amount of coating applied to the substrate web is metered by the depth of an all-over engraved pattern in the application roll. This process is frequently modified by interposing a resilient offset roll between the engraved roll and the web. (Printing ink manual, 5th edn. Leach RH, Pierce RJ, Hickman EP, Mackenzie MJ, Smith HG (eds). Blueprint, New York, 1993).

Gravure Printing n Depressions in an engraved printing cylinder or plate are filled with ink, the excess on raised portions being wiped off by a doctor blade. Ink remaining in the depressions is deposited on the plastic film as it passes between the gravure roll and a resilient backup roll. (Printing ink manual, 5th edn. Leach RH, Pierce RJ, Hickman EP, Mackenzie MJ, Smith HG (eds). Blueprint, New York, 1993).

Gravure Tissue n Sheet of paper coated with dichromated gelatin but containing no pigment. (Printing ink manual, 5th edn. Leach RH, Pierce RJ, Hickman EP, Mackenzie MJ, Smith HG (eds). Blueprint, New York, 1993).

Gray n (Gy) The SI unit of absorbed radiation dose, defined as the absorption of 1 J/kg of body mass, i.e., Gy = 1 J/kg. The former unit, the *rad*, now deprecated but still being used widely in the US, equals 0.01 Gy.

Gray Blast n (**Commercial Blast**) See ▶ Nace No. 3.

Gray Fabric See ▶ Greige Fabric.

Gray Scale n An archromatic scale ranging from black through a series of successively lighter grays to white. Such a series may be made up of steps which appear to be equally distant from one another (such as the Munsell Value Scale) or may be arranged according to some other criteria such as a geometric progression based on lightness. Such scales may be used to describe the relative amount of difference between two similar colors. (Colour physics for industry, 2nd edn. McDonald, Roderick, Society of Dyers and Colourists, West Yorkshire, England, 1997).

Grazing Angle n Angle of light very nearly parallel to the surface of a material; used to describe incident and/or reflected angles.

Grease \ˈgrēs\ n [ME *grese*, fr. OF *craisse*, *graisse*, fr. (assumed) VL *crassia*, fr. L *crassus* fat] (13c) A semisolid lubricant composed of emulsified petroleum oils and soluble hydrocarbon soaps.

Grease Forming See ▶ Mechanical Grease Forming.

Grease Paint (1) Nondrying and nonoxidizing coating for void spaces in ships which can be easily removed for control of the steel sheets. (2) Theatrical make-up.

Grease Proof Inks and Coatings n Inks (and coatings) that are resistant to the action of fats, oils and greases.

Greasiness See ▶ Spewing.

Greasing n Sensitization of the non-printing areas of a lithographic plate resulting in undesirable adhesion of the ink to these areas. *Also called Scumming*.

Green B, Naphthol n Iron lake of 1-nitroso-2-naphthol-3-carboxylic acid.

Green B, Pigment n Pigment produced by treating 1-nitroso-2-naphthol with a ferrous salt.

Green, Brunswick See ▶ Chrome Green.

Green, Chrome See ▶ Chrome Green.

Green Cinnabar See ▶ Chromium Oxide Green.

Green Earth Naturally occurring earth pigment which is primarily an iron silicate. Used to make lime green. See ▶ Lime. Syn: ▶ Terre Verte.

Green, Emerald See ▶ Emerald Green.

Green Gold See Niclel Azo Yellow.

Green Lake n Mixture of Prussian blue and Yellow Lake.

Green, Lime n Term applied to those green pigments which are employed for their fastness to alkali (lime). For a considerable time it consisted of the green produced from green earth but the term is now applied to other types, e.g., reduced Pigment Green B.

Green Lumber *n* Lumber which has not been dried or seasoned.

Greenockite \ gre-nə-ˌkīt\ *n* [Charles M. Cathcart, Lord *Greenock* † 1859 English soldier] (1844) Naturally occurring cadmium sulfide mineral.

Green, Paris *n* Pigment consisting essentially or copper acetoarsenite. It is sometimes used in anti-fouling compositions, but its main use is in insecticides.

Green Rouge See Chromium Oxide Green.

Green Stock *n* Raw or uncured plastic or rubber stock which is ready for vulcanization. The term is not applied to crude rubber, synthetic rubber, or resin that has not been compounded.

Green Strength *n* Of a perform or incompletely cured molding, the ability to withstand handling without distorting.

Green Tack *n* A term used in fabric bonding for the preliminary bond created in the first stage of curing by the wet adhesive process. At this point, the bond is not fully cured and hence is "green."

Greentone Ink Blue See ▶ Iron Blue.

Green Vitriol \- vi-trē-əl\ $FeSO_4 \cdot 7H_2O$. Crystalline ferrous sulfate used in the manufacture of iron blue.

Green, Zinc *n* Range of greens made by mixing zinc chromate with iron blue. They are less bright and weaker than the Brunswick greens, but find application where lead-free greens are required.

Greige Fabric \ grā(zh)-\ *n* [F *grège* raw (of silk), fr. It *greggio*] (1926) An unfinished fabric just off the loom or knitting machine.

Grenadine \ ˌgre-nə-ˈdēn, ˈgre-nə-ˌ\ *n* [F, fr. *grenade* coarse silk fabric, pomegranate] (1852) (1) A fine, loosely woven fabric in leno weave made with dyed filling yarns and having a clipped dobby design. (2) A silk cord constructed by twisting together several twisted strands.

Grex (1) A unit of linear density equal to the weight in grams of 10 km of yarn, filament, fiber, or other textile strand. (2) The system of yarn numbering based on the use of grex units. (Kadolph SJJ, Langford AL (2001) Textiles. Pearson Education, New York) Also see ▶ Yarn Number.

Grex Number *n* Like *cut* and *denier*, a deprecated measure of lineal density of yarns and fibers, the weight in grams of 10 km of the yarn. A grex number of 1 corresponds to an SI lineal density of 10^{-7} kg/m or 0.1 tex. See also ▶ Cut, ▶ Denier, and ▶ Tex.

Grey Fabric See ▶ Greige Fabric.

Grid \ grid\ *n* [back-formation fr. *gridiron*] (1839) Channel-shaped, mold-supporting members.

Griffith Theory *n* The idea that, in brittle fracture, crack growth occurs, causing fracture, when the rate of decrease of stored elastic energy equals or exceeds the rate of creation of new fracture-surface energy. The Griffith equation states that, in thin sheets, tensile strength is given by: $\sigma = 2\gamma E/(\pi c)^{0.5}$, where $\gamma =$ the fracture-surface energy per unit area, $E =$ the elastic modulus, and $c =$ half the crack length.

Grind \ grīnd\ *v* [ME, fr. OE *grindan*; akin to L *frendere* to crush, grind] (before 12c) (1) To reduce the size of breakable particles by impact or by squeezing them between relatively moving surfaces. (2) To remove material from a part by contact with an abrasive, to change either the part's shape or its finish.

Grinding *n* (1) Process by which pigment particles are reduced in size, mechanically. (2) Its use to describe pigment dispersion in mill bases is deprecated.

Grinding Media Charge See Mill Charge.

Grinding Medium *n* Any medium which is used as a means of dispersing pigments or fillers by grinding or dispersing.

Grinding Slab *n* A flat piece, usually of glass or stone, on which color is ground from a coarse to a finely divided state, frequently with the medium that is to blend it as a paint.

Grinding-Type Resin *n* A vinyl resin that must be ground to effect dispersion in plastisols or organosols.

Grinning *n* (1) A flaw in fabric, especially a ribbed fabric, that occurs when warp threads show through the covering filling threads or when the threads have slipped leaving open spaces on either side. (2) A condition that occurs when the carpet backing shows through the pile. (3) A printing term referring to either poor cover where the background shade shows through the print, or to the "two-tone" appearance of a shade printed with incompatible dyes. (Printing ink manual, 5th edn. Leach RH, Pierce RJ, Hickman EP, Mackenzie MJ, SmithHG (eds). Blueprint, New York, 1993).

Grinning Through *n* Showing through of the underlying surface due to the inadequate opacity of a paint film which has been applied to it.

Gripper Loams Shuttleless looms. These looms employ a projectile with a jaw that grips the end of the filling yarn during the insertion of the pick.

Grisaille \gri- zī, - zā(ə)l\ *n* [F, fr. *gris* gray, fr. MF] (1848) Decorative painting in which objects are rendered in tones of one color, usually gray, often applied on glass, and intended to give the effect of sculptured relief. (Merriam-Webster's collegiate dictionary, 11th edn. Merriam-Webster, Springfield, MA, 2004).

Grit Blasting *n* A mold-finishing process in which abrasive particles are blown against mold surfaces in order to produce a controlled degree of roughness. The process is often used on molds for blow molding to assist air escape near the end of the blow, and on other types of molds to produce a desired texture in the product. (Rosato's plastics encyclopedia and dictionary. Rosato DV (ed). Hanser-Gardner, New York, 1992) See ▶ Blast Cleaning.

Grit Number See ▶ Mesh Number.

Gritty \ˈgri-tē\ *adj* (1598) In a coating, roughness resembling grains of sand. (Weismantal GF (1981) Paint handbook. McGraw-Hill, New York).

Grooved Barrel *n* In an extruder or injection molder, a barrel that has shallow grooves over the first few screw diameters from the feed opening. The groves may be parallel to the machine axis or helical with direction opposite that of the screw. Grooves have been shown to improve feeding of difficult-to-feed materials and to reduce energy consumption per unit of throughput, but can cause overfeeding and attendant problems with screws and materials for which the grooves and screw were not designed. (Weismantal GF (1981) Paint handbook. McGraw-Hill, New York).

Grosgrain \ˈgrō-ˌgrān\ *n* [F *gros grain* coarse texture] (1869) A heavy fabric with prominent ribs, grosgrain has a dressy appearance and is used in ribbons, vestments, and ceremonial cloths. (Fairchild's dictionary of textiles. Tortora PG (ed). Fairchild Books, New York, 1997).

Ground *n* (1) Surface applied to a canvas or other support, upon which a picture is painted. (2) Base on which organic agents are precipitated to form lakes. (3) Any surface which is or will be painted. (Gair A (1991) Artist's manual. Chronicle Books LLC, San Francisco, CA; Mayer R, Sheehan S (1991) Artist's handbook of materials and techniques. Viking Adult, New York).

Ground Color *n* A term describing the plain background color against which a design is created. (Mayer R, Sheehan S (1991) Artist's handbook of materials and techniques, Viking Adult, New York).

Groundnut Oil Another name for arachis or peanut oil. See ▶ Peanut Oil.

Ground Oyster Shells See ▶ Calcium Carbonate.

Ground Perlite See ▶ Perlite.

Ground Silica See ▶ Silica, ▶ Crystalline.

Ground State *n* (1926) The state of lowest energy of a particle.

Ground Wire *n* A wire attached between apparatus, containers and instruments and an electrical ground to dissipate an electric or electrostatic charge.

Group \ˈgrüp\ *n* {*often attributive*} [F *groupe*, fr. It *gruppo*, of Germanic origin; akin to OHGr *kropf* craw] (1686) A vertical column of elements in the periodic table, sometimes called a family of elements. (Whitten KW, Davis RE, Davis E, Peck LM, Stanley GG (2003) General chemistry. Brookes/Cole, New York).

Growth See Secondary Creep.

GRP *n* Abbreviation for "▶ Glass-Reinforced Plastic". See ▶ Reinforced Plastic.

Guaiacol *n* 2-Methoxyphenol, is a natural organic compound with the molecular formula C7H8O2. It is the monomethyl ether of catechol. It is a yellowish oil drived from guaiacum or wood creosote. It is used medicinally as an expectorant, antiseptic, and local anesthetic. In coating technology, it is used for anti-skinning properties. (www.wikipedia.org, Asphalt science and technology. Usmani AM (ed). Marcel Dekker, New York, 1997; Paint/coatings dictionary, Federation of Societies for Coatings Technology, Philadelphia, Blue Bell, PA, 1978)

Guanidine Aldehyde Resins *n* Reaction products of guanidine aldehydes.

Guar Gum \ˈgwär ˈgən\ *n* (1950) The ground endosperms of *Cyanopsis tetra gonolobus*. The free-flowing powder is soluble in cold or hot water and can be gelled with borax. It is used as a thickening agent or protective colloid. (Industrial gums: Polysaccharides and their derivatives. Whistler JN, BeMiller JN (eds). Elsevier Science and Technology Books, New York, 1992).

Guide Bar *n* A mechanism on a warp-knitting machine that directs warp threads to the latch needles.

Guide Coat *n* Coat similar to the finish or color coat but of a different color to assure good coverage.

Guide Eye *n* In filament winding, a moving metal or ceramic loop (eye) through which the fiber passes as it flows from creel to mandrel.

Guide Pin *n* (dowel pin) In compression, transfer, and injection molding, usually one of two or more hardened steel pins that maintain proper alignment of the mold halves a they open and close.

Guide-Pin Bushing *n* (dowel-pin bushing) In molding, a hardened bushing that receives and guides the leader pin, controlling alignment of the mold halves as the mold closes. (Strong AB (2000) Plastics materials and processing. Prentice Hall, Columbus, OH).

Guides Fittings of various shapes for controlling the path of a threadline. (Fairchild's dictionary of textiles. Tortora PG (ed). Fairchild Books, New York, 1997).

Guignet's Green See ▶ Chrome-Oxide Green and ▶ Hydrated Chromium Oxide.

Guillotine \ gi-lə-ˌtēn\ *n* [F, fr. Joseph *Guillotin* † 1814 F physician] (1793) Cutting device that consists of a single blade that descends between guides for chopping fibers, plastic strands, etc. (Fairchild's dictionary of textiles. Tortora PG (ed). Fairchild Books, New York, 1997).

Guimet's Blue *n* Synthetic ultramarine blue.

Gum \ gəm\ *n* [ME *gomme*, fr. MF, fr. L *cummi*, gummi, fr. Gk *kommi*, fr. Egyptian *qmyt*] (14c) (Adhesives) Any of a class of colloidal substances, exuded by or prepared from plants, sticky when moist, composed of complex carbohydrates and organic acids, which are soluble or swell in water. The term gum is sometimes used loosely to denote various materials that exhibit gummy characteristics under certain conditions, for example. Gum balata, gum benzoin, and gum asphaltum. Gums are included by some in the category of natural resins. (Handbook of adhesives. Skeist I (ed). Van Nostrand Reinhold, New York, 1990, 1977, 1962).

Gum *n* (Lithography) A water-soluble colloid such as gum arabic, cellulose gum, etc., used for coating a lithographic plate to make the non-image areas ink-repellent and to preserve the plate for future use. (Printing ink manual, 5th edn. Leach RH, Pierce RJ, Hickman EP, Mackenzie MJ, Smith HG (eds). Blueprint, New York, 1993).

Gum Acacia See ▶ Acacia Gum.

Gum Accroides See ▶ Accroides.

Gum Animi See ▶ Animi.

Gum Benzoin See Benzoin.

Gum Bloom *n* A defect in a painted surface, appearing as a lack of gloss or a haze resulting from the use of incorrect reducer.

Gum Karaya See ▶ Karaya Gum.

Gum Mastic *n* A natural resin used for picture varnish to protect oil paintings, water colors and the like.

Gumming *n* (1) Clogging of sandpaper by a coating. (2) Addition of a gum resin to a coating.

Gummy *n* British term for brush drag.

Gum, Natural *n* A carbohydrate high polymer that is insoluble in alcohol and other organic solvents, but generally soluble or dispersible in water. Natural gums are hydrophilic polysaccharides composed of monosaccharide units joined by glycosidic bonds. They occur as exudations from various trees and shrubs in tropical areas or as phycocolloids (algae), and differ from natural resins in both chemical composition and solubility properties. Some contain acidic components and others are neutral. Their chief use is as protective colloids and emulsifying agents in food products and pharmaceuticals; as sizing for textiles; and in electrolytic deposition of metals. Examples are: Arabic, tragacanth, guar, karaya. The word "gum", often used as an adjective, seems to acquire a different meaning from the noun. For example, the resinous products obtained from pine pitch (produced by the parenchyma cells of softwoods) are conventionally called "gum turpentine" and "gum rosin". There are also such "gum reins" as gum benzoin, gum camphor, and others. The so called ester gum is a semi synthetic reaction product of rosin and a polyhydric alcohol. All these are actually resinous products having properties quite different from hose of natural gums. Still further complicating the matter is the common application of the word "gum" to such plant lattices as chicle and natural rubber, which are different from both carbohydrate gums and resins. (Langenheim JH (2003) Plant resins: chemistry, evolution ecology and ethnobotany. Timber, Portland, OR; Industrial gums: Polysaccharides and their derivatives. Whistler JN, BeMiller JN (eds). Elsevier Science and Technology Books, New York, 1992).

Gum Rosin *n* (1712) Rosin which is produced from the gum turpentine obtained by tapping living pine trees, as distinguished from the wood rosin obtained from the stump and branches of dead trees.

Gum Rubber *n* (pure gum) Raw, unvulcanized rubber recovered from the latex of the *hevea* tree or from polymerization of synthetic rubber. Gum runner has almost no useful properties prior to being vulcanized. (Langenheim JH (2003) Plant resins: Chemistry, evolution and ethnobotany. Timber, Portland, OR; Industrial gums: Polysaccharides and their derivatives. Whistler JN, BeMiller JN (eds). Elsevier Science and Technology Books, New York, 1992).

Gums *n* Although rarely used in the plastics industry, this term is used in kindred industries to include materials that can be dissolved in water to produce viscous or mucilaginous solutions. Thus, water-soluble polymers such as polyvinyl pyrrolidone, polyvinyl alcohol, ethylene oxide resins, polyacrylic acid, and polyacrylamide are regarded as gums. (Langenheim JH (2003) Plant resins: Chemistry, evolution ecology and ethnobotany. Timber, Portland, OR; Industrial gums: Polysaccharides and their derivatives. Whistler JN, BeMiller JN (eds). Elsevier Science and Technology Books, New York, 1992).

Gum Sandarac \-ˈsan-də-ˌrak\ Soft resin obtained from *Tetraclinis articulate*, used in the manufacture of paints, linoleum, and oil cloth. *Also known as Juniper Gum.* (Langenheim JH (2003) Plant resins: Chemistry, evolution ecology and ethnobotany. Timber, Portland, OR; Industrial gums: Polysaccharides and their derivatives. Whistler JN, BeMiller JN (eds). Elsevier Science and Technology Books, New York, 1992).

Gum Spirits See ▶ Turpentine.

Gum Thus *n* Botanically, the oleoresin from trees of *Boswellia* species native to Arabia and Somaliland. As applied to the naval stores industry, the term refers to the crystallized pine oleoresin or "scarpe" collected from scarified "faces" of trees being worked for turpentine. *Also known as Olibanum and Frankioncense.* (Langenheim JH (2003) Plant resins: Chemistry, evolution ecology and ethnobotany. Timber, Portland, OR; Industrial gums: Polysaccharides and their derivatives. Whistler JN, BeMiller JN (eds). Elsevier Science and Technology Books, New York, 1992).

Gum Turpentine *n* (1926) See ▶ Turpentine.

Guncotton \-ˈkä-tᵊn\ *n* (1846) A highly inflammable and explosive form of ▶ Cellulose Nitrate made by digesting clean cotton in a mixture of one part nitric acid and three parts sulfuric acid, the latter acting as a catalyst and scavenger of the water generated in the reaction. As it ages in contact with air, dry guncotton becomes gradually more unstable and dangerous to handle. (Goldberg DE (2003) Fundamentals of chemistry. McGraw-Hill Science/Engineering/Math, New York).

Gunk \ˈgeŋk\ *n* [fr. *Gunk*, trademark for a cleaning solvent] (1943) A slang term for ▶ Premix.

Gunk Molding See ▶ Premix Molding.

GUP *n* Glass-fiber-reinforced polyester resins.

Gusset \ˈgə-sət\ *n* [ME, piece of armor covering the joints in a suit of armor, fr. MF *gousset*] (ca. 1570) A tuck in the side of a bag, usually made in symmetrical pairs in both paper and plastic-film bags, that permit the bag to assume nearly rectangular-boxy form when opened. Gussets may be formed in the tubular ▶ Blown Film from which bags are made just before the film enters the pinch rolls.

Gutta-Percha \ˌgə-tə-ˈpər-chə\ *n* [Malay *gètah-pèrcha*, fr. *gètah* sap, latex + *pèrcha* scrap, rag] (1845) (GP, PI, *trans*-1,4-polyisoprene) A rubber-related, polymeric substance extracted from the milky sap of leaves and bark of certain trees belonging to the family *Sapotaceae*, genera *Palaquium* and *Payena*, plants native to Malaysia. Its mer has the same empirical formula as natural rubber, its *cis* isomer. Gutta-percha is a tough, horny substance at room temperature, but becomes soft and tacky when warmed to 100°C. In the past it was used in compounds for golf-ball covers, electrical insulation, cutlery handles, and machinery belting, and as a stiffening agent in natural rubber, but has been replaced in many of these applications by synthetics. See also ▶ Balata. (Langenheim JH (2003) Plant resins: Chemistry, evolution ecology and ethnobotany. Timber, Portland, OR).

Gutta-Percha, Synthetic See ▶ Polyisoprene.

Gutter \ˈgə-tər\ *n* [ME *goter*, fr. MF *goutiere* fr. *goute* drop, fr. L *gutta*] (14c) A channel at the eaves for conveying away rain water. (Harris CM (2005) Dictionary of architecture and construction. McGraw-Hill, New York).

Gutzeit Test *n* Method for testing and estimating small quantities of arsenic and antimony in other materials. Details are given in the *British Pharmacopoeia*.

Gym Finish *n* Composition of oils and resins in a solvent; designed to produce a high gloss, tough surface finish on wood floors. (Weismantal GF (1981) Paint handbook. McGraw-Hill, New York).

Gynocardia Oil *n* A drying oil, smelling like linseed oil, and obtained from *Gynocardia odorata*, which grows in India and Pakistan. Its component acids include isolinolenic, linolenic, linoleic, oleic and palmitric acids. It has a sp gr of 0.925/25°C, iodine value of 153–160. (Paint: Pigment, drying oils, polymers, resins, naval stores, cellulosics esters, and ink vehicles, vol 3. American Society for Testing and Material, 2001).

Gypsum \ˈjip-səm\ *n* [L, fr. Gk *gypsos*] (14c) $CaSO_4 \cdot 2H_2O$. Natural crystalline calcium sulfate used as an extender in paint manufacture. See ▶ Calcium Sulfate. (Hibbard MJ (2001) Mineralogy. McGraw-Hill, New York; Solomon DH, Hawthorne DG (1991) Chemistry of pigments and fillers. Krieger, New York) Also known as *Mineral White*.

Gypsum, Calcined *n* $CaSO_4 \cdot \frac{1}{2}G_2O$. Gypsum partially dehydrated by heat. See ▶ Plaster of Paris.

Gypsum Cement See ▶ Plaster of Paris.

Gypsum Plaster See ▶ Cement Plaster.

Gypsum Wallboard *n* A sheet or slab having an incombustible core, essentially gypsum, surfaced with paper suitable to receive decoration. Also known as *Plasterboard*. (Harris CM (2005) Dictionary of architecture and construction. McGraw-Hill).

H

h *n* \ˈāch\ (1) SI abbreviation for Hour and for the prefix Hecto-. (2) Also H, the symbol for Channel Depth of an extruder screw.

H (1) The chemical symbol for the element hydrogen. (2) Abbreviation for the SI unit of magnetic inductance, the ▶ Henry. (3) Symbol for Enthalpy. (CRC handbook of chemistry and physics. Lide DR (ed). CRC, Boca Raton, FL, 2004 Version)

Habit \ˈha-bət\ *n* [ME, fr. OF, fr. L *habitus* condition, character, from *habēre* to have, hold] (13c) The general shape of a crystal. The same descriptive terms are used in each of the six crystal systems, e.g., rods, plates, tablets, needles, flakes, etc. (Hibbard MJ (2001) Mineralogy. McGraw-Hill, New York)

Hackberry Tree Seed Oil \ˈhak-ˌber-ē-\ Obtained from the kernels of the tree, *Celtis occidentalis*. Its main constituent fatty acid is linoleic acid, present to the extent of over 70%. It has a sp gr of 0.9204/25°C, refractive index of 1.4794/25°C, saponification value of 191, and an iodine value of 150.

Hackles \ˈha-kəl\ *n* [ME *hakell*; akin to OHGr *hāko* hook] (15c) Thin, needle-like or sliver-like protrusions (ranging from 3 to 6 μm) found on steel plates which have been blasted with steel or grit. (Anderson TL (2005) Fracture mechanics: Fundamentals and applications. CRC, Boca Raton, FL)

Haco Oil *n* Specifically processed whale oil, of German origin, with good drying properties.

Hagen-Poiseuille Equation *n* (Poiseuille equation) The equation of steady, laminar, Newtonian flow through circular tubes:

$$Q = \pi R^4 \Delta P/(8\eta L)$$

where Q = the volumetric flow rate, R and L are the tube radius and length, ΔP = the pressure drop (including any gravity head) in the direction of low, and η = the fluid viscosity. With the roles of Q and η interchanged, this is the basic equation of capillary viscometry. Any *consistent* system of units may be used. This important equation was first derived theoretically in 1839 by G. Hagen and, a year later, inferred from experimental measurements by J.L. Poiseuille. In a laminar flow through a circular tube, a simple force balance shows that the shear stress at the wall, τ_w, = $\Delta P\ R/(2\ L)$. By Newton's law of viscosity (see ▶ Viscosity), the shear rate at the wall, γ_w, must equal to the shear stress divided by the viscosity. Solving the above equation for $\Delta P\ R/(2\ \eta\ L)$, one obtains $\gamma_w = 4\ Q/(\pi\ R^3)$. By applying the Rabinowitsch Correction to this expression for Newtonian shear rate, one can get the true shear rate at the wall for a non-Newtonian liquid. An often seen, equivalent, but slightly different form for the Newtonian shear rate at a tube wall is 8 V/D, where V = the average fluid velocity = $Q/(\pi\ D^2/4)$ and $D = 2R$. (Goodwin JW, Goodwin J, Hughes RW (2000) Rheology for chemists. Royal Society of Chemistry, August; Patton TC (1979) Paint flow and pigment dispersion: A rheological approach to coating and ink technology. Wiley, New York; Parfitt GD (1969) Dispersion of powders in liquids. Elsevier, New York; Van Wazer, Lyons, Kim, Colwell (1963) Viscosity and flow measurement. Lyons Kim, Colwell (eds). Interscience, New York)

Hair-Cracking See ▶ Cracking.

Hairy See ▶ Fuzziness.

Half-Life *n* (1907) The period of time necessary for one-half of a reactant to be consumed in reaction or process. Used to measure the rate of radioactive decay of disintegration. The time lapse during which a radioactive mass losses one half of its radioactivity.

Half-Reaction An equation for an oxidation or a reduction which specifically indicates the electrons lost or gained.

Halftone \ˈhaf-ˌtōn\ *n* (1651) A printed image composed of dots of varying frequency (number per square inches), size or density, producing tonal gradations. The term is also applied to the process, "plates" and ink used to produce this image.

Halftone Screens *n* Transparent plates ruled diagonally with opaque lines at right angles to one another, the distance between them and their thickness being approximately equal.

Halide \ˈha-ˌlīd\ *n* (1876) Any compound including one (or more) of the halogen elements (F, Cl, Br, I) in its −1 valence state. The term halide is often used to indicate that any of the four principal halogens may be interchangeably present.

Hall Effect When a steady current is flowing in a steady magnetic field, electromotive forces are developed which are all right angles both to the magnetic force and to the current and are proportional to the product of the intensity of the current, the magnetic force and the sine of the angle between the directions of these quantities.

Halocarbon Plastic \ˈha-lə-ˌkär-bən-\ *n* (1950) A term listed by ASTM (D 883) to mean a polymer containing only carbon and one or more halogens. The primary members of the family are the Chlorofluorocarbon and Fluorocarbon Resins.

Halo Effect *n* (ca. 1928) Piling up of ink at the edges of printed letters and dots. Also colored or sometimes uncolored areas adjacent to them, caused by the spread of colored or uncolored vehicles. (Printing ink manual, 5th edn. Leach RH, Pierce RJ, Hickman EP, Mackenzie MJ, Smith HG (eds). Blueprint, New York, 1993)

Halogen \ˈha-lə-jən\ *n* [Sw, fr. *hal-* + *-gen*] (1842) The elements of group 7a of the periodic table; fluorine, chlorine, bromine, and iodine (F, Cl, Br, I). The fifth member, astatine, is rare, radioactive, and unstable, with a half-life less than 9 h, so it is never seen in commerce.

Halogenated Solvents *n* The solvents containing halogen (usually chlorine) have improved solvency compared with the hydrocarbons from which they are derived and in addition flammability is reduced. Some of these are highly toxic, and precautions must be taken to avoid inhalation of their vapors.

Halogenation *n* (1882) Process or reaction wherein any members of the halogen group are introduced into an organic compound, either by simple addition or substitution.

Halogen/Phosphorus Flame Retardant *n* Chlorine- and bromine-substituted esters of phosphoric acid or phosphonic acid, used mostly in polyurethane foams to impart fire resistance.

Hammer Finish \ˈha-mər-\ A paint finish which appears to have been applied over hammered metal; produced by the use of nonleafing metallic pigment, plus tinting pigments, which are mixed in a special binder.

Hammer Mill *n* (1610) Type of mill used for pulverizing dry materials in which the disintegration is caused by the action of a number of small hammers.

Hammett Equation \ˈha-met-\ An equation used in organic chemistry to predict reactions for aromatic compounds.

$$\log \frac{K}{K_o} = \rho\sigma \quad \text{or} \quad \log \frac{K}{K_o} = \rho\sigma$$

Where K or k refers to the reaction of a *m-* or *p-* substituted phenyl compound (e.g., ionization of a substituted benzoic acid) and K_o or k_o refers to the same reaction of the unsubstituted compound (e.g., ionization of benzoic acid), the substituent group constant ρ is a number (+ or −) indicating the relative electron withdrawing or electron-releasing effect of a particular substituent, the reaction constant σ is a number (+ or −) indicating the relative need of a particular reaction for electron withdrawal or electron release; a linear free energy relationship since it is based on the fact that a linear relationship exists between free energy change and the effect exerted by a substituent. (Smith MB, March J (2001) Advanced organic chemistry, 5th edn. Wiley, New York; Morrison RT, Boyd RN (1992) Organic chemistry, 6th edn. Prentice Hall, Englewood Cliffs, NJ)

Hand \ˈhand\ *n* {*often attributive*} ME, fr. OE, akin to OHGr *hant* hand] (before 12c) The feel of film, fabric, or coated fabric − its flexibility, smoothness, and softness − as judged by the touch of a person.

Hand-Blocked Print *n* A fabric that has been printed by hand with wooden or linoleum blocks. Also see ▶ Printing.

Hand-Blocking See ▶ Block Printing.

Hand Layup *n* Method of forming reinforced plastics articles comprising the steps of placing a web of the reinforcement, which may or may not be preimpregnated with a resin, in a mold or over a form and applying fluid resin to impregnate and/or coat the reinforcement, followed by curing of the resin and extraction of the cured article from the mold. See ▶ Layup Molding.

Handle See ▶ Hand.

Hand Mold *n* A mold that is removed from the press after each shot for extraction of the molded article; generally used only for short runs and experimental moldings.

Hand-Screening See ▶ Silk-Screening.

Hanger Wire *n* The wire employed to suspend the acoustical ceiling from the existing structure (wood joists, steel bar joists, steel beams, concrete slabs, etc. (Harris CM (2005) Dictionary of architecture and construction. McGraw-Hill, New York)

Hang Pick *n* A pick that is caught on a warp yarn knot for a short distance which produces a triangular hole in the fabric. Hang picks usually result from knots that are tied incorrectly, shuttle tension that is too loose, or harness that is timed too early.

Hang Shot See ▶ Hang Pick.

Hangup *n* (1) Stray bits of extrudate that cling to the face of the die and eventually have to be cleaned off. (2) Failure of feed material in a hopper to fall out of the hopper, caused sometimes by arching when the feed opening is only a few times larger than the pellet, but more often by clumping together of sticky particles.

Hank \ˈhaŋk\ *n* [ME, of Scand origin; akin to ON *hǫnk* hank; akin to OE *hangian* to hand] (14) A skein of yarn. (2) A standard length of slubbing, roving, or yarn. The length is specified by the yarn numbering system in use; e.g., cotton hanks have a length of 840 yards. (3) A term applied to slubbing or roving that indicates the yarn number (count); e.g., a 1.5 hank roving.

Hank Shellac Form of bleached shellac which, instead of being ground and dried, is pulled into hanks. Contains up to 25% moisture.

Hansa Yellow \ˈhan(t)sə-\ *n* Range of yellow pigment dyestuffs which are characterized by clean tones and good fastness to light in masstones but poor fastness to light in pale tints. Heat resistance is also poor and therefore these pigments should not be used in baking finishes. They are produced by coupling diazotized bases with acetoacetanilide or its substitution products.

Hanus Iodine Number *n* Method claimed to measure the complete unsaturation of both conjugated and nonconjugated substances. The method employs as a reagent a solution of iodine and bromine in glacial acetic acid.

Hardboard \ˈhärd-ˌbȯrd\ *n* (1925) A generic term for a smooth, grainless panel manufactured primarily from interfelted lignocellulosic fibers (usually wood), consolidated under heat and pressure in a hot-press to a density of 31 lb/ft³ (specific gravity 0.50) or greater, and to which other materials may have been added during manufacture to improve certain properties. In its most common form, it is smooth on one side and has a light waffle-like texture on the reverse side. Used for interior panels or durable siding. It is also available with a smooth surface on both sides, and with a variety of embossed or textured finishes, as well as a perforated form that accepts hooks and fixtures, for converting any wall into a storage area.

Hard Clay *n* Sedimentary rocks composed mainly of fine clay mineral material without natural plasticity or any compacted or indurated clay.

Hard Dry See ▶ Dry-Hard.

Hardener \ˈhärd-nər\ *n* (1611) Additive (crosslinking agent, resin or other modifier) used to promote or control the hardening or curing reaction of a coating, plastic, adhesive or resin system. (Odian GC (2004) Principles of polymerization. Wiley, New York) See ▶ Curing Agent, ▶ Catalyst, and ▶ Crosslinking. {G Härter m, F durcisseur m, S endurecedor m, I indutiore m}

Hard-Facing Alloy *n* Any of several metals applied by welding to the contact surfaces (screw-flight tips) to reduce the rate of wear. Colmonoys® are nickel-based, containing boron, chromium, iron, and silicon. Stellites® are cobalt-based, with chromium, and tungsten. They are usually applied into a machined helical groove prior to cutting the main channel, then finish-ground afterward.

Hard Ferrite See ▶ Ferrite.

Hard Fiber *n* Stiff, elongated fibers obtained from leaves or stems of plants. Coarse and stiff, they are used in matting and industrial products. (Elsevier's textile dictionary. Vincenti R (ed). Elsevier Science and Technology Books, New York, 1994)

Hard Gloss Paint *n* A high-gloss enamel, formulated with a hard-drying resin vehicle.

Hard Gums *n* Resins exuded from living plants centuries ago. Some of the hard gums are Amber, Zanzibar, Kauri, Pontianak, Manila, and Congo.

Hardinge Mill *n* Conical type of ball mill used for grinding rock and similar materials.

Hard Lac Resin *n* Resin resulting from shellac when treated by fractional solvent extraction to remove the soft resin component. Hard lac resin is claimed to give films which are harder, and more durable, than those of ordinary shellac. In addition, hard lac resin polymerizes more easily when heated and has better water resistance. (Weismantal GF (1981) Paint handbook. McGraw-Hill, New York)

Hardness \-nəs\ *n* (before 12c) (1) The resistance to local deformation. (2) Ability of a coating film, as distinct from its substrate, to resist cutting, indentation, or penetration by a hard object. An arbitrary scale of hardness is based upon ten selected minerals. For metals the diameter of the indentation made by a hardened steel sphere (Brinnell) or the height of rebound of a small drop hammer (Shore Scleroscope) serve to measure hardness. See ▶ Barcol Hardness, ▶ Brinell Hardness, ▶ Durometer, ▶ Indentation Hardness, ▶ Knoop Hardness Number, ▶ Knoop Microhard-Ness, ▶ Modulus of Elasticity, ▶ Rockwell Hardness, ▶ Scratch Hardness, and ▶ Vickers Hardness. {G Härte f, F dureté f, S dureza f, I durezza f}

Hardness Scale See ▶ Scale, ▶ Knoop Hardness Number, ▶ Pencil Hardness, and ▶ Rocker Hardness Tester.

Hard Putty *n* (1) Linseed oil putty hardened with litharge or basic lead carbonate. (2) A quick setting putty containing water and plaster of Paris.

Hard Rubber *n* The material obtained by heating a highly unsaturated diene rubber with a high percentage of sulfur. The first hard rubber was Ebonite, made from natural rubber, and black, as its name implies. Similar products have been made from several of the synthetics. The sulfur is mostly contained in 3- and 4-carbon rings with a few –C-S-C–crosslinks.

Hard Size *n* A condition found in areas of fabric where the warp contains an excessive quantity of sizing (Elsevier's textile dictionary. Vincenti R (ed). Elsevier Science and Technology Books, New York, 1994)

Hard Stopping See ▶ Hard Putty.

Hard-Surfacing See Hard-Facing Alloy and Case Hardening.

Hardware *n* (ca. 1515) The physical parts of mechanical devices and electronic control systems.

Hardwoods *n* (1568) Generally, the botanical group of trees that have broad leaves, in contrast to the conifers or softwoods. The term has no reference to the actual hardness of the wood. Angiosperms is the botanical name for hardwoods.

Harmonic Motion *n* (1867) Simple Harmonic Motion and Angular Harmonic Motion.

Harness \här-nəs\ *n* [ME *herneis* baggage, gear, fr. OF] (14c) A frame holding the heddles in position in the loom during weaving. (Elsevier's textile dictionary. Vincenti R (ed). Elsevier Science and Technology Books, New York, 1994)

Harness Chain *n* A mechanism used to control the vertical movements of the harness, or shaft, on a loom.

Harsh Fiber *n* Fiber that is rough or coarse to the touch, but not fused or bonded filaments. (Elsevier's textile dictionary. Vincenti R (ed). Elsevier Science and Technology Books, New York, 1994)

Hashab *n* Name applied to a fine pale grade of gum acacia.

Haul-Off *n* In sheet extrusion, the three-roll stand and cooling-conveyor assembly that polishes and cools the molten sheet emerging from the die, often extended in scope to include the rubber pull rolls and winder. The term has analogous meaning for other extruded products, such as pipe. See, too, Sheet Line.

Hazardous Substance *n* A substance which, by reason of being explosive, flammable, poisonous, corrosive, oxidizing, or otherwise harmful, is likely to cause death or injury. Element or compound which, when discharged in any quantity, presents an imminent and substantial danger to the public health or welfare.

Haze *n* Haze and luminous transmittance are tests to measure the light transmittancy of different types and grades of plastic materials; Haze is defined in ASTM D 1003 as the percentage of transmitted light passing through the specimen which deviates more than 2.5° from the incident beam by forward scattering; luminous transmittance is defined as the ratio of transmitted to incident light. The cloudy or turbid appearance of an otherwise transparent specimen caused by light scattered from within the specimen or from its surfaces. (www.astm.org, Paint and coating testing manual. Koleske JV (ed). American Society for Testing and Materials, 1995; Paint testing manual: Physical and chemical examination of paints, varnishes, lacquers, and colors – STP 500. American Society for Testing and Materials, 1973)

Hazen Color Scale *n* See ▶ Apha Color Scale.

Hazy (1) A relatively small amount of nonsettling, finely dispersed matter which is not visibly homogeneous with the mass of the liquid specified, even through the liquid is transparent and transmits most of the light incident upon it. (2) Displaying haze.

HDPE *n* Abbreviation for High-Density Polyethylene. See ▶ Polyethylene.

HDT See ▶ Heat Deflection Temperature.

Head *n* (1) In any extrusion operation, the delivery end of the extruder, usually fitted with a hinged gate that may contain a breaker plate and screen pack, to which the adapter and die are attached. (2) In blow molding, the entire apparatus by which the molten plastic is shaped into a tubular parison. This may include an adapter, a parison die, and a melt accumulator.

Head End *n* (1) The beginning of a new piece of fabric in the loom that bears appropriate identification. (2) A small sample of fabric that may be submitted to a customer for approval.

Headers *n* Double wood pieces supporting joists in a floor, or double wood members placed on edge over windows and doors to transfer the roof and floor weight to the studs.

Head Pressure *n* In extrusion, the pressure of the melt at the delivery, or head end of the screw, typically as signaled by a Pressure Transducer mounted in the barrel opposite the tip of the screw. Head pressure is one of the most important indicators of the state of any extrusion operation.

Head-to-Head Polymer *n* A configuration whereas the functional groups are on adjacent carbon atoms on a polymer chain. A polymer in which the monomeric units are alternately reversed produced from the monomer, $CH_2=CHR$. (Odian GC (2004) Principles of polymerization. Wiley, New York)

Head-to-Tail Polymer *n* A structure of the type

$$\sim\sim CH_2CHX - CHXCH_2 \sim\sim$$

in a vinyl or related type of polymer. During polymerization, it results from a growing chain $\sim\sim CH_2C^*HX$ adding on a monomer ($CH_2=CHX$) to give $-CH_2CHX-CHXCH_2^*$, the substituted carbon atom being designated the head. (Odian GC (2004) Principles of polymerization. Wiley, New York) {G Kopf-Schwang-Polymerisation f, F polymérisation têtebêche, polymérisation f, S polimerización cabeza a cola, polimerización f, I polimerizzazione testacoda, polimerizzazione f}

Heartwood \-wúd\ *n* (1810) The wood extending from the pith to the sapwood, the cells of which no longer participate in the life processes of the tree. Heartwood may be infiltrated with gums, resins and other materials which usually make it darker and more decay resistant that sapwood.

Heat \ˈhēt\ *n* [ME *hete*, fr. OE *hîte*, *hîtu*; akin to OE *hāt* hot] (before 12c) Energy which is transit from a hot to a cold object because of the temperature difference, to become warm or hot. Heat is a form of energy associated with the motion of atoms, molecules and other particles matter is composed of. It can be created by chemical reactions (such as burning) nuclear reactions (such as fusion reactions taking place inside the Sun) electromagnetic dissipation (as in electric stoves) or mechanical dissipation (such as friction). Heat can be transferred between objects by radiation, conduction and convection. Temperature is used to indicate the level of elementary movement associated with heat. Heat can only be transferred between objects, or areas within an object, that have different temperatures. By common knowledge, the term heat has been used in connection with the warmth, or hotness, of surrounding objects. The concept that warm objects "contain heat" is not uncommon. During its 350 year development, the science of thermodynamics had established a physical quantity named temperature to quantify the level of "warmth", whereas heat (also improperly called heat change) was defined as a transient form of energy that quantifies the spontaneous transfer of thermal energy due to a temperature difference or gradient. (Smith JM, Abbott MM, Abbot M, Van Ness HC, Van Ness HC (2004) Introduction to chemical engineering thermodynamics. McGraw-Hill, New York). The SI unit for heat is the joule; an alternative unit still in use in the United States and other countries is the British thermal unit. (CRC handbook of chemistry and physics. Lide DR (ed). CRC, Boca Raton, FL, 2004) The amount of heat exchanged by an object when its temperature varies by one degree is called heat capacity. Heat capacity is specific to each and every object. When referred to a quantity unit (such as mass or moles), the heat exchanged per degree is termed specific heat, and depends primarily on the composition and physical state (phase) of objects. Fuels generate predictable amounts of heat when burned; this heat is known as heating value and is expressed per unit of quantity. Upon transitioning from one phase to another, pure substances can exchange heat without their temperature suffering any change. The amount of heat exchanged during a phase change is known as latent heat and depends primarily on the substance and the initial and final phase. Heat is a process quantity, as opposed to being a state quantity, and is to thermal energy as work is to mechanical energy. Heat flows between regions that are not in thermal equilibrium with each other; it spontaneously flows from areas of high temperature to areas of low temperature. All objects (matter) have a certain amount of internal energy, a state quantity that is related to the random motion of their atoms or molecules. When two bodies of different temperature come into thermal contact, they will exchange internal energy until the temperature is equalized (that is, until they reach thermal equilibrium). The amount of energy transferred is the amount of heat exchanged. It is a common misconception to confuse heat with internal energy: heat is related to the change in internal energy and the work performed by the system. The term heat is used to describe the flow of energy, while the term internal energy is used to describe the energy itself. Understanding this difference is a necessary part of understanding the first law of thermodynamics. Infrared radiation is often linked to heat, since objects at room temperature or above will emit radiation mostly concentrated in the mid-infrared band (see ▶ black body) (See also Hudson RD Jr (1969) Infrared system engineering. Wiley, New York; and www.wikipedia.org)

Heat Bodied *n* Thickening or polymerizing a drying oil by heat processing.

Heat-Bodied Oil See ▶ Oil.

Heat Capacity *n* (specific heat) The amount of heat required to raise the temperature of a unit mass of a substance one degree. In the SI system, the unit of heat capacity is J/(kg K), but kJ/(kg K) or J/g K are often more convenient. Conversions from older units are: 1 cal/(g °C) = 1 Btu/(lb °F) = 4.186 J/(g K). Most neat resins have heat capacities (averaged from room temperature to about 100°C) between 0.92 J/(g K) for polychlorotrifluoroethylene and 2.9 for polyolefins. (The heat capacity of water, one of the highest of all materials, is 4.18 J/(g K) at room temperature.) A term loosely used as a synonymous with heat capacity but not truly so is ▶ Specific Heat.

Heat Cleaning *n* (1) A batch or continuous process in which sizing on glass fibers is vaporized off. (2) Cleaning residual polymer from small extruder screws, dies or other small parts by immersing them in fused salts such as sodium nitrate.

Heat Content *n* Syn: ▶ Enthalpy.

Heat-Convertible Resins *n* Thermosetting resins which on controlled heating become infusible and insoluble. This phenomenon is associated with crosslinking of the resin molecules.

Heat Deflection Temperature *n* The temperature at which a material specimen (standard bar) is deflected by a certain degree under specified load. At this temperature, a material achieves a specific modulus which is defined by the applied stress and the sample geometry. Also called heat distortion temperature, heat distortion point, heat deflection point, deflection temperature under load, DTUL, tensile heat distortion temperature, HDT. See also ▶ ISO 75.

Heat Distortion *n* The degree of deformation of a plastic material under load for a short time at elevated temperature.

Heat Distortion Point See ▶ Heat Deflection Temperature.

Heat-Distortion Point *n* The former name, now deprecated and fading from use, of ▶ Deflection Temperature.

Heat Distortion Temperature See ▶ Heat Deflection Temperature.

Heated-Tool Welding See ▶ Hot-Plate Welding.

Heat Effect *n* The heat in calories developed in a circuit by an electric current of *I* amperes flowing through a resistance of *R* ohms, with a difference of potential *E* volts for a time *t* seconds,

$$H = \frac{RI^2 t}{4.18} = \frac{EIt}{4.18}.$$

(Handbook of chemistry and physics, 52nd edn. Weast RC (ed). The Chemical Rubber, Boca Raton, FL)

Heat Equivalent, or Latent Heat, of Fusion *n* The quantity of heat necessary to change 1 g of solid to a liquid with no temperature change. Dimensions $[l^2 t^{-2}]$.

Heater Band *n* An electrical heating unit shaped to fit extruder barrels, adapters, die surfaces, etc, for maintaining high temperatures of the banded items and furnishing heat to the plastics within them. Nichrome wires within the bands are embedded in insulation such as magnesium oxide and their ends are welded to screw-type terminals at the end of the band. The ends of cylindrical bands are brought together by Monel bolts to tighten them snugly against the surface to be heated. Service life is greatly increased by operating them at 80% or less of rated wattage, the usual practice. See also ▶ Cast-in Heater.

Heat Forming See ▶ Thermoforming.

Heather Yarn **he-thər-\ *n* A term describing mottled or mélange-type yarns.

Heating Cylinder *n* (heating chamber) In elderly injection molders, that part of the machine in which the cold plastic pellets were heated to the molten condition before injection. Until the late 1950s, heating cylinders for injection machines were static devices equipped internally with *torpedoes* or *spreaders*, sometimes heated separately, that caused the charge to be distributed in a thin annulus with the cylinder. This hastened the heating but was still slow and could degrade the plastic. Today, all new commercial machines are screw-injection types, accomplishing the heating mainly by the frictional action of the screw. This heating and melting process is not only faster but produces a melt of more uniform temperature, less likely to have been scorched. (Strong AB (2000) Plastics materials and processing. Prentice Hall, Columbus, OH)

Heat Mark *n* An extremely shallow depression or groove in the surface of a plastic article, visible because of a sharply defined rim or a roughened surface. See also ▶ Sink Mark.

Heat of Combustion *n* The amount of heat evolved by the combustion of a unit mass, usually 1 g-molecular weight, of the substance (kJ/mol). The former unit of heat of combustion, deprecated now but still widely used, was the kcal/mol, equal to 4.186 kJ/mol. Heats of combustion, which are tabulated for many pure compounds, are an important segment of the database of physical chemistry and chemical engineering, for it is from sums and differences in heats of combustion that most heats of reaction are estimated. (Ready RG (1996) Thermodynamics. Pleum, New York)

Heat of Formation *n*, ΔH_f The change in enthalpy for a reaction in which one mole of a compound in formed from its uncombined elements (*enthalpy of formation*). (Ready RG (1996) Thermodynamics. Pleum, New York)

Heat of Fusion *n* (enthalpy of fusion) The quantity of heat needed to melt a unit mass of a solid at constant temperature (J/kg or J/mol). Because all meltable plastics consist of broad mixtures of homologous molecules having different molecular weights, and because melting points of homologs increase with molecular weight, it isn't possible to melt a plastic at a constant temperature. Instead, melting occurs over a range of temperature. With crystalline resins, such as polyethylene, the range may be relatively small, about 20°C, and it is really only for such resins that "heat of fusion" is meaningful and measurable. To estimate the heat of fusion, enthalpy (J/g) is measured beginning at a temperature well below that at which melting begins and carried on to temperatures well above that at which the plastic has completely melted. Plots (or equations) of enthalpy versus temperature for both ranges are extrapolated to the center of the melting range, the "melting point." The difference between the higher and lower enthalpies at that point is taken as the heat of fusion of the plastic. (Ready RG (1996) Thermodynamics. Pleum, New York)

Heat of Polymerization (enthalpy of polymerization) The difference between the enthalpy of one mole of monomer and the enthalpy of the products of the polymerization reaction. Addition polymerizations are exothermic, values ranging from about 35 to 100 kJ/mol, and removal of this heat is an important aspect of reactor design. The reported value of an enthalpy of polymerization may be referred either to the temperature at which the polymerization is usually carried out or to a standard temperature, such as 25°C or 18°C. (Ready RG (1996) Thermodynamics. Pleum, New York)

Heat Quantity n The cgs unit of heat is the *calorie*, the quantity of heat necessary to change the temperature of 1 g of water from 3.5°C to 4.5°C (called a small calorie). If the temperature change involved is from 14.5°C to 15.5°C, the unit is the normal calorie. The mean calorie is $\frac{1}{100}$ the quantity of heat necessary to raise 1 g of water from 0°C to 100°C. The large calorie is equal to 1,000 small calories. The British thermal unit is the heat required to raise the temperature of 1 lb of water at its maximum density, 1°F. It is equal to about 252 cal. Dimensions of energy [$m\ l^2 t^{-2}$]. (CRC handbook of chemistry and physics. Lide DR (ed). CRC, Boca Raton, FL, 2004 Version; Ready RG (1996) Thermodynamics. Plenum, New York)

Heat-Reactive Phenolic Resin n Phenol-formaldehyde resin which has been manufactured under alkaline conditions and which reacts with oils, rosin, or other unsaturated components during varnish making, through the free methylol groups in its structure along with further condensation of the methylol groups in the resin.

Heat Resistance n A property of certain fibers or yarns whereby they resist degradation at high temperature. Heat resistance may be an inherent property of the fiber-forming polymer or it may be imparted by additives or treatment during manufacture. Also see ▶ Heat Stabilized.

Heat-Resistant Finishes Finishes designed to resist deterioration on continuous or intermittent exposure to a predetermined elevated temperature.

Heat Seal n A method of uniting two or more surfaces by fusion, either of the coatings or the base materials, under controlled conditions of temperature, pressure, and dwell time.

Heat Sealing n The process of joining two or more thermoplastic films or sheets by heating areas in contact with each other to the temperature at which fusion occurs, usually aided by pressure. When the heat is applied by dies or rotating wheels maintained at a constant temperature, the process is called *thermal sealing*. In *melt-bead sealing*, a narrow strand of molten polymer is extruded along one surface, trailed by a wheel that presses the two surfaces together. In *impulse sealing*, heat is applied by resistance elements that are applied to the work when relatively cool, then are rapidly heated. Simultaneous sealing and cutting can be performed in this way. *Dielectric sealing* is accomplished with polar materials by inducing heat within the films by means of radio-frequency waves. When heating is performed with ultrasonic vibrations, the process is called *ultrasonic sealing*. See also ▶ Welding.

Heat-Seal Strength n With heat-sealed flexible films, the force required to pull apart a heat-sealed joint divided by the joined area tested. The strength of a heat seal is sometimes expressed as a percentage of the film's tear strength or tensile strength.

Heat-Sensitive Paints n Coatings which change color at a specific temperature.

Heat Sensitivity n The tendency of a plastic to undergo changes in properties, color, or even to degrade at elevated temperatures. Severity of change is always a matter of both temperature and time. ASTM D 794 describes the procedures to be used in determining the permanent effects on plastics of elevated-temperature exposure.

Heat Set Inks n Letterpress and lithographic inks which dry under the action of heat by evaporation of their high boiling solvent.

Heat-Shrinkable Film n A film that is stretched and oriented while it is being cooled so that later, when used in packaging, it will, upon being rewarmed, shrink tightly around the package contents. Blown film made from plasticized PVC is the largest-volume shrink film. Heat-shrinkable tubing of several polymers if widely used in the electronics industry to protect bundles of wiring. ASTM D 2671 (Section 10.01) describes a method for testing such tubing.

Heat Sink n (1936) A device for the absorption or transfer of heat away from a critical part or assembly.

Heat Stability n The resistance to change in color or other properties as a result of heat encountered by a plastic compound or article either during processing or in service. Such resistance may be enhanced by the incorporation of a stabilizer.

Heat Stability Test *n* An accelerated test used to predict viscosity stability of coatings with time. The test generally involves measuring viscosity change after the paint has been heat-aged about a week to 10 days at 50–60°C. An excessive increase in viscosity is unacceptable.

Heat Stabilizers *n* A stabilizer additive to reduce or eliminate thermal degradation or its effects. Most commonly used with chlorinated polymers which are particularly thermally sensitive due to their tendency to dehydrochlorinate.

Heat Test *n* Test employed for assessing the purity of tung and similar rapidly polymerizing oils. The test involves heating the oil at a specified temperature and noting the time required for gelation to occur. Syn: ▶ Browne Heat Test. See ▶ Gel Time.

Heat Transfer *n* The movement of energy as heat from a hotter body to a cooler body. The three basic mechanisms of heat transfer are radiation, conduction, and convection. Radiation heating occurs when heat passes from the emitting body to the receiving body through a medium, such as air, that is not warmed. Conduction heating is the flow of heat from a hot region to a cooler one in either single homogeneous substances or two substances in close contact with each other. Convection is the transfer of heat by flow of a fluid, either a gas or a liquid, and either by natural currents caused by differences in density or by forced movement caused by a fan, pump, or stirrer. All three modes of transfer are important in plastics processing. {*heat transfer coefficient* G Härte f, F dureté f, S dureza f, I durezza f}

Heat-Transfer Medium See ▶ Thermal Fluid.

Heat Transfer Printing *n* A method, whereby a printed image is transferred from a carrier to a receiving substrate by the use of heat. In the process, as currently performed, the ink is made up of sublimable dyes in conventional ink vehicles, the carrier is paper, and the receiving substrate is a synthetic fabric. *Also known as Thermal Printing.*

Heavy-Bodied *n* Having a thick consistency or high viscosity.

Heavy Bodied Inks *n* Inks of a high viscosity or stiff consistency.

Heavy-Centered Pattern *n* Spray pattern having most paint in center, less at edges.

Heavy End *n* (1) The higher boiling fraction in distillation. (2) See ▶ Coarse Thread.

Heavy Filling See ▶ Coarse Thread.

Heavy Metals *n* (1974) Metallic elements with high molecular weights generally toxic to plant and animal life. Examples: mercury, chromium, arsenic, lead, etc.

Heavy Pick See ▶ Coarse Thread.

Heavy Solvent Naphtha *n* High-boiling coal-tar naphtha generally distilling below 190°C (90/190°C is a usual grade).

Heavy Spar See ▶ Barium Sulfate.

Hecto- *n* {*combining form*} [F, irreg. fr. Gk *hekaton*] (h) The SI prefix meaning × 100.

Heddle \ˈhe-dəl\ *n* [prob. alt. of ME *helde*, fr. OE *hefeld*; akin to ON *hafald* heddle, OE *hebban* to lift] (1513) A cord, round steel wire, or thin flat steel strip with a loop or eye near the center through which one or more warp threads pass on the loom so that the thread movement may be controlled in weaving. The heddles are held at both ends by the harness frame. They control the weave pattern and shed as the harnesses are raised and lowered during weaving.

Hegman Gauge See ▶ Fineness of Grind Gauge.

Hehner Number (value) A number expressing the percentage (i.e., g/100 g) of water-insoluble fatty acids in an oil or fat.

Heisenberg Theory of Atomic Structure \ˈhī-zən-bərg-\ *n* The currently accepted view of the structure of atoms, formulated by Heisenberg in 1934, according to which the atomic nuclei are built of nucleons, which may be protons or neutrons, while the extranuclear shells consist of electrons only. If m and m' are given in grams, and r in centimeters, F will be in dynes if $k = 6.670 \times 10^{-8}$.

$$F = k\frac{mm'}{r^2}$$

(Goldberg DE (2003) Fundamentals of chemistry. McGraw-Hill Science/Engineering/Math, New York)

Heisenberg Uncertainty Principle *n* [Gr. Werner *Heisenberg*] (1939) The product of the uncertainty in the position of a particle times that in its momentum is a constant. It is impossible to determine simultaneously both the position and the momentum of a particle exactly. (Goldberg DE (2003) Fundamentals of chemistry. McGraw-Hill Science/Engineering/Math, New York)

Helical Screw Feeder \ˈhe-li-kəl-\ *n* See ▶ Crammer-Feeder, ▶ Screw Conveyor.

Helical Transition *n* In the transition zone (compression zone) of an extruder screw, the root surface of the screw describing an advancing spiral surface of increasing radius. See also ▶ Conical Transition.

Helical Winding *n* A winding in which the filament or band advances along a spiral path, not necessarily at a constant angle except in the case of a cylindrical article.

Helio Fast Rubine 4BL Lake *n* (58055) This pigment is marketed as an alumina hydrate lake; the parent toner is hard and horny when dried and, consequently, two difficult to disperse, It is black in masstone but yields a purpose tint on reduction with white. Its good lightfastness in dark shades, high color intensity, good chemical and bleed resistance make Helio Fast Rubine 4BL lake particularly suitable for shading full shade durable maroon finishes. On the other hand, its poor lightfastness in very weak pastel shades plus its poor bake resistance at high temperatures make it unsuitable as a white toning agent for high bake exterior white finishes.

Helio-Klischograph *n* A method of engraving gravure cylinders by the use of an electronic scanning system which transmits a signal modulated by the density of a positive copy to a diamond cutting head effecting the engraving mechanically rather than by chemical etching. (Printing ink manual, 5th edn. Leach RH, Pierce RJ, Hickman EP, Mackenzie MJ, Smith HG (eds). Blueprint, New York, 1993)

Helix Angle \hē-liks-\ *n* (1) Of an extruder screw, the lead angle of the flight with respect to the screw axis, the angle whose tangent = $t/(\pi D)$, where t = the axial distance the flight advances per turn (lead) and D = the major (or nominal) screw diameter. (2) In filament winding, the angle between the axis of rotation and the filament at its point of contact with the winding.

Hellige System or Comparator *n* Standard instrument for color determination of liquids such as varnishes, oils, etc., based on comparisons which are made against a series of discs of colored glass, and which are carefully graded.

Hematite \hē-mə-ˌtīt\ *n* (1540) Fe_2O_3. Natural oxide of iron. See ▶ Ferric Oxide.

$$O^{--}$$
$$O^{--} \quad Fe^{+++} \quad O^{--}$$
$$Fe^{+++}$$

Hemiacetal \he-mē-ˈa-sə-ˌtal\ *n* (1893) Any class of compounds characterized by the grouping C(OH)(OR) where R is an alkyl group and usually formed as intermediates in the preparation of acetals from aldehydes or ketones. (Morrison RT, Boyd RN (1992) Organic chemistry, 6th edn. Prentice Hall, Englewood Cliffs, NJ)

Hemiacetal Group *n* Functional groups derived from carbonyl groups by addition of one molecule of an alcohol, of the general structure shown: -C-OH-OR-. (Morrison RT, Boyd RN (1992) Organic chemistry, 6th edn. Prentice Hall, Englewood Cliffs, NJ)

Hemicelluloses \ˌhe-mi-ˈsel-yə-ˌlōs\ *n* [ISV] (1891) The principal noncellulosic polysaccharides in wood. Wood contains 28–35% hemicelluloses, the balance being cellulose and lignin. (Morrison RT, Boyd RN (1992) Organic chemistry, 6th edn. Prentice Hall, Englewood Cliffs, NJ)

Hemicolloid \-ˈkä-ˌlóid\ *n* Colloidal particle of specified size, both as regards number of molecules in association, and length of chain. This includes substances involving between 20 and 100 molecules, and possessing an overall chain length not exceeding 25 nm.

Hemihydrate \-ˈhī-ˌdrāt\ *n* (ca. 1901) A hydrate which contains one-half molecule of water to one molecule of the compound; the most common such material is partially dehydrated gypsum (plaster of Paris).

Hemimorphic \-ˈmór-fik\ *adj* [ISV] (ca. 1959) A crystal with unlike faces at opposite ends of the same axis.

Hemipolymer *n* A readily soluble polymer with molecular weights between 1,000 and 10,000.

Hemisymmetric \-sə-ˈme-trik\ *n* A crystal on which one-half the faces of at least one form are missing.

Hemp \ˈhemp\ *n* [ME, fr. OE *hinep*; akin to OHGr *hanaf* hemp, Gk *kannabis*] (before 12c) A coarse, durable bast fiber of *Cannabis sativa* found all over the world. Used primarily for twines, cordage, halyards, and tarred riggings.

Hempseed Oil *n* A semidrying oil obtained from the seeds of *Cannabis sativa*, which grows in India, Manchuria, China, Japan, and some parts of Europe. It is usually classed with soybean, poppyseed, sunflower, and walnut oils, and has an iodine value of approximately 160, arising in the main from the presence of practically 50% of linoleic acid and 25% of linolenic acid. It has some application in manufacture of paints and varnishes.

Henderson-Hasselbalch Equation *n* A formula relating the pH value of a solution to the pK_a value of the acid in the solution and the ratio of the acid and the conjugate base concentrations: $pH = pK_a + \log([A-]/[HA])$, where [A−] is the concentration of the conjugate base and [HA] is the concentration of the protonated acid. For the bicarbonate buffer system in blood,

$$pH = pK + \log([HCO_3-]/[CO_2])$$

(Whitten KW, Davis RE, Davis E, Peck LM, Stanley GG (2003) General chemistry. Brookes/Cole, New York)

Henry \ˈhen-rē\ *n* [for Joseph *Henry*] (H) The SI unit of electric inductance, the inductance of a closed circuit in which an electromotive force of 1 V is produced when the electric current in the circuit varies uniformly at a rate of 1 A/s. The magnetic flux in this situation will be just 1 weber (Wb). Thus, 1 H = 1 V s/A = 1 Wb/A.

(Handbook of chemistry and physics, 52nd edn. Weast RC (ed). The Chemical Rubber, Boca Raton, FL)

Henry's Law *n* [for W. *Henry*] The mass of a slightly soluble gas that dissolves in a definite mass of a liquid at a given temperature is directly proportional to the partial pressure of that gas. This law holds only for gases that do not chemically react with the solvent. (Watson P (1997) Physical chemistry. Wiley, New York; Handbook of chemistry and physics, 52nd edn. Weast RC (ed). The Chemical Rubber, Boca Raton, FL)

Heptanol-2 *n* Another name for methyl amyl carbinol. It has a bp of 160.4°C, vp of 0.9 mm Hg/20°C, and a sp gr of 0.8187/20°C.

Heptyl Acetate *n* Bp, 191°C; sp gr, 0.874.

***n*-Heptyl *n*-Decyl Phthalate** *n* $C_7H_{15}OOCCF_6H_4COOC_{10}H_{21}$. A general purpose plasticizer for PVC and several other thermoplastics, with lower volatility and better low-temperature performance than ▶ Dioctyl Phthalate. It is excellent for use in vinyl plastisols.

***n*-Heptyl *n*-Dinonyl Trimellitate** *n* An aromatic plasticizer similar to Trioctyl Trimellitate with better low-temperature performance and lower volatility.

***n*-Heptyl *n*-Nonyl Adipate** *n* A plasticizer for PVC similar to Dioctyl Adipate but with better low-temperature performance and lower volatility. It is also approved for use in contact with foods.

Herbicide \ ▪(h)ər-bə- ▪sīd\ *n* [L *herba* + ISV *–cide*] (1899) An agent used to destroy or inhibit plant growth.

Herringbone \ ▪her-iŋ- ▪bōn\ *n* {*often attributive*} (1659) A broken twill weave characterized by a balanced zigzag effect produced by having the rib run first to the right and then to the left for an equal number of threads.

HERTZ \ ▪hərts\ *n* [Heinrich R. *Hertz*] (ca. 1928) (Hz) Cycles per second of any periodic phenomenon. The term and its multiple, megahertz (MHz), were in the past used most frequently for waves in the radio-frequency range, but the hertz is now the SI unit of frequency.

Hessian \ ▪he-shən\ *n* {*chiefly British*} (1729) A name for burlap used in the United Kingdom, India, and parts or Europe. Also see ▶ Burlap.

Hess's Law of Constant Heat Summation *n* The amount of heat generated by a chemical reaction is the same whether the reaction takes place in one step or in several steps, or all chemical reactions which start with the same original substances and end with the same final substances liberate the same amounts of heat, irrespective of the process by which the final state is reached. (Ready RG (1996) Thermodynamics. Pleum, New York)

Heterochain Polymers *n* Polymers which contain more than one type of atom in the polymer chain.

Heterocyclic Compound *n* A compound whose molecule includes at least one ring that contains one or more elements other than carbon and hydrogen. A simple example is *pyridine*, C_5H_5N, a benzene molecule in which one carbon is replaced by nitrogen.

Heterocyclic Monomers *n* Monomer whose molecule includes at least one ring that contains one or more elements other than carbon and hydrogen. A simple example is pyridine, C_5H_5N, a benzene molecule in which one carbon is replaced by nitrogen.

Heterofillament Also called *Heterofil.* See ▶ Bicomponent Fibers.

Heterogeneous \ ▪he-tə-rə- ▪jē-nē-əs\ *adj* [ML *heterogenus*, fr. Gk *heterogenēs*, fr. *heter-* + *genos* kind] (1630) Composed of two or more phases.

Heteropolymerization *n* A special case of addition polymerization involving the combination of two dissimilar unsaturated monomers, the product being a *heteropolymer*.

Hevea Rubber See ▶ Rubber.

Hexabromide Value \-▮brō-▮mīd ▮val(▮)yü\ *n* Hexabromide value is a measure of the amount of acids of the linolenic types which are present in a mixture of natural fatty acids. Bromine, under suitable conditions, forms addition compounds through addition at the unsaturated linkages. The hexabromides are separated from the dibromides and tetrabromides by reason of the insolubility of the former in cold either. The hexabromide value is the percentage of hexabromide obtainable from a given material.

Hexabromobiphenyl \-▮brō-(▮)mō-(▮)bī-▮fe-n°l\ *n* [2,4,6-$(Br)_3C_6H_2$-]$_2$. A flame retardant suitable for use in thermosetting resins and thermoplastics such as acrylonitrile-butadiene-styrene resin, nylons, polycarbonate, polyolefins, PVC, polyphenylene oxide, and polystyrene-acrylonitrile. It is insoluble in water, heat-stable, and furnishes a high bromine content in the end product.

Hexachloroethane *n* (carbon hexachloride, perchloroethane) A substitute for camphor in celluloid manufacture.

Hexachlorophene *n* $(C_6HCl_3OH)_2CH_2$. A white, essentially odorless, free-flowing powder widely used as a bacteriostat in many thermoplastics including vinyls, polyolefins, acrylics and polystyrene.

Hexafluoroacetone Trihydrate *n* A solvent cement, active at room temperatures, for bonding acetal resin articles to themselves and to other polymers such as nylon, acrylonitrile-butadiene-styrene, styrene-acrylonitrile, polyester, cellulosics, and natural or synthetic rubber. It is also a toxic irritant, so it must be handled with care.

Hexagonal \hek-▮sa-gə-n°l\ *adj* (1571) (1) Of a closed plane figure, having six sides. (2) Having the properties of a *regular* hexagon, i.e., all six sides and all six angles equal. (3) One of the six crystal systems, in which there are four principal axes, three in one plane at 120° to each other, the fourth perpendicular to the others. Atomic spacing is equal along the planar axes, different along the fourth. (Hibbard MJ (2001) Mineralogy. McGraw-Hill, New York)

Hexahydrophenol *n* Syn: ▶ Cyclohexanol.

Hexahydrophthalic Anhyride (HHPPA) $C_6H_{10}(CO)_2O$. A curing agent for epoxy resins and an intermediate for the production of alkyd resins.

Hexalin See ▶ Cyclohexanol.

Hexalin Acetate See Cyclohexyl Acetate.

Hexamethylene \▮hek-sə-▮me-thə-▮lēn\ Syn: ▶ Cyclohexane.

Hexamethylene Adipamide (nylon 6/6) A nylon made by condensing hexamethylene diamine with adipic acid. See ▶ Nylon 6/6.

Hexamethylene Diamine \-▮dī-ə-▮mēn\ (1,6-diaminohexane) A colorless solid in leaflet form, which, when condensed with adipic acid, forms nylon 6/6, It has also been used to cure epoxy resins, especially for coatings and usually in a modified form, for example as an epoxy resin adduct.

Hexamethylene-1,6-Diisocyanate (HDI) A colorless liquid, the first aliphatic diisocyanates to be used commercially in the production of urethanes. When used with certain metal catalysts, it produces urethane polymers with good resistance to discoloration, hydrolysis, and thermal degradation.

Hexamethylenetetramine (Hexamine) \-ˈte-trə-ˌmēn\ n [ISV hexa- + methylene + tetra- + amine] (1888) (HMT, methenamine, urotropine) $(CH_2)_6-N_4$. A bicyclic compound, the reaction product of ammonia and formaldehyde. It is used as a basic catalyst and accelerator for phenolic and urea resins, and a solid, catalytic-type curing agent (hardener) for epoxies.

Hexamethylphosphoric Triamide $[N(CH_3)_2]_3PO$. A pale, water-soluble liquid used as an ultraviolet absorber in PVC compounds.

Hexane \ˈhek-ˌsān\ n [ISV] (1877) C_6H_{14}. A straight-chain hydrocarbon, extracted from petroleum or natural gas. Commercial grades contain other hydrocarbons such as cyclohexane, methyl cyclopentane, and benzene. Hexanes are used as catalyst-carrying solvents in the polymerization of olefins and elastomers. Sp gr, 0.660; bp, 69°C; solidification range −95° to −100°C; flp, −22°C (−70°F).

Hexanedioic Acid Syn: ▶ Adipic Acid.

Hexanol $CH_3(CH_2)_4CH_2OH$. High-boiling solvent (144–156°C) with a vp of 4 mmHg at 30°C; sp gr, 0.8186.

Hexone See ▶ Methyl Isobutyl Ketone.

Hexyl- \ˈhek-səl\ n [ISV] (1869) The straight-chain radical of hexane, C_6H_{13}–.

Hexyl Acetate $CH_3COO(CH_2)_5·CH_3$. Cellulose nitrate, cellulose acetate-butyrate, polyvinyl acetate, polystyrene, phenolics, alkyds, and coumarone-indene resins. It has a bp of 169°C, sp gr, of 0.890, and vp of 5 mmHg/20°C.

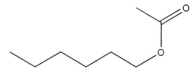

Hexyl Ether High boiling solvent. It has a bp of 226°C, sp gr of 0.794/20°C, vp of 0.07 mmHg/20°C, and flp of 77°C (170°F).

n-**Hexylethyl** *n*-**Decyl Phthalate** (NODP) See ▶ *n*-Octyl *n*-Decyl Phthalate.

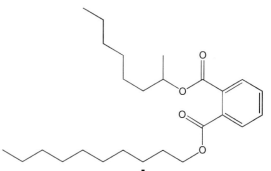

Hexyl Methacrylate \-ˌme-ˈtha-krə-ˌlāt\ n A monomer used in producing acrylic resins.

HF n Abbreviation for High-Frequency.
HF Preheating See ▶ Dielectric Heating.

Hickies \ˈhi-kēs\ *n* (ca. 1915) Defects in a print appearing as specks surrounded by an unprinted "halo".

Hide Finishes *n* Designed for use on leather. They are based on pigmented nitrocellulose, and completely cover the surface.

Hiding, Complete *n* The state of uniform application of paint at such a thickness that, when dry, the substrate is no longer affecting the color, i.e., thickness such that increasing the thickness even slightly does change the *perceived* color. Complete hiding is frequently judged by application of a uniform film over a white and black substrate such as a Morest Chart. It is sometimes defined as the state when a contrast ratio of 0.98 is obtained (although this is seldom complete hiding as judged visually). Visual complete hiding is not necessarily spectrophotometric complete hiding. (Paint and coating testing manual. Koleske JV (ed). American Society for Testing and Materials, 1995) See ▶ Contrast Ratio.

Hiding, Incomplete *n* The state of application of a coating at such a thickness that the color of the substrate has an effect on the color which the coating would exhibit if it were applied at complete hiding. Increasing the thickness of the coating thus changes the perceived color. The relative degree of incomplete hiding is frequently described by the contrast ratio. (Paint and coating testing manual. Koleske JV (ed). American Society for Testing and Materials, 1995) See ▶ Contrast Ratio and ▶ Complete.

Hiding Power *n* The ability of a paint, ink, etc., to hide or obscure a surface over which it has been applied uniformly. When expressed numerically, it is generally in terms of the number of square feet over which a gallon of paint, or pound of pigment, as used, can be uniformly spread to produce a specified contrast ratio. Hiding power is usually expressed in square feet per gallon or square meters per liter. (Paint and coating testing manual. Koleske JV (ed). American Society for Testing and Materials, 1995) See ▶ Contrast Ratio.

Hiding Power Chart *n* (1) A paper chart, partially black and partially white. A paint film applied over such a chart gives a visual evaluation of the degree of hiding. If a measurement of the reflectance of the film over the black portion is divided by the measurement of the reflectance over the white portion, the contrast ratio (opacity) is obtained. (2) A chart prepared calculation of the Kubelka-Munk equation for a white substrate of specified reflectance for various reflectances at complete hiding ($R\infty$) and for various reflectances over black, various contrast ratios, scattering-thickness values (SX). From such a chart, using measurements of reflectances over black and white, or measurements over black and of $R\infty$, changes in thickness, pigment volume concentration, $R\infty$ for obtaining a desired contrast ratio (degree of hiding) or $R\infty$, an be determined. Such charts are frequently called Judd Charts. (Paint and coating testing manual. Koleske JV (ed). American Society for Testing and Materials, 1995)

Hiding, Spectrophotometric, Complete *n* The state of uniform application of a paint film which, when dry, exhibits identical spectrophotometric curves over a perfect white of 100% reflectance and a perfect black of 0% reflectance. (Paint and coating testing manual. Koleske JV (ed). American Society for Testing and Materials, 1995) In the Kulblka-Munk equations, spectrophotometric hiding is defined as $R\infty$ at all wavelengths.

Hiding, Spectrophotometric, Incomplete *n* The state of a uniform paint film in which the spectrophotometric curves measured over white and black areas are not identical; addition of more paint is required to achieve complete spectrophotometric hiding or identity of reflectance curves of the film over white and black substrates. Films which appear visually to be at complete hiding are not necessarily at complete spectrophotometric hiding.

Hi-Flash Naphtha See ▶ Naphthas.

High-Boiling Solvent *n* A solvent with an initial boiling point above 150°C (302°F).

High Build *n* Proceeding thick, dry films per coat. See ▶ Build.

High-Build Coating *n* A coating composed of a series of uniform, tile-like films which are applied in thicknesses (minimum 5 mils) greater than those normally associated with paint films and thinner than those normally applied with a trowel. See also ▶ Tile-Like Coating.

High-Bulk Yarn See ▶ Textured Yarns.

High Density *n* (1960) A term to describe a material with heavier than normal weight per unit volume. Also see ▶ Density.

High-Density Overlay *n* An overlay consisting of paper that is impregnated with a thermosetting resin and then applied to plywood; provides a smooth, hard, wear-resistant surface for high-quality concrete formwork and decking.

High-Density Polyethylene *n* (HDPE) This term is generally considered to include polyethylenes ranging in density from about 0.94 to 0.96 g/cm^3 and higher. Whereas the molecules in low-density polyethylenes are branched and linked in random fashion, those in the higher-density polyethylenes are linked in longer chains with fewer side branches, resulting in higher-modulus materials with greater strength, hardness, and chemical resistance, and higher softening temperatures. See also ▶ Polyethylene.

Higher Fatty Acid See ▶ Fats.

High-Frequency *n* (1892) Pertaining to the part of the electromagnetic spectrum between 3 and 200 MHz, employed in plastics welding, sealing and preheating operations. Frequencies of 30 MHz and below are the most used.

High-Frequency Heating *n* Syn: ▶ Dielectric Heating.

High-Frequency Welding *n* A method of welding thermoplastic articles in which the surfaces to be joined are heated by contact with electrodes of a high-frequency electrical generator.

High Gloss See ▶ Gloss. *Also called Full Gloss.*

High-Impact Polystyrene *n* (HIPS) Polystyrene whose impact strength has been elevated by the incorporation of rubber particles. The best grades, produced by polymerization of styrene containing dissolved rubber, have better impact resistance than those made by missing rubber and polystyrene together. Whereas crystal PS has notched-Izod impact between 0.13 and 0.24 J/cm, HIPS resins range between 0.37 to 2.1 J/cm.

High-Intensity Mixer *n* A type of mixer consisting of a large bowl with a high-speed rotor in its bottom, used for producing dry blends of PVC with liquid and powder additives, also for producing other powder blends. Though there is considerable heating of the charge during mixing, the main constituent is not melted and the mix is discharged in powder form. Compare with Internal Mixer.

Highlight \\ˈhī-ˌlīt\\ *n* (ca. 1889) The light or open areas of a halftone print.

Highlighting *n* In antiquing, accenting raised areas on furniture with a lighter tone than the crevices. Produced by rubbing harder. Emphasizing the impression of relief by making certain parts of a finished surface lighter than the general color of that surface.

High-Load Melt Index *n* (1) The rate of flow of a molten resin through an orifice 2.096 mm in diameter and 8.000 mm long at 190°C when subjected to a pressure difference of 2.982 MPa. Also known as the *flow-behavior index*. See ▶ Power Law (Shah V (1998) Handbook of plastics testing technology. Wiley, New York)

Highloft *n* General term for a fiber structure containing more air than fiber. Specifically, a lofty, low-density nonwoven structure that is used for applications such as fiberfill, insulation, health care, personal protection, and cleaning material.

High-Low Profile *n* A pile construction characterized by the presence of two or more pile heights. High-low pile carpets sometimes combine looped and cut surface yarns.

High Modulus *n* A term that refers to a material with a higher than normal resistance to deformation. Also see ▶ Modulus.

High-Molecular-Weight High-Density Polyethylene *n* (HMWHDPE) See ▶ Polyethylene. {G Niederdruck-Polyethylen n; Polyethylen hoher Ditche, Polyethylen n, F polyéthylène haute densité (PEHD), polyéthylène m S polietileno de alta densidad, polietilieno m, I polietilene alta densità (PE a.d.), polietilene m}.

High Molecular Weight Low Density Polyethylene *n* Thermoplastic with improved abrasion and stress crack resistance and impact strength, but poor processibility and reduced tensile strength. *Also called HMWLDPE.*

High-Performance Composite See ▶ Advanced Composite.

High-Performance Plastic *n* (advanced plastic) Any neat, filled, or reinforced resin, thermoplastic or thermoset, that maintains stable dimensions and mechanical properties above 100°C. Some writers consider many of the ▶ Engineering Plastics to be high-performance plastics and the distinction is blurred. Others have limited the term to specialty thermoplastics (generally high-priced) such as polyimide, polytetrafluoroethylene, polyphenylene oxide and sulfide, liquid-crystal polymers, ultra-high molecular-weight polyethylene, etc. See also ▶ Advanced Resin.

High-Performance Polyethylene Fiber *n* A strong, stiff fiber made from ultra-high-molecular-weight polyethylene by gel-spinning and drawing to a highly oriented state. Produced in the US by Allied-Signal as Spectra® fiber.

High pH Finish *n* A finish, basic in nature rather than acid or neutral, that is applied to yarn or fiber.

High Polymer *n* (1942) A polymer with molecules of high molecular weight, sometimes arbitrarily designated as greater than 10,000. All materials commonly regarded as plastics are high polymers, but not all high polymers are plastics. See ▶ Polymer.

High-Pressure Laminate *n* A laminate molded and cured at pressures not lower than 6.895 MPa (1,000 psi), and more commonly in the range of 8–14 MPa. See also ▶ Decorative Board and ▶ Laminate.

High-Pressure Molding *n* A method of molding or laminating in which the pressure used is greater than 1.4 MPA (200 psi) (ASTM D 883).

High-Pressure Powder Molding *n* Some polymers in powder form can be molded by high-pressure compaction at room temperature followed by heating to complete sintering, curing or polymerization reactions. The process is limited to fairly simple shapes, and to polymers that do not release vapors when heated.

It has been most successful with semicrystalline polymers that can be post-heated for a sufficient time at a temperature within the crystalline endotherm of the polymer. Examples of such polymers are polyphenylene oxide, polytetrafluoroethylene, and ultra-high-molecular-weight polyethylene. (Strong AB (2000) Plastics materials and processing. Prentice Hall, Columbus, OH)

High-Pressure Spot *n* A defect in reinforced plastics: an area containing very little resin, usually due to an excess of reinforcing material.

High-Shrink Staple *n* Staple with a higher degree of potential shrinkage than regular staple of the same generic fiber. When blended with regular staple and treated (in yarn or fabric form) to induce shrinkage, it produces a high degree of bulk in the product.

High-Temperature Dyeing See ▶ Dyeing.

High-Temperature Plasticizer *n* Any plasticizer that imparts higher than the normal resistance to high temperatures to plastics compounds in which it is incorporated. An example is di-tridecyl phthalate, which permits vinyl compounds to be used at temperatures up to 136°C.

High Tenacity *n* A term to describe a material with a higher than normal tensile strength. Also see ▶ Tenacity.

High-Visibility Paints *n* Generally applied to paints containing fluorescent pigments or to paints having high reflectivity because of the use of retroreflective pigments in them. See ▶ Fluorescent Pigments and ▶ Retroreflective.

High-Visibility Pigments See ▶ Fluorescent Dyes.

Hindered Isocyanate \- ˈī-sō- ˈsī-ə- ˈnāt\ *n* See ▶ Isocyanate Generator.

HIPS *n* Abbreviation for High-Impact Polystyrene.

Hiroe Grade of pale East India resin exported from Macassar.

Hitch-Back See ▶ Sticker (1).

HLB Value See ▶ Hydrophile-Lipophile Balance Value.

HMT *n* Abbreviation for Hexamethylenetetramine.

HMWLDPE See ▶ High Molecular Weight Low Density Polyethylene.

HMWPE *n* Abbreviation for High-Molecular-Weight Polyethy-Lene. See ▶ Polyethylene.

Hob \ˈhäb\ *n* [ME *hobbe*, fr. *Hobbe*, nickname for Robert] A master model of hardened steel that is pressed into a block of softer metal to form a number of identical mold cavities. The hobbed cavities are inserted into recesses in a steel mold base and are connected by a runner system. (2, *v*) To form a mold cavity by forcing a hardened steel *hob*, having the inverse shape of the cavity into a soft metal block (that may subsequently be hardened).

Hold-Down Groove *n* A small groove cut into the side wall of a mold to assist in retaining the molding in the mold as it opens.

Hold Out *v* (1585) Ability (or property) to prevent soaking into the substrate.

Hold Paint *n* Marine paint for holds of ships.

Hole \ˈhōl\ *n* [ME, fr. OE *hol* (fr. neuter of *hol*, adj, hollow) & *holh*; akin to OHGr *hol*, *adj*, hollow; perhaps akin to OE *helan* to conceal] (before 12c) The absence of an electron in a semiconductor.

Holes *n* (Tow) In tow opening processes, partial or complete filament breakage within a confined spread of tow, usually circular or oval in shape. Not to be confused with splitting or partial crimp deregistration, which are linear.

Holiday Detector *n* Device for detection of pinholes or voids.

Holidays Application defect whereby small areas are left uncoated. Syn: ▶ Misses, ▶ Skips, ▶ Voids, ▶ Discontinuities, and ▶ Vacations.

Holland \ˈhä-lənd\ *n* {*often capitalized*} [ME *holand*, fr. *Holand*, county in the Netherlands, fr. MD *Holland*] (14 c) See ▶ Shadecloth.

Holland Finish *n* A glazed or unglazed finish containing oil and a filling material. The finish is applied to cotton fabrics to make them opaque or semiopaque. The resultant fabric resembles a beetled linen fabric called Holland fabric.

Hollow Filament Fibers *n* Manufactured, continuous filament fibers, having voids created by introduction of air or other gas in the polymer solution or by melt spinning through specially designed spinnerets.

Holomicroscopy \-mī-ˈkräs-kə-pē\ *n* A three-dimensional photomicrographic process utilizing laser beams and time-differential interferometry to study miscrocopic particles in three dimension. This technology has potential for studying crystalline structures in polymeric materials. (Loveland RP (1981) Photomicrograsphy. Krieger, New York; Rhodes G (1999) Crystallography made crystal clear: A guide for users of macromolecular models. Elsevier Science and Technology Books, New York)

Homespun \- ˈspən\ *adj* (1591) Coarse plain-weave fabric of uneven yarns that have a handspun appearance.

Homochain Polymers *n* A polymer containing only one type of atom in the polymer chain.

Homogeneous \- ˈjē-nē-əs\ *adj* [ML *homogeneus*, *homogenus*, fr. Gk *homogenēs*, fr. *hom-* + *genos* kind] (1641) Composed of a single phase.

Homogeneous Polymerization *n* Polymerization in which a homopolymer is fanned.

Homogenizer \hō-ˈmä-jə-ˌnīz-ər\ *n* (1886) A machine used to break up agglomerates and disperse elemental particles in fluids, consisting of a positive-displacement pump capable of attaining very high pressure, an orifice through which the material is forced at high velocity, and an impact ring on which the stream impinges, Homogenizers are used in the preparation of monomer emulsions and coating compounds, and for dispersing pigments and plasticizers into resin. The term is not restricted to one type of machine, and is used to describe agitators and masticators as well as high-speed rotor-stator type precision machines with close clearances. (Perry's chemical engineer's handbook, 7th edn. Perry RH, Green DW (eds). McGraw-Hill, New York, 1997) See also ▶ Colloid Mill.

Homologous Series \hō-ˈmä-lə-gəs\ *n* A family of organic compounds that have identical functionality with each succeeding member having one more $–CH_2–$ group in its molecule that the preceding member. An example is the series *methanol* (CH_3OH), *ethanol* (CH_3CH_2OH), *propanol* ($C_3H_7–OH$), *butanol* (C_4H_9OH), etc. (Morrison RT, Boyd RN (1992) Organic chemistry, 6th edn. Prentice Hall, Englewood Cliffs, NJ)

Homologous Temperature *n* the ratio of the absolute temperature of a material sample to the absolute temperature at which the material melts.

Homonuclear \ˌhō-mə-ˈnü-klē-ər\ *adj* (1930) Possessing nuclei of atoms of the same element.

Homopolymer *n* (1946) High polymers consisting of molecules that contain (neglecting the ends, branch junctions, and other minor irregularities) either a single type of unit or two or more chemically different types in regular sequence. (Odian GC (2004) Principles of polymerization. Wiley, New York) See ▶ Copolymer and ▶ Polymer.

Homopolymerization *n* A polymer consisting of a single species of monomer, as polyadenylic acid or polyglutamic acid. (Odian GC (2004) Principles of polymerization. Wiley, New York)

Honan \ˈhə-ˌnän\ *n* A pongee-type fabric of the very best Chinese wild silk. Honan is sometimes woven with blue edges.

Honeycomb \ˈhə-nē-ˌkōm\ *n* (before 12c) Manufactured product consisting of sheet metal or resin-impregnated sheet material (paper, fibrous glass, etc.) that has been formed into a network of open-ended, hexagonal cells, each cell's walls being shared with its immediate neighbors. Honeycombs are used as cores for sandwich constructions.

Honeycomb Cores *n* Plastic impregnated woven or non-woven fiber fabric material that serves as a core in sandwich construction.

Honeycombing *n* (1) Checks often visible at the surface, that occurs in the interior of a piece of wood, usually along the wood drays. (2) Lack of vertical film integrity; formation of cell structure; voids.

Hookean Elasticity *n* (ideal elasticity) Stress-strain behavior in which stress and strain are directly proportional, in accordance with Hooke's Law. (Serway RA, Faugh JS, Bennett CV (2005) College physics. Thomas, New York)

Hookean Spring *n* A concept visualized as a coil spring whose extension is proportional to the applied load, useful by analogy in modeling the viscoelastic behavior of polymers. (Serway RA, Faugh JS, Bennett CV (2005) College physics. Thomas, New York)

Hooke Model See ▶ Hookean Spring.

Hooke's Law *n* [Robert *Hooke*] (1853) The observation, by Robert Hooke, that, in a body placed in tension, the fractional increase in length is proportional to the applied load divided by the body's cross-sectional area perpendicular to the load. It was later extended to shear and compressive loading. Thomas Young later contributed the idea of the modulus and formulated Hooke's law as $e = \sigma/E$ where σ = the applied stress, E = the modulus of elasticity of the material, and e = the relative elongation, i.e., the change in lengthy divided by the original length. Few plastics conform to Hooke's law beyond deformations greater than one or two percent. (Serway RA, Faugh JS, Bennett CV (2005) College physics. Thomas, New York)

Hook Reed See ▶ Reed.

Hoop Stress *c* The circumferential stress in a cylindrical body, such as a pipe, that is subjected to internal (or external) pressure. Hoop stress (σ) is given as $\sigma = PD/(2t)$ where P = the pressure, D = the mean diameter of the cylinder, $(D_o + D_i)/2$, and t = the wall thickness = $(D_o - D_i)/2$. In pipes under internal pressure, the hoop stress is twice the lengthwise stress, so dominates the processes of creep and failure. Hoop stress is an often used term when designing and testing plastic pipe and vessels. (Engineering plastics and composites. Pittance JC (ed). SAM International, Materials Park, OH, 1990; Perry's Chemical Engineer's Handbook, 7th edn. Perry RH, Green DW (eds). McGraw-Hill, New York, 1997; Chung DD (1994) Carbon fiber compositers. Elsevier Science and Technology Books, New York)

Hopper \ˈhä-pər\ *n* (13c) In extrusion injection molding, the bin mounted over the feed opening that holds a supply of molding material. The hopper may be intermittently filled or continuously fed (see ▶ Hopper Loader). Feeding from the hopper is ordinarily by gravity, but it may be aided by vibrators, stirrers, or screw feeders. In some setups the hopper feeds into a metering device such as a Gravimetric Feeder or Vibratory Feeder. that meters the rate at which the feedstock enters the extruder or molder.

Hopper Dryer *n* A hopper through which hot air flows upward, drying and heating the feedstock. To improve extraction of moisture from the plastic the air may be passed through a desiccant prior to entering the hopper. The attendant preheating also reduces the amount of power the extruder must furnish per unit mass of material fed and can significantly increase the extruder output. (Strong AB (2000) Plastics materials and processing. Prentice Hall, Columbus, OH)

Hopper Loader *n* (hopper filler) A device for automatically feeding molding powders to hoppers of extruders and injection molders, and maintaining the level of feedstock in the hopper. The functions of drying and blending color concentrates with the feedstock are also sometimes accomplished by loaders. There are two general types: mechanical and pneumatic. (Strong AB (2000) Plastics materials and processing. Prentice Hall, Columbus, OH)

Hopsacking \ˈhäp-ˌsak-\ *n* [ME *hopsak* sack for hops, fr. *hoppe* hop + *sak* sack] (1888) A coarse, open, basket-weave fabric that gets its name from the plain-weave fabric of jute or hemp used for sacking in which hops are gathered.

Horizontal Line See ▶ Ring.

Hot Air Shrinkage *n* Generally, the reduction in the dimensions of a fabric, yarn, or fiber induced by exposure to dry heat. Specifically, a fundamental property of fibers.

Hotbench Test *n* A method of determining gelation properties of plastisols, employing a bar or plate whose temperature rises from one side to the other ("temperature-gradient plate"). The sample is spread on the bar and from the positions at which various changes occur in its state, the temperatures of those changes may be estimated.

Hot-Gas Welding *n* A welding process for plastics analogous to that used for metals, except that a stream of hot gas is used instead of a flame. Welding guns for plastics consist of a blower behind a heating element, much like a hair drier, and a nozzle that focuses the stream at the weld zone. Either air, or, better, dry nitrogen is used. The heated gas is directed at the joint that has been prepared for welding, while a rod of the plastic being welded is applied to the heated zone and melted into the groove.

Hot Head Press *n* A pressing machine capable of generating high temperatures and pressures. Used for pressing and processing permanent-press fabrics.

Hot-Leaf Stamping See ▶ Hot Stamping.

Hot-Manifold Mold *n* An injection mold equipped with an internal heater located in the center of the melt stream in the manifold and nozzle system. This type of mold was developed for thermally sensitive resins to provide gentler heating and avoid the decomposition problems experienced with external heating techniques because of excessive temperature differences.

Hot-Melt \ˈhät-ˌmelt\ *n* (1939) A fast-drying nonvolatile adhesive applied hot in the molten state. (Handbook of adhesives. Skeist I (ed). Van Nostrand Reinhold, New York, 1990, 1977, 1962)

Hot-Melt Adhesive *n* A thermoplastic adhesive applied to the surfaces to be joined in the molten state, then allowed to cool, usually under pressure. They are convenient, require no drying, and leave no voids in the joint. But they can creep and generally have lower strength than standard adhesives. ASTM has developed a number of tests for hot-melt adhesives, most of them in Section 15.06 of the Annual Book of Standards. (Handbook of adhesives. Skeist I (ed). Van Nostrand Reinhold, New York, 1990, 1977, 1962)

Hotmelt Coatings *n* Compositions which liquify readily on heating and are applied to various surfaces in molten condition.

Hot Oil Expression *n* Method of obtaining vegetable oils from seeds by subjecting the oil-bearing seeds to pressure in a suitable press. To facilitate expression of the oil and/or to increase the yield, the seeds are carefully and uniformly preheated before expression. Oils obtained from this process are known as hot-expressed oils.

Hot-Plate Welding *n* (hot-tool welding) Two plastic surfaces to be joined are first held lightly against a heated metal surface, which may be coated with polytetrafluoroethylene to prevent sticking, until the surface layers have melted. The surfaces are then quickly joined and held under light pressure until the joint has cooled. See also ▶ Thermoband Welding and ▶ Welding.

Hot-Pressing *n* The pressure-forming, between heated platens, of plywood, laminates, particleboard, fiberboard, etc.; usually requires thermosetting resins and heat for curing.

Hot-Runner Mold *n* (insulated-runner mold) An injection mold for thermoplastics in which the runners and sometimes the secondary sprues are insulated from the chilled cavity plate so that they remain hot during the entire cycle. The plastic in the runners remains molten and is not ejected with the molded part, thus avoiding the normal handling, grinding, and reprocessing of sprues and runners.

Hot-Setting Adhesive See ▶ Adhesive.

Hot-Short *adj* Inelastic, not stretchable, and easily broken in tension when hot.

Hot Spray *n* Process in which paint is heated prior to spraying to reduce viscosity so that higher solids may be applied.

Hot Spraying *n* Spraying of hot lacquers or paints, the viscosities of which have been reduced to spraying consistency by means of heat instead of by the addition of volatile solvents. By such a process it is possible to apply materials with higher solid contents and therefore better build.

Hot Stage *n* A microscope stage equipped with heating and cooling at controllable rates, enabling one to observe changes in morphology with temperature, such as spherulite growth in polymers as they are cooled from the melt.

Hot Stamping *n* (roll-leaf stamping, gold-leaf stamping) A method of marking plastics in which a special pigmented, dyed, or metallized foil is pressed against the plastic article by a heated die, welding selected areas of the foil to the article. The term also includes a process of impressing inked type into the material when the type is heated.

Hot Surface *n* (1) An alkaline surface such as plaster, concrete, etc. (2) An abnormally absorbent surface (British).

Hot-Tack Strength *n* In heat sealing, the strength of the seal just at the end of the dwell when the die halves part (clearly more a concept than a measurable property).

Hot-Tip-Gate Molding *n* An injection-molding technique used for thin, large-area articles molded in a single cavity. In conventional molding, a sprue connects the machine nozzle with runners leading to each gated cavity. In hot-tip-gate molding, the sprue is eliminated and material is injected directly from the nozzle through a heated bushing that serves as the sprue and gate for the individual cavity. Advantages are faster molding cycles, less material waste and reprocessing, fewer post-molding operations, and reduced sink marks and flow lines. (Strong AB (2000) Plastics materials and processing. Prentice Hall, Columbus, OH)

Hot Wire Cutter *n* A process for splitting some types of plastic foam blocks into smaller pieces whereas heated wires slowly melt through the block as they are fed into the wires by gravity or conveyor belt.

Houndstooth \ˈhaún(d)z-ˌtüth-\ *n* (1937) A term describing a medium-sized broken-check effect; the check is actually a four pointed star.

Housekeeping \ˈhaús-ˌkēp-piŋ\ *n* (1550) Cleanliness, neatness, and orderliness of an area, with the designation of a proper place for everything and everything in its proper place.

House Paint *n* Coating designed for use on large exterior surfaces of a building; generally of lower gloss than the coating used on trim areas. (Coatings technology handbook. Tracton AA (ed). Taylor & Francis, New York, 2005; Wicks ZN, Jones FN, Pappas (1999) Organic coatings science and technology, 2nd edn. Wiley-Interscience, New York)

HT-1 A type of nylon made from phenylenediamine and iosphthalic or terephthalic acid, with good high-temperature properties. (Whittington's dictionary of plastics. Carley, James F (ed). Technomic, 1993)

Hubl Solution *n* Reagent used for the determination of Hubl iodine values, involving iodine and mercuric chloride.

Huckaback \ˈhə-kə-ˌbak\ *n* (1690) A heavy, serviceable toweling made with slackly twisted filling yarns to aid absorption. The cloth has a honeycomb effect. (Fairchild's dictionary of textiles. Tortora PG (ed). Fairchild Books, New York, 1997)

Hue \ˈhyü\ *n* [ME *hewe*, fr. OE *hīw*; akin to ON *h ȳ* plant down, Gothic *hiwi* form] (before 12c) The specific quality distinguishing one color from other, such as yellow, red, or blue. (Color, character, dominant wavelength) Blue, green, red, etc. White, black and grays possess no hue.

Huemann's Blue *n* Synthetic ultramarine blue. See ▶ Ultramarine Blue.

Hue, Munsell See Munsell Hue.

Huggins Constant *n* (k′) The slope coefficient of the Huggins Equation found to be constant for a series of homologous polymers of different molecular weights dissolved in a particular solvent. (Huggins ML (1958) Physical chemistry of high polymers. Wiley, New York)

Huggins Equation *n* In dilute polymer solutions $\eta_{sp}/c = [\eta] + k'[\eta]^2 c$, where η_{sp} and $[\eta]$ are the specific and intrinsic viscosities, c = the concentration in g/dL, and k' = the Huggins constant; one of several equations (Schulz, Blasche, Huggins and Kraemer) to extracted values $c \longrightarrow 0$ or infinite dilution of solute. By making viscosity measurements at several concentrations, plotting η_{sp}/c versus c, and extrapolating the line obtained to $c = 0$, the intrinsic viscosity can be evaluated. Then k' = the slope of the line divided by $[\eta]^2$. (Kamide K, dobashi T (2000) Physical chemistry of polymer solutions. Elsevier, New York; Huggins ML (1958) Physical chemistry of high polymers. Wiley, New York) See also ▶ Dilute-Solution Viscosity, ▶ Staudinger Index, and ▶ Mark-Houwink Equation.

Hull Paint *n* Ships' hull paint, bottom paint.

Humectant \hyü-▪mek-tənt\ *n* [L *humectant-*, *humectans*, pp of *humectare* to moisten, fr. *humectus* moist, fr. *humēre* to be moist] (ca. 1857) A substance, e.g., sorbitol that promotes retention of moisture. (Merriam-Webster's collegiate dictionary, 11th edn. Merriam-Webster, Springfield, MA, 2004) Humectants have been used in antistatic coatings for plastics. (Bart J (2005) Additives in polymers: Industrial analysis and applications. Wiley, New York)

Humidity \hyü-▪mi-də-tē\ *n* (15c) (1) The amount of a volatile compound, normally a liquid at the prevailing ambient conditions, present as its vapor in a gas. (2) Specifically, the amount of water vapor present in air, with absolute humidity expressed as the mass of water per unit mass of dry air. See ▶ Relative Humidity.

Humidity, Absolute *n* (1867) Mass of water vapor present in unit volume of the atmosphere, usually measured as grams per cubic meter. It may also be expressed in terms of the actual pressure of the water vapor present.

Humidity Blush See ▶ Blushing.

Humidity Ratio *n* In a mixture of water vapor and air, the weight of water vapor per unit weight of dry air. *Also called Specify Humidity.*

Humidity, Relative *n* (1820) The ratio of the pressure of water vapor present to the pressure of saturated water vapor at the same temperature. The ratio is generally expressed as a percentage or as a decimal fraction.

Humidity, Specific See ▶ Humidity Ratio.

Hund's Rule The rule which states that two electrons tend to remain unpaired and in separate orbitals of the same energy, rather than paired in the same orbital.

Hungary Blue \▪həŋ-g(ə-)rē\ *n* See ▶ Cobalt Blue.

Hungry Surface *n* Surface, the absorptive powers of which have not been fully satisfied by the coats of paint applied to it, usually resulting in a patchy film. Also known as *Starved Surface.*

Hunter Color Difference Equation *n* Generally refers to the color difference equation used for calculating color differences on certain color difference meters (especially those made by Hunter Labs and Gardner Instruments). It is a simplified equation to approximate the NBS equation:

$$\Delta E = [(\Delta a_L)2 + (\Delta b_L)2 + (\Delta L)^2]^{1/2}$$

where
$L = 10Y^{1/2}$
$a_L = 17.5 (1.02X - Y)/Y^{1/2}$
$b_L = 7.0 (Y - 0.847Z)/Y^{1/2}$ and the deltas are obtained as the difference of the sample minus the standard. Positive a or Δ a indicates redness, negative indicates greenness; plus b or Δ indicates yellowness, minus indicates blueness. Other instruments may use the following version:

$$\Delta E = [(\Delta a_R)^2 + (\Delta b_R)^2 + (\Delta L)^2]^{1/2}$$

where
$R = Y$
$a_R = 2.75f_y (1.02X - Y)$
$b_R = 0.70f_y (Y - 0.847Z)$
$f_y = 0.50 (21 + 0.2Y)/(1 + 0.2Y)$. The particular equation used must be specified. (Colour physics for industry, 2nd edn. McDonald, Roderick, Society of Dyers and Colourists, West Yorkshire, England, 1997; Billmeyer FW, Saltzman M (1966) Principles of color technology. Wiley, New York)

Huygens' Theory of Light \▪hī-gənz-\ *n* This theory states that light is a disturbance traveling through some medium such as the ether. Thus light is due to wave motion in ether. Every vibrating point on the wave-front is regarded as the center of a new disturbance. These secondary disturbances traveling with equal velocity, are enveloped by a surface identical in its properties with the surface from which the secondary disturbances start and this surface forms the new wave-front. (Serway RA, Faugh JS, Bennett CV (2005) College physics. Thomas, New York; Saleh BEA, Teich MC (1991) Fundamentals of photonics. Wiley, New York)

Hybrid Composite \▪hī-brəd\ *n* Advanced composite with a combination of different high-strength continuous filaments in the matrix. Also, composite in which continuous and staple fibers are used in the same matrix.

Hybrid Fabric *n* Fabric for composite manufacture in which two or more different yarns are used in the fabric construction. This provides design flexibility to meet performance requirements and controls cost by permitting some lower priced fibers to be used.

Hybrid Orbital *n* An atomic orbital formed by combining or mixing two or more ground-state atomic orbitals.

Hybrid Yarn *n* In aerospace textiles, a yarn having more than one component. Also see ▶ Commingled Yarn.

Hydantoin Epoxy Resin See ▶ Epoxy Resin.

Hydrate \▪hī-▪drāt\ *n* (1802) Compound in which molecules of water, as such, are present. Many crystalline salts contain what is described as water of crystallization. Each molecule of salt is associated with one or more molecules of water. These salts are known as hydrated salts.

Hydrated Alumina \-ə▪lü-mə-nə\ *n* Syn: ▶ Alumina Trihydrate.

Hydrated Aluminum Silicate See ▶ Aluminum Silicate (Clay).

Hydrated Chromium Oxide n $Cr_2O_3 \cdot 2H_2O$. Pigment Green 18 (77289). Permanent green pigment with excellent lightfastness but lower temperature resistance than chromium oxide and less tinting strength. Density, 2.9–3.7 g/cm³ (24.2–30.8 lb/gal); O.A., 110. Syn: ▶ Guignet's Green, ▶ French Veronese Green, ▶ Emerald Oxide of Chromium, ▶ Mittler's Green, ▶ Pelletier's Green, ▶ Emerald Green, and ▶ Smeraldino.

Hydrated Lime See ▶ Calcium Hydroxide.

Hydrated Magnesium Aluminum Silicate n $3MgO \cdot 1.5Al_2O_3 \cdot 8SiO_2 \cdot 9H_2O$. Natural adsorptive clay with a primary constituent of attapulgite clay which has rodlike particle shape. Density, 2.36 g/cm³ (19.7 lb/gal); O.A., 80–190, particle size, 0.13 μm. Syn: ▶ Fuller's Earth, and ▶ Attapulgite.

Hydration \hī-ˈdrā-shən\ n (1850) The addition of water to another substance. The water may react with the other material, as in the hydration of lime; may be taken up in a mixed chemical/physical affinity (absorption), as in 6/6 nylon and numerous other plastics, in time reaching an equilibrium value for the prevailing conditions; or may simply be drawn into pores of the material by capillary action (wetting). See, too, ▶ Solvation.

Hydraulic Entanglement See ▶ Hydroentangling.

Hydraulic Spraying n Spraying by hydraulic pressure. See ▶ Airless Spraying.

Hydric \ˈhī-drik\ adj (1926) Containing or relating to, hydrogen in combination.

Hydride \ˈhī-ˌdrīd\ n (1869) A compound containing hydrogen in the – 1 oxidation state.

Hydroabietyl Alcohol Obtained by hydrogenation of Abietic acid (a triple-ring, aromatic acid extracted from rosin) and used as a plasticizer for PVC and some cellulosic resins.

Hydroblasting n Cleaning with high-pressure water jet.

Hydrocarbon \ˈhī-drō-ˌkär-bən\ n (1826) An organic compound (as acetylene or butane) containing only carbon and hydrogen and often occurring in petroleum, natural gas, coal, and bitumens.

Hydrocarbon Plastics n Plastics based on resins composed of hydrogen and carbon alone. NOTE – Oxygen and other elements may be present as impurities in the resins, less than 1% by wt.

Hydrocarbon Resins n Polymers consisting solely of hydrogen and carbon, specifically denoting, within the trade, those resins which are hard and friable, which give good adhesive strength, and lack the cohesive strength typical of elastomeric polymers.

Hydroentangling n Process for forming a fabric by mechanically wrapping and knotting fibers in a web through the use of high-velocity jets or curtains of water.

Hydroextractor See ▶ Centrifuge.

Hydrogel n A three-dimensional network of a hydrophilic polymer, generally covalently or ionically crosslinked. The most widely used example is poly (hydroxyethyl methacrylate), especially in medical applications such as implants, blood bags, and syringes.

Hydrogenated Methyl Abietate n $C_{19}H_{31}COOCH_3$. A derivative of Abietic acid, which is extracted from pine rosin, used as a plasticizer for cellulose nitrate, ethyl cellulose, acrylic and vinyl resins, and polystyrene.

Hydrogenated Naphthalenes n Two commercial solvents are included in this description, namely tetrahydronaphthalene and decahydronaphthalene. The former is a partly hydrogenated naphthalene, whereas the latter is completely hydrogenated.

Hydrogenated Naphthas n Series of solvents obtained by subjecting the sulfur dioxide extracts of petroleum to hydrogenation.

Hydrogenated Rosin n Modified rosin obtained by the hydrogenation of rosin at high pressures in the presence of a catalyst.

Hydrogenated Rubber n Rubber reacted in solution under pressure, in the presence of a suitable catalyst, to yield hydrogenated rubber, in which the unsaturation of natural rubber has been substantially satisfied.

Hydrogenation \hī-ˌdrä-jə-ˈnā-shən\ n (1809) The process of passing hydrogen into an unsaturated chemical in the presence of a catalyst to convert the material to a more saturated state (i.e., containing more combined hydrogen).

Hydrogenation of Oils n Method of treatment employed to convert unsaturated fatty acids or oils into saturated types.

Hydrogen Bonds *n* (1923) A very strong attraction between a hydrogen atom which is attached to an electronegative atom, and an electronegative atom which is usually on another molecule.

—O—H⋯O(H)	0.28 ± 0.01	H bond formed in water
—O—H⋯O—C	0.28 ± 0.01	Bonding of water to other molecules often involves these
N—H⋯O(H)	0.29 ± 0.01	
N—H⋯O—C	0.29 ± 0.01	Very important in protein and nucleic acid structures
N—H⋯N	0.31 ± 0.01	
N—H⋯S	0.37	Relative rare; weaker than above

*Defined as distance from center of donor atom to center of acceptor atom. for example, in the N – H ⋯ O = C bond it is the N – O distance.

The conventional view of the water molecule   The actual view of the water molecule

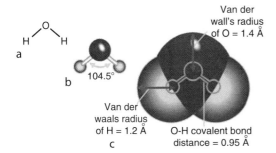

a
b 104.5°
c Van der waals radius of H = 1.2 Å
O-H covalent bond distance = 0.95 Å
Van der wall's radius of O = 1.4 Å

Hydrogen Bromide *n* (1885) HBr. A colorless irritating gas that fumes in moist air and yields hydrobromic acid when dissolved in water.

$$H\text{——}Br$$

Hydrogen Chloride *n* (1869) HCl. A colorless pungent poisonous gas that fumes in moist air and yields hydrochloric acid when dissolved in water.

$$H\text{——}Cl$$

Hydrogen Cyanide *n* (1882) A poisonous usually gaseous compound HCN that has the odor of bitter almonds.

$$\equiv\!\!\text{N}$$

Hydrogen Equivalent *n* The number of replaceable hydrogen atoms in one molecule of a substance, or the number of atoms of hydrogen with which one molecule could react.

Hydrogen Fluoride *n* (1885) A colorless corrosive gaseous compound HF that yields hydrofluoric acid when dissolved in water.

$$H\text{——}F$$

Hydrogen Iodide *n* (1885) An acrid colorless gas HI that fumes in moist air and yields hydriodic acid when dissolved in water.

$$H\text{——}I$$

Hydrogen Ion *n* (1896) The cation H^+ of acids consisting of a hydrogen atom whose electron has been transferred to the anion of the acid.

Hydrogen Ion Concentration *n* (**pH**) The concentration of hydrogen ions in solution when the concentration is expressed as gram-ionic weights per liter. A convenient form of expressing hydrogen ion concentration is in terms of the negative logarithm of this concentration. The negative logarithm of the hydrogen ion concentration is called pH. The significance of pH is still in dispute. Water at 25°C has a concentration of H ion of 10^{-7} and of OH ion of 10^{-7} moles/L. Thus, the pH of water is 7 at 24°C. A greater accuracy is obtained if one substitutes the thermodynamic activity of the union for its concentration. See ▶ pH.

Hydrogen Peroxide *n* (1872) H_2O_2. An unstable compound used as an oxidizing and bleaching agent.

$$HO\text{——}OH$$

Hydrogen Plastics *n* (1) Plastics based on polymers made from monomers containing only carbon and hydrogen. (2) In the plastics industry, hydrocarbon resins are considered to be those thermoplastic resins of low molecular weight made from relatively impure monomers that are derived from coal-tar fractions, cracked-petroleum distillates, and turpentine. The family includes ▶ Coumarone-Indene Resins, cyclopentadiene resins, petroleum resins, terpene resins, and many others of little commercial importance. Having little strength, most hydrocarbon resins are rarely used alone. Their primary applications are as binders in asphalt flooring, processing aids in elastomers and polyolefins, and coating additives. (Ash M, Ash I (1982–1988) Encyclopedia of plastics polymers, and resins, vols I–III. Chemical Publishing, New York)

Hydrogen Value *n* Method of determining hydrogen absorption for unsaturated oils or similar compounds. Hydrogen is reacted with unsaturated oils, and iodine values are calculated from the amount of hydrogen absorbed.

Hydrolysis \hī-ˈdrä-lə-səs\ *n* [NL] (1880) Disruptive reaction consisting of splitting a compound into two parts, one of which combines with the H ion of water and the other combines with the OH ion of water. For example, when vegetable oils or fats are split, the esters are broken down into an avid component and an alcohol component, by taking unto themselves the H and OH ions of water. When alkaline reagents are present, the acids do not appear as such at the end of the reaction, but as the corresponding alkali metal salts. (Whitten KW, Davis RE, Davis E, Peck LM, Stanley GG (2003) General chemistry. Brookes/Cole, New York)

Hydrolysis (Hydrolytic) Constant *n*, K_h The equilibrium constant for a hydrolysis equilibrium.

Hydrolytic Degradation *n* Any breakdown of a plastic in which reaction with water or water vapor (steam) plays a role.

Hydrometer \hī-ˈdrä-mə-tər\ *n* (1675) An instrument that senses and indicates the density of a liquid in order to measure the specific gravity of a liquid or solution. A simple type consists of a glass tube with a bulb at its bottom, fine lead shot in the bulb, and a graduated scale within the tube. The amount of shot is adjusted before the top of the tube is sealed so that, when the hydrometer is floated in a liquid of density within its calibrated range, the scale reads the density of the liquids at the meniscus.

Hydronium Ion \hī-ˈdrō-nē-əm ˈī-ən\ *n* The hydrated hydrogen ion; represented by H_3O^+. (Whitten KW, Davis RE, Davis E, Peck LM, Stanley GG (2003) General chemistry. Brookes/Cole, New York) Also known as the *Oxonium Ion*.

Hydroperoxides \hī-ˈdrō-pə-ˈräk-ˌsīd\ *n* (1937) Compounds of the type R-O-OH, where R is an alkyl group. Useful as free radial initiators for free radical polymerization. (Morrison RT, Boyd RN (1992) Organic chemistry, 6th edn. Prentice Hall, Englewood Cliffs, NJ)

Hydrophile See ▶ Emulsoid.

Hydrophile-Lipophile Balance (HLB) Value *n* An empirical value (arbitrary units of 0–20) which is based on the premise that all surfactants combine hydrophilic and lipophilic groups in one molecule and that the proportion between the weight percentages of these two groups for nonionic surfactants is an indication of the behavior that may be expected from the product. (McCutcheon's emulsifiers and detergenst (2000): North American edition, vol 1. McCutcheon Division of McCutcheon, 2000; Gooch JW (2002) Emulsification and polymerization of alkyd resins. Kluwer Academic/Plenum, New York)

Hydrophilic \ˌhī-drə-ˈfi-lik\ *adj* [NL *hydrophilus*, fr. Gk *hydr-* + *-philos* –philous] (1916) (1) A "water-loving", i.e., material having an affinity for, and readily wetted by water (Goldberg DE (2003) Fundamentals of chemistry. McGraw-Hill Science/Engineering/Math, New York). (2) In lithography, referring to that property of a substance that makes it more receptive to water and fountain solutions than to oils and inks.

Hydrophobic (Chemistry) \-ˈfō-bik\ *adj* (1807) A substance which does not absorb or exhibit affinity for water; nonwettable. (Goldberg DE (2003) Fundamentals of chemistry. McGraw-Hill Science/Engineering/Math, New York)

Hydrophobic *n* (Lithography) Hydrophobic refers to that property of a substance that makes it more receptive to oils and inks than to water and fountain solutions. Also called *Lipophilic*.

Hydroquinone \-kwi-ˈnōn\ *n* [ISV] (ca. 1872) (*p*-dihydroxybenzene, hydroquinol, *p*-hydroxyphenol, quinol) $C_6H_4(OH)_2$. A white crystalline material derived from aniline, used, as are many of its derivatives, as an inhibitor of free-radical polymerization in unsaturated polyester resins and in monomers such as vinyl acetate. Hydroquinone is almost colorless and can retain its inhibitory action even in the presence of oxygen. (Odian GC (2004) Principles of polymerization. Wiley, New York)

Hydroquinone Di(β-Hydroxyethyl) Ether *n* (HQEE) A white solid material used as a reactant in the preparation of polyesters, polyolefins, and polyurethanes. As a chain extender in urethane prepolymers, HQEE increases to 150°C the high-temperature resistance of parts molded from the prepolymer.

Hydroscopic *adj* Having the ability to absorb moisture from the atmosphere. All fibers have this property in varying degrees. (Complete textile glossary. Celanese Acetate LLC, Three Park Avenue, New York, 2000; Goldberg DE (2003) Fundamentals of chemistry. McGraw-Hill Science/Engineering/Math, New York)

Hydrosol \ˈhī-drə-ˌsäl\ *n* [*hydr-* + *sol* (fr. *solution*] (1864) (1) In physical chemistry, a colloidal suspension in water. (2) In the plastics industry, a suspension of resin such as PVC or nylon in water, not necessarily of colloidal nature. See also ▶ Latex.

Hydrostatic \-ˈsta-tik\ *adj* [prob. fr. NL *hydrostaticus*, fr. *hydra-* + *staticus* static] (1666) Of or relating to fluids at rest or to the pressures they exert or transmit.

Hydrostatic Design Basis *n* (HDB) One of a series of 10^5-h or 50-year strength values and forming the basis for choosing the wall thickness of pipe for various diameters and service pressures and temperatures. (American Society for Testing and Materials, 100 Barr Harbor Drive, West Conshohocken, PA 19428-2959, www.astm.org)

Hydrostatic Design Stress *n* The estimated sustained hoop stress that can exist in a pipe at expected service conditions and over the life of the pipe with a high degree of certainty that failure will not occur. See preceding entry.

Hydrostatic Pressure *n* At a distance *h* from the surface of a liquid of density *d*,

$$P = hdg.$$

The total force on an area *A* due to hydrostatic pressure,

$$F = PA = Ahdg.$$

Force in dynes and pressure in dynes per square centimeter will be given if *h* is in centimeter, *d* in grams per cubic centimeter and *g* in centimeter per square seconds. (Munson BR, Young DF, Okiishi TH (2005) Fundamentals of fluid mechanics. Wiley, New York)

Hydrostatic Strength *n* The hoop, stress calculated from the following equation, at which a pipe fails because of rising internal pressure, usually in about 1 min. The equation is: $\sigma = P\,D_m/(2\,t)$, in which P = the internal (gauge) pressure at rupture, D_m = the initial mean diameter (= outer diameter − *t*), and *t* = the initial wall thickness. This strength is usually less than the tensile strength of the pipe material as determined in a standard tensile test because of the presence in the pipe of longitudinal stress equal to half the hoop stress. The pressure at which the failure occurs is known as the *quickburst pressure*. (Munson BR, Young DF, Okiishi TH (2005) Fundamentals of fluid mechanics. Wiley, New York)

Hydroterpins *n* Hydrogenation products derived from turpentine. They differ from the parent material in that they possess greater stability, i.e., less tendency to oxidation, higher boiling ranges, and slower evaporation rates.

Hydrous \ˈhī-drəs\ *adj* (1826) Containing water usually in chemical association (as in hydrates).

Hydrous Alumina *n* Syn: ▶ Alumina Trihydrate.

Hydrous Oxides *n* Poorly characterized compounds formed from the combination of certain oxides with water or by precipitation from the addition of base to certain ions in aqueous solution.

Hydrox-, Hydroxy- A chemical prefix denoting the presence of the –OH group in a compound.

Hydroxide \hī-ˈdräk-ˌsīd\ *n* [ISV] (1851) Hydrated metallic oxide; a base; a compound which will give hydroxyl (OH) ions in solution.

(2-Hydroxpropyl) Methyl Cellulose *n* A mixed ether of cellulose containing both hydropropyl and methyl groups.

2(2′-Hydroxy-3,5-Ditertiarybutylphenyl)-7-Chlorobenzotriazol *n* An off-white, nontoxic, crystalline powder with high thermal stability, used as an ultraviolet absorber for polyolefins, PVC, polyurethanes, polyamides, and polyesters.

Hydroxyethyl Acetamide See ▶ N-Acetyl Ethanolamine.

Hydroxyethyl Cellulose *n* Any of a family of polymeric ethers formed by reacting alkali cellulose with ethylene oxide. Water solubility and applications depend on the degree of substitution of hydroxyenthyl groups for –OH groups, and include textile sizes, adhesives, thickeners, and stabilizers for vinyl polymers.

Hydroxyethyl Cellulose Resins *n* Prepared by reaction of alkali cellulose with ethylene oxide, which then forms long polyoxyethylene chains by continued reaction with the hydroxyethyl groups.

Hydroxyethylmethyl Methacrylate *n* A monomer that polymerizes to a hydrophilic polymer that is rigid when dry but when saturated with water becomes a soft, clear material (Hydron®). Applications include masonry coatings, soft contact lens, and other biomedical devices.

Hydroxyl \hī-ˈdräk-səl\ *n* [*hydr-* + *ox-* + *-yl*] (1869) The chemical group or ion OH that consists of one atom of hydrogen and one of oxygen and is neutral or positively charged.

$$OH_2\bullet$$

Hydroxyl End Group *n* A polymer chain-terminating (–OH) group.

Hydroxyl Group *n* –OH. The monovalent group characteristic of hydroxides and alcohols.

2-Hydroxyl-4-Methoxy-5-Sulfonbenzophenone *n* An ultraviolet absorber for thermoplastics.

Hydroxyl Value *n* A measurement of hydroxyl groups (–OH) in an organic material (American Society for Testing and Materials, 100 Barr Harbor Drive, West Conshohocken, PA 19428-2959, www.astm.org,). In plasticizers, the hydroxyl value includes –OH groups present in any free unesterified alcohol as well as those of the plasticizer molecule itself. In some plasticizers, large hydroxy values signal that the plasticizer may become incompatible on aging. In urethane technology, hydroxyl number is an important factor in the selection of polyols to achieve desired characteristics in elastomers and foams (Plasticizer's data base. Wypych G (ed). Noyes, New York, 2003).

2-Hydroxy-4-Methoxybenzophenone *n* An ultraviolet absorber for numerous thermoplastics.

2(2′-Hydroxy-4′-Methylphenyl) Benzotriazole *n* ("Tinuvin P") A nontoxic crystalline powder with high thermal stability, an ultraviolet absorber for polystyrene, acrylics, PVC, polyesters, and polycarbonates.

2-Hydroxy-4-*n*-Octoxybenzophenone *n* $C_{21}H_{26}O_3$. A pale yellow powder, an ultraviolet absorber for PVC, and several other plastics. It is compatible withy highly plasticized vinyls and has a very low order of toxicity.

Hydroxypropylglycerin *n* A pale-straw-colored liquid used as a plasticizer for cellulosic resins and as an intermediate for alkyd and polyester resins.

Hydroxypropyl Methacrylate *n* (HPMA) A reactive monomer copolymerizable with a wide variety of acrylic and vinyl monomers, used for thermosetting resins and surface coatings.

Hygrometer \hī-ˈgrä-mə-tər\ *n* [prob. fr. F *hygromètre*, fr. *hygr-* + *-mètre* –meter] (1670) An instrument that senses and indicates the relative humidity (of moisture) in air.

Hygroscopic \ˌhī-grə-ˈskä-pik\ *adj* [*hygroscope*, an instrument showing changes in humidity + ^{1}-*ic*, fr.] (1790) Having a strong tendency to absorb moisture from the air. Some resins are hygroscopic, and therefore usually require drying before being extruded or molded. See ▶ Hydroscopic.

Hypalon^R Hypalon® chlorosulfonated polyethylene, manufactured by DuPont Performance Elastomers. A versatile, high-performance elastomer used extensively in the automotive and general rubber industries.

Hyperon \ˈhī-pə-ˌrän\ *n* [prob. fr. *hyper-* + ^{2}-*on*] (1953) Any article with mass intermediate between that of the neutron and the deuteron. See ▶ Meson. (Giambattista A, Richardson R, Richardson RC, Richardson B (2003) College physics. McGraw Hill Science/Engineering/Math, New York)

Hypogoeic Acid *n* $C_{15}H_{29}COOH$. Unsaturated fatty acid found in certain vegetable oils. It has a mp of 33°C, and a bp of 236°C/15 mmHg.

Hypohalous Compound in which a hydroxyl group is combined with a halogen atom.

Hysteresis \ˌhis-tə-ˈrē-səs\ *n* [NL, fr. Gk *hysterēsis* shortcoming, fr. *hysterein* to be late, fall short, fr. *hysteros* later] (1881) (1) The lagging of the physical effect on a body behind its cause, particularly during repeated cycling. Hysteresis is accompanied by dissipation of energy as heat. If a fatigue specimen is cycled at too high a frequency/or stress, the resulting heat buildup can seriously bias the test and even soften the specimen (Shah V (1998) Handbook of plastics testing technology. Wiley, New York). (2) The influence of prior history that causes the repetition of a process or its reversal to take a different path from the original. In tensile loading of plastics past the proportional limit but short of failure, the path (stress vs. strain) followed upon unloading will generally be at lower stresses than were recorded during loading at the same strains. Open or closed loops may be observed. (Handbook of physical polymer testing, vol 50. Brown R (ed). Marcel Dekker, New York, 1999)

Hysteresis Loop n Flow curve for a dilatent (shear thining) material obtained by measurements on a rotational viscometer showing for each value of rate of shear, two values of shearing stress, one for an increasing rate of shear and the other for a decreasing rate of shear. A hysteresis loop characterizes a dilatant material. If no hysteresis loop is obtained, the material is nondilatant. The area within the loop is a measure of dilatancy (increasing shear thining with increasing shear stress). Dilatancy should not be confused with thixoptropy (shear thining at constant shear stress over time) (Munson BR, Young DF, Okiishi TH (2005) Fundamentals of fluid mechanics. Wiley, New York; Patton TC (1979) Paint flow and pigment dispersion: A rheological approach to coating and ink technology. Wiley, New York; Van Wazer, Lyons, Kim, Colwell (1963) Viscosity and flow measurement. Lyons, Kim, Colwell (eds). Interscience, New York)

Hytrel® n DuPont trade name for a family of copolyester elastomers. Typical reactants from which the elastomers are derived are terephthalic acid, polytetramethylene glycol, and 1,4-butanediol. Powders and pellets are available for extrusion and molding. The products are highly resilient, have good flex-fatigue life at low and high temperatures, and are resistant to oils and chemicals. Some grades termed *segmented copolyesters* are excellent modifiers for PVC, improving processability and imparting resistance to abrasion, impact, and fungi.

Hz n SI abbreviation for Hertz. (CRC handbook of chemistry and physics. Lide DR (ed). CRC, Boca Raton, FL, 2004 Version)

I

i \ˈī\ *n* (1) (also I) Symbol for electric current. (2) Symbol for $\sqrt{-1}$, the coefficient of "imaginary", i.e., out-of-phase, components in complex quantities.

(I) *n* (1) Chemical symbol for the element iodine (Whitten, K. W., et al., *General Chemistry*, Brookes/Cole, New York, 2003). (2) The symbol for electric current (Giambattista, A., et al., *College Physics*, McGraw-Hill Science/Engineering/Math, New York, 2003). (3) The symbol for moment of inertia (Giambattista, A., et al., *College Physics*, McGraw-Hill Science/Engineering/Math, New York, 2003).

Iatrochemistry *n* [Greek *iatros*, "doctor" + chemistry] Eighteenth-century chemistry was derived from and remained unsovled with questions of mechanics, light and heat as well as iatrochemistry and notions of medical therapy and the interaction between substances and the formations of new substances. This period was also the time of alchemy (approximately 1550–1750 A.D.). Iatrochemistry attempted to explain life and physiological processes using the knowledge of physics at that time. From the root word was derived: iatrogenic [Greek *latros doctor* + -genic] meaning "caused by a doctor" – a symptom or illus brought on unintentionally by something that a doctor does or says. (www.britannica.com and Microsoft Encarta 2004).

Iceland Moss \ˈī-slən(d)-, -ˌslan(d)-\ *n* (1805) Lichen grown in Iceland, Norway and Sweden, which yields a mucilaginous extract, used in sizes and the like. Syn: ► Cetraria.

Iceland Spar *n* (1771) Natural, transparent, double refracting, crystalline form of calcium carbonate. (Hibbard, M. J., *Mineralogy*, McGraw-Hill, New York, 2001).

Ichnography *n* (1) Art of making decorative drawings by the use of compass and rule. (2) The term also refers to the art of tracing plans and illustrations.

ICI *n* (1) Older abbreviation for International Commission on Illumination, (Use international abbreviation CIE).

Ideal Copolymerization *n* A copolymerization in which, in the simple binary case, the product of the monomer reactivity ratios r_A and r_B for monomers A and B ($r_A r_B$) equals unity. Thus each type of growing center ∼∼A^* and ∼∼B^* shows the same preference for adding on or the other monomer during propagation. (Odian, G. C., *Principles of Polymerization*, Wiley, New York, 2004).

Ideal Elastomer *n* (ideal rubber) An elastomer which on deformation at constant, e.g., by stretching by application of a force f, satisfies the conditions

$$(\partial H/\partial l)_{T,P} = 0 \text{ and } f = -T(\partial S/\partial l)_{T,P}$$

where *l* is the length, *T* is the temperature, *P* is the pressure, *S* is the entropy and *H* is the enthalpy. (*Physical Properties of Polymers Handbook*, Mark, J. E., ed., Springer, New York, 1996).

Ideal Gas *n* A gas whose behavior is described by the ideal-gas law, $PV = nRT$. (Goldberg, D. E., *Fundamentals of Chemistry*, McGraw-Hill Science/Engineering/Math, New York, 2003).

Ideal-Gas Law *n* The combination of Charles' and Boyle's laws, usually stated in the form $P \cdot V = n \cdot R \cdot T$, where P = the absolute pressure, V = the gas volume, n = the number of moles of gas present, T = the absolute temperature, and R = the universal molar gas (or energy) constant. R has many numerical values, to be consistent with the units chosen for the variables. In SI, R = 8.31439K/(mol·K). (Goldberg, D. E., *Fundamentals of Chemistry*, McGraw-Hill Science/Engineering/Math, New York, 2003).

Ideal Liquid *n* This term has several meanings in physical chemistry and hydrodynamics. (Munson, B. R., et al., *Fundamentals of Fluid Mechanics*, Wiley, New York, 2005; Goldberg, D. E., *Fundamentals of Chemistry*, McGraw-Hill Science/Engineering/Math, New York, 2003). The one relevant to plastics is ► Newtonian Liquid.

Ideal Solid *n* A material that obeys ► Hooke's Law. (Serway, R. A., et al., *College Physics*, Thomas, New York, 2005).

Ideal Solution *n* Solutions that exhibit no change of internal energy on mixing and complete uniformity of cohesive forces. The behavior of ideal solutions is described by Raoult's law over all ranges of temperature and concentration.

Ignition \ig-ˈni-shən\ *n* (1612) The beginning of burning. (Babrauskas, V., *Ignition Handbook*, Fie Science, New York, 2003; Goldberg, D. E., *Fundamentals of Chemistry*, McGraw-Hill Science/Engineering/Math, New York, 2003).

Ignition Loss See ASTM D2584 at Flammability.

Ignition Resistant Chemical Additives See ► Flame Retardant.

Ignition Time See ► Flammability. (Babrauskas, V., *Ignition Handbook*, Fie Science, New York, 2003).

IIR *n* Butyl rubber.

Illuminance \i-ˈlü-mə-nən(t)s\ *n* (ca. 1938) The luminous flux incident per unit area of a surface. (*CRC Handbook of Chemistry and Physics*, Lide, D. R., ed., CRC, Boca Raton, FL, 2004 Version).

Illuminant \-nənt\ *n* (1644) Mathematical description of the relative spectral power distribution of a real or imaginary light source, that is, the relative energy emitted by a source at each wavelength in its emission spectrum. Popularly used synonymously with the term "light source" or "lamp." Such usage is not recommended. (*CRC Handbook of Chemistry and Physics*, Lide, D. R., ed., CRC, Boca Raton, FL, 2004 Version).

Illuminant A *n* (**CIE**) Incandescent illumination, yellow orange in color, with a correlated color temperature of 2.856K. It is defined in the wavelength range of 380–770 nm. See ▶ Correlated Color Temperature.

Illuminant B *n* (**CIE**) An approximation of mean noon sunlight, approximately neutral in hue, with a correlated color temperature of 4.874 K. It is defined in the wavelength range of 380–770 nm. See ▶ Correlated Color Temperature.

Illuminant C *n* (**CIE**) An approximation of average daylight, bluish, with a correlated color temperature of 6.774 K. It is defined in the wavelength range of 380–770 nm. See ▶ Correlated Color Temperature.

Illuminant D *n* (**CIE**) Daylight illuminants, defined from 300 to 830 nm, the UV portion 300–380 nm being necessary to describe correctly colors which contain fluorescent dyes or pigments. They are designated as D with a subscript to describe the correlated color temperature: D_{65} having a correlated color temperature of 6.504 K, close to that of Illuminant C, is the most commonly used. They are based on actual measurements of the spectral distribution of daylight.

Illuminating \-ˌnā-tiŋ\ *vt* [ME, fr. L *illuminatus*, pp of *illuminare*, fr. *in* + *luminare* to light up, fr. *lumen-*, *lumen* light] (15c) Hand decoration of books, as done in medieval times, with drawings and miniature paintings. Also ornamentation and embellishments, usually in red, blue and gold, which were sometimes added to initial letters and borders. (Giambattista, A., et al., *College Physics*, McGraw-Hill Science/Engineering/Math, New York, 2003).

Illumination \i-ˌlü-mə-ˈnā-shən\ *n* (14c) The illumination on any surface is measured by the luminous flux incident on unit area. The units in use are: the *lux*, (abbreviation lx) one lumen per square meter; the *photo*, (abbreviation ph) one lumen per square centimeter; the *Footcandle*, (abbreviation fc) one lumen per square foot. (Giambattista, A., et al., *College Physics*, McGraw-Hill Science/Engineering/Math, New York, 2003).

Illumination, Diffuse *n* Nonspecular illumination or nondirect incident light source projected onto the object of observation. Indirect or reflected rays of light are often used as diffuse sources. Also, a scattering lens over a light source will diffuse the rays to reduce specular light.

Ilmenite \ˈil-nə-ˌnīt\ *n* [Gr *Ilmenit*, fr. *Ilmen* range, Ural Mts., Russia] (ca. 1827) $FeO \cdot TiO_2$. Natural iron titanate. It is an iron-black mineral and is one of the main sources of titanium dioxide.

Ilmenite Black *n* Opaque black pigment, prepared from the mineral of the same name. (Hibbard, M. J., *Mineralogy*, McGraw-Hill, New York, 2001).

Image, Real *n* The image which would register on a ground glass screen or photographic film placed in that plane. (Moller, K. D., *Optics*, Springer, New York, 2003).

Image, Virtual *n* This image would appear by construction to be in a given plane, but a ground glass screen or photographic film placed in that plane would show no image. (Moller, K. D., *Optics*, Springer, New York, 2003).

Imbibition \ˌim-bə-ˈbi-shən\ *n* (15c) Swelling of a solid, semisolid or gel by absorption of a liquid.

Imitation Gold Ink *n* A simulated gold ink that uses aluminum powder to produce the metallic luster.

Immediate Elastic Deformation *n* Recoverable deformation that is essentially independent of time, i.e., occurring in (a time approaching) zero time and recoverable in (a time approaching) zero time after removal of the applied load.

Immediate Set *n* The deformation found by measurement immediately after the removal of the load causing the deformation in a short-time test.

Immersion Objective *n* A microscope objective that is used with a liquid of refractive index greater than 1.00 between the specimen and objective, and usually between the specimen and substage condenser. Immersion objectives are used when a numerical aperture greater than 1.00 is desired, since this cannot be achieved with a dry objective. (Moller, K. D., *Optics*, Springer, New York, 2003).

Immiscible \(ˌ)i(m)-ˈmi-sə-bəl\ *adj* (1671) Not miscible. Any liquid which will not mix with another liquid, in which case it forms two separate layers or exhibits cloudiness or turbidity. Compare ▶ Incompatible.

Impact Adhesive *n* See ▶ Contact Adhesive.

Impact Energy *n* The energy required to break a specimen, equal to the difference between the energy in the striking member of the impact apparatus at the instant of impact and the energy remaining after complete fracture of the specimen. Also called impact strength. See also ▶ ASTM D256, ASTM D3763.

Impact Modifiers A general term for any additive, usually an elastomer or plastic of different type, incorporated in a plastic compound to improve the ▶ Impact Resistance of finished articles. The improvement is customarily assessed by performance of test specimens in standard tests.

Impact Polystyrene *n* An alternative name for high impact polystyrene; usually polystyrene modified with thermoplastic elastomer to provide a tougher material. See ▶ High-Impact Polystyrene.

Impact Property Tests *n* Names and designations of the methods for impact testing of materials. Also called impact tests. See also ▶ Impact Toughness.

Impact PS See ▶ Impact Polystyrene.

Impact Resistance *n* (1) Ability of a coating to resist a sudden blow. Ability to resist deformation from impact. (2) The relative durability of plastics article to fracture under stresses applied at high speeds. A widely used ASTM impact test, www.astm.org, employs the Izod pendulum striker swung from a fixed height to strike a specimen in the form of a notched bar mounted vertically as a cantilever beam. The Charpy tester, an alternative in D 256, uses a specimen in the form of a horizontal beam supported at both ends. ASTM lists some several different impact tests for plastics and plastics products. See also ▶ Brittleness Temperature, Drop-Weight Test, Free-Falling-Dart Test, and ▶ Tensile-Impact Test.

Impact Strength *n* The quantitative measure of the ability of a material to withstand shock loading in a standard test. For plastics the test is usually either the Izod or Charpy test described in ASTM (www.astm,org), and the result is calculated as the energy expended (work done) in breaking a specimen, divided by its width or thickness. Specimens of both tests are usually notched on the wide opposite that where they are struck, though the notch position may be reversed and unnotched specimens may be tested. In SI the convenient reporting unit is J/cm of notch width, or, for unnotched specimens, J/cm. To provide a much simpler stress distribution free of notch effects, researchers, developed the ▶ Tensile-Impact Test. (Shah, V., *Handbook of Plastics Testing Technology*, Wiley, New York, 1998).

Impact Test *n* Used to assess the adhesion and flexibility of applied coatings. A weighted plunger is allowed to fall from a specified height onto the front (direct) or back (reverse) of a coated panel. The extent of the damage caused by the impact is used as a basis for assessing the adhesion and flexibility. (*Paint and Coating Testing Manual (Gardner-Sward Handbook)* MNL 17, 14th Ed., ASTM, Conshohocken, PA, 1995).

Impact Toughness *n* property of a material indicating its ability to absorb energy of a high-speed impact by plastic deformation rather than crack or fracture. See also ▶ Impact Property Tests.

Impasto \im-ˈpas-(ˌ)tō, -ˈpäs-\ *n* [It, fr. *impastare*] (1784) This term implied a thick application (paste, impasted) of paint or coloring material. Particularly thick or heavy application of paint. The thinly painted and impasted areas of an oil painting provide texture. (Gair, A., *Artist's Manual*, Chronicle Books LLC, San Francisco, 1996).

Imperial Gallon See ▶ Gallon, Imperial.

Impingement Mixing *n* Very intense, rapid and thorough mixing accomplished by causing two fast-moving liquid streams or sprays to meet within a confined space. Typically the streams are resin and curing agent or resin streams containing catalyst in one and accelerator in the other. The technique is used in reaction-injection molding, in sprayup of reinforced plastics, and in foam-in-place molding.

Impregnated Cloth *n* A cloth impregnated with resin, varnish shellac, etc.

Impregnated Fabric *n* A fabric in which the interstices between the yarns are completely filled, as compared to sized or coated material where the interstices are not completely filled. Not included in the definition is a woven fabric constructed from impregnated yarns, rather than one impregnated after weaving.

Impregnating Insulating Varnish *n* First coating insulating varnish applied to electrical equipment by means of vacuum-pressure impregnation, or by dipping. Its main purpose is to fill up the voids between the windings and elsewhere. The name distinguishes it from finishing insulating varnishes, which are used primarily to give an effective seal to the outside of windings, to prevent ingress of moisture.

Impregnation \im-ˈpreg-ˌnāt, ˈim-ˌ\ *vt* [LL *impraegnatus*, pp of *impraegnare*, fr. L *in-* + *praegnas* pregnant] (1605) The process of thoroughly soaking a material of an open or porous nature with a resin. When webs or shapes of reinforcing fibers are impregnated with a thermosetting resin in the B stage or with a thermoplastic, and such webs are intended for subsequent shaping or laminating, the masses are called ▶ Sheet-Molding Compounds or ▶ Prepreg. The main difference between impregnation and ▶ Encapsulation is that in encapsulation an outer protective coating is formed with little or no penetration of the resin into the article, whereas in impregnation there is little or no protective coating.

Impression \im-ˈpre-shən\ *n* (14c) (1) The printing pressure necessary for ink transfer. (2) Also refers to a single print.

Impression Cylinder *n* The cylinder on a printing press that holds the material being printed against the printing plate, cylinder, or blanket.

Impsonite Naturally occurring asphaltum, found in Oklahoma.

Impulse Sealing *n* (thermal-impulse sealing) The process of joining thermoplastic sheets or films by pressing them between elements equipped to provide a pulse of intense heat to the sealing area for a very short time, followed immediately by cooling. The heating element may be a length of thin resistance wire such as nichrome, or a bar heated by a high-frequency electric field and cored for water cooling. See also ▶ Heat Sealing.

Incandescent \ˌin-kən-ˈdes-sᵊnt\ *adj* [prob fr. F, fr. L *incandescent-. incandescens*, pre. part. of *incandescere* to become hot, fr. *in-* + *candescere* to become hot, fr. *candēre* to glow] (1794) Luminous in the yellow-to-white range because of being at a high temperature (1,250–1,550°C).

Inching *v* (1599) A very low rate of mold closing used in many molding operations and often automatically included in the cycle, for the last millimeter or so before the mold faces meet.

Incident Angle *n* (1628) The angle between the incident ray and a normal to the surface at the point of incidence. It is the same as "entrance angle" in SAE automotive nomenclature.

Incident Ray *n* The center line of the light beam incident upon a surface.

Inclusion \in-ˈklü-zhən\ *n* [L *inclusion-, inclusio*, fr. *includere*] (1600) (1) A foreign body or impurity phase in a solid. (2) Presence of foreign material in the finished material.

Inclusion Complex See ▶ Adduct.

Incompatibility \ˌin-kəm-ˌpa-tə-ˈbi-lə-tē\ *n* (1611) The effect when two or more components of a composition do not blend properly to produce a uniform or homogeneous mixture. It can be characterized by gelatin, curding, precipitation, cloudiness, seediness, land loss in gloss of blooming on resulting film.

Incompatible \ˌin-kəm-ˈpa-tə-ˈbəl\ *adj* [ME, fr. MF & ML; MF. fr. ML incompatibilis, fr. L *in-* + ML *compatibilis* compatible] (15c) A term that describes two materials, such as two resins, or a resin and plasticizer, that are incapable of forming a solution of a two-phase blend and that tend to separate after being mixed.

Incompatible Polyblends *n* Polymers that do not alloy to form stable compositions.

Incomplete Hiding See ▶ Hiding, Incomplete.

Indanthrone Blue *n* A member of a group of vat dye pigments. Relative to phthalocyanine blue, it is considerably redder in hue and superior in bronze resistance; lightfastness is excellent. It shows no tendency toward crystal growth in conventional paint thinners or solvents.

Indelible Ink \in-ˈde-lə-bəl-\ Compositions which are made water- and alkali-fast. Used on cloth to withstand laundering.

Indene *n* C_9H_8. Colorless liquid with a sp gr of 1.006 (20/4°C), mp of 3.5°C, bp of 182°C, refractive index of 1.5726, and a flp of 78.3°C. Insoluble in water, soluble in most organic solvents, rapidly absorbs oxygen from the air, forms polymers by exposure to air and sunlight. Derivation: Contained in the fraction of crude coal tar distillates which boils from 176°C to 182°C. Uses: Preparation of synthetic resins. See ▶ Coumarone-Indene Resins.

Indene Resins *n* A colorless, liquid hydrocarbon, C_9H_8, obtained from coal tar by fractional distillation: used in synthesizing resins. See ▶ Coumarone-Indene Resins.

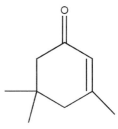

Indentation Hardness *n* Resistance to penetration by an indenter. The hardness of a material as determined by either the size of an indentation made by an indenting tool under a fixed load, or the load needed to produce penetration of an indenter to a predetermined depth. The instruments commonly used with plastics are the Shore Durometer (indenter A for soft resins and elastomers, D for hard materials) described in ASTM D 2240, and the Barcol Impressor, ASTM D 2583. In D 2240, the authors say, "No simple relationship exists between indentation hardness determined by this method and any fundamental property of the material tested. For specification purposes it is recommended that Test Method D 1415 (section 09.01) be used for soft materials and method A of D 530 or Test D 785 be used for hard materials." D 530 and D 785 use the Rockwell test in which a ball of suitable diameter is pressed into the sample with known force, and the indentation depth is noted. A similar test format is used in D 1415. See ▶ Knoop Hardness Number and ▶ Pfund Hardness Number.

Index of Metamerism \-mə-ˈta-mə-ˌri-zəm\ An index to describe the degree of metamerism existing between two samples. There is no standard accepted index. However, a frequently computed index is the color difference, using a specified equation, which exists under a second illuminant when a pair of colors is an exact match under a primary or first illuminant.

Index of Refraction *n* (1829) (refractive index) The ratio of the velocity of light in a vacuum to its velocity in the transparent material of interest. It generally varies with the wavelength of the light, being higher at the short wavelengths, also with temperature. When a light beam passes from a less dense medium to a denser medium at an acute angle θ_1 with the interfacial normal, it will be bend closer to the normal in the more dense medium, defining the angle of refraction, θ_2. The index of refraction is given by sin θ_1/sin θ_2. This relationship is the basis for measurement of refractive index with an instrument such as the Abbé refractometer (www.astm.org) for transparent plastics). Index of refraction is useful for identification of plastics, as well as minerals and liquids, and is involved in many quantitative analytical methods. (Serway, R. A., et al., *College Physics*, Thomas, New York, 2005; Moller, K. D., *Optics*, Springer, New York, 2003; *Handbook of Chemistry and Physics*, Weast, R. C., ed., The Chemical Rubber, Boca Raton, FL, 52nd Ed.).

India Ink *n* {*often cap 1st* I} 1665) A liquid ink made of lampblack mixed with a binding material and water. *Also called Chinese Ink.*

Indian Ochre \-ˈō-kər\ See ▶ Iron Oxides, Natural.

Indian Red *n* (ca. 1753) Natural, Pigment Red 101 (77491). Synthetic Pigment Red 101 (77491). Crystalline copperas (ferrous sulfate in crystal form) is converted by calcinations to ferric oxide, producing a range of light to dark red oxide pigments. The dark shades are known as Indian red. Original Indian reds were a variety of natural red iron oxides obtained from India.

Indian Red Ochre See ▶ Iron Oxides, Natural.

Indian Yellow *n* Purree. A yellow pigment containing the magnesium salt of euxanthic acid.

India Rubber *n* {*often cap I*} (1790) See ▶ Rubber, Natural.

Indicator \ˈin-də-ˌkā-tər\ *n* (1666) A conjugate acid-base pair in which at least one of the pair is highly colored.

Indicators *n* Substances which change from one color to another when the hydrogen ion concentration reaches a certain value, different for each indicator.

Indicatrix *n* A three-dimensional construction of the optical relationships in the crystal. Radii proportional in length to refractive index values or velocities may be used to represent principal index directions of the crystal.

Indicia \in-ˈdi-sh(ē)ə\ *n plural* [L, plural of *indicium* sign, fr. *indicare*] (ca. 1626) Any markings such as symbols, lettering, small pictures, etc., applied to a plastic article to identify it.

Indigo \ˈin-di-ˌgō\ *n* [It dialect, fr. L *indicum*, fr. Gk. *indikon*, fr. neuter of *indikos* Indic, fr. *Indos* India] (1555) $C_{16}H_{10}N_2O_2$ (molecular weight, 262.26). Dark blue, transparent dye produced either by fermentation from the genus, *Indigofera*, indigenous to the Far East, or synthetically.

Indophenol Test \ˌin-dō-ˈfē-ˌnōl-\ *n* [ISV] See Gibbs Indophenol Test.

Induced Decomposition Reaction *n* Chain transfer in a free radical polymerization by reaction of a growing active center with an undissociated initiator molecule, especially with a peroxide.

Induced Electromotive Force *n* In a circuit is proportional to the rate of change of magnetic flux through the circuit

$$E = \frac{d\phi}{dt}$$

where $d\phi$ is the change of magnetic flux in a time dt. The induced current will be given by

$$I = \frac{d\phi}{Rdt}$$

where R is the resistance of the circuit. (*Handbook of Chemistry and Physics*, Weast, R. C., ed., The Chemical Rubber, Boca Raton, FL, 52nd Ed.).

Inductance \in-ˈdək-tən(t)s\ *n* (1886) The change in magnetic field due to the variation of a current in a conducting circuit causing an induced counter electromotive force in the circuit itself. The phenomenon is known as *self-induction*. If an electromotive force is induced in a neighboring circuit the term mutual injection is used. Inductance may thus be distinguished as self- or mutual and is measured by the electromotive force produced in a conductor by unit rate of variation of the current. Units of inductance are the centimeter (absolute electromagnetic) and the ▶ Henry, which is equal to 10^9 cm of inductance. The ▶ *Henry* is that inductance in which an induced electromotive force of

1 V is produced when the inducing current is changed at the rate of 1 A/s. Thus, 1 H = 1 V/(A/s) = 1 V·s/A = 1 Wb/A. Dimensions, $[e^{-1}l^{-1}t^2]$; $[\mu l]$. (*Handbook of Chemistry and Physics*, Weast, R. C., ed., The Chemical Rubber, Boca Raton, FL, 52nd Ed.).

Induction \-shən\ *n* (14c) Any change in the intensity or direction of a magnetic field causes an electromotive force in any conductor in the field. The induced electromotive force generates an induced current if the conductor forms a closed circuit. (*Handbook of Chemistry and Physics*, Weast, R. C., ed., The Chemical Rubber, Boca Raton, FL, 52nd Ed.).

Induction Heating *n* (1919) A method of heating electrically-conductive materials, usually metallic parts, by placing the part or material is an alternating magnetic field generated by passing an alternating current through a primary coil. The alternating magnetic field induces eddy currents in the piece that generate hysteretic heat. Plastics, being poor conductors, cannot be heated directly by induction, but the process is used indirectly in welding of plastics and has been used to heat extruders and dies. (*Handbook of Chemistry and Physics*, Weast, R. C., ed., The Chemical Rubber, Boca Raton, FL, 52nd Ed.).

Induction Period *n* The time period before a chemical reaction (e.g., polymerization) proceeds that usually includes the production of free radicals or ions; rise in temperature and other reaction conditions before an initiation of polymerization occurs; other necessary periods of time preceding a chemical reaction between one or more chemical species prior to production of a different species. (Odian, G. C., *Principles of Polymerization*, Wiley, New York, 2004).

Induction Period *n* (**of Drying**) That period prior to set-to-touch in the drying process where there is but a small change in consistency of the binder and the natural inhibitors of the oil are removed by oxidation.

Induction Welding *n* A method of welding thermoplastic materials by placing a conductive metal insert between two plastic surfaces to be joined, applying pressure to hold the surfaces together, heating the metallic insert in an alternating magnetic field until the adjacent plastic is softened and welded, then killing the field and cooling the joint.

Induline *n* Range of blue or black dyestuffs belonging to the safranine group. They can be prepared in oil, spirit, and water soluble forms. They are used as dyes for leather and in lake manufacture.

Induline Bases *n* Substances prepared from amidoazobenzene aniline and aniline hydrochloride.

Industrial Fabric *n* A broad term for fabrics used for non-apparel and non-decorative uses. They fall into several classes: (1) a broad group including fabrics employed in industrial processes (e.g., filtering, polishing, and absorption), (2) fabrics combined with other materials to produce a different type of product (e.g., rubberized fabric for hose, belting, and tires; fabric combined with synthetic resins to be used for timing gears and electrical machinery parts; coated or enameled fabrics for automobile tops and book bindings; and fabrics impregnated with adhesive and dielectric compounds for application in the electrical industry), and (3) fabrics incorporated directly in a finished product (e.g., sails, tarpaulins, tents, awnings, and specialty belts for agricultural machinery, airplanes, and conveyors). Fabrics developed for industrial uses cover a wide variety of widths, weights, and constructions and are attained, in many cases, only after painstaking research and experiment. Cotton and manufactured fibers are important fibers in this group, but virtually all textile fibers have industrial uses. The names mechanical fabrics or technical fabrics sometimes have been applied to certain industrial fabrics.

Industrial Finishes or Coatings *n* Coatings which are applied to factory-made articles (before or after fabrication), usually with the help of special techniques for applying and drying – as opposed to trade sales paints. See ▶ Architectural Coatings.

Industrial Maintenance Paints *n* High performance coatings which are formulated for the purpose of heavy abrasion, water immersion, chemical, corrosion, temperature, electrical, or solvent resistance. See also ▶ Maintenance Paints.

Industrial Methylated Spirits See ▶ Methylated Spirit.

Industrial Talc *n* $Mg_3Si_4O_{10}(OH)_2$. A mineral product varying in composition from that approaching the theoretical formula of talc (which is non-fibrous), to mixtures of talc and other naturally associated minerals, some of which may be fibrous as defined in ASTM D 2946 and/or asbestos. See ▶ Asbestos; Magnesium Silicate, Fibrous.

Industrial Talc, Asbestos-free *n* Industrial talc of which less than two particles per 100 particles are asbestos, where "asbestos fiber" is defined as being both by a fiber by ASTM D 2946 and one of the asbestiform varieties of serpentine, riebeckite, cummingtonite (which are chrysotile, crocidolite and amosite, respectively), anthopyllite, tremolite, or actinolite. The non-asbestiform varieties of these same materials are not asbestos.

Industrial Waste Water *n* Water discharged from an industrial process as a result of formation or utilization in that process.

Industrial Water *n* Water (including its impurities) used directly or indirectly in industrial processes.

Inert \i-ˈnərt\ *adj* [L *inert-, iners* unskilled, idle, fr. *in-* + *art-, ars* skill] (1647) The term applied to various extended pigments such as asbestine, barites, silica, calcium sulfate, mica, talc, etc. In general, they have poor hiding power but they are inert from a chemical and physical standpoint. While they contribute some desirable properties to paint, they are primarily used to lower the cost.

Inert Additive *n* A material added to a plastic compound to improve properties, but which does not react chemically with any other constituents of the composition.

Inert Base *n* A paint base which does not provide hiding, color, or drying properties. Its main function is to provide solid, usually at low cost.

Inert Gas *n* (1902) Any of a group of rare gases that include helium, neon, argon, krypton, xenon, and sometimes radon and that exhibit heat stability and extremely low reaction rates. *Also known as Nobel Gas.*

Inertia \i-ˈnər-shə, -shē-ə\ *n* [NL, fr. L, lack of skill, fr. *inert-, iners*] (1713) The resistance shown by all matter to any attempt to change its state of motion.

Inert Pair *n* The pair of *s* electrons in an atom with a valence shell containing at least one *p* electron.

Inert Pigment *n* (1) A pigment which remains relatively inactive or chemically unchanged in paints under stated conditions. The term has little significance unless the conditions are stated. (2) Frequently used for extender.

Inflammable \in-ˈfla-mə-bəl\ *adj* [F, fr. ML *inflammabilis*, fr. L *inflammare*] (1605) (Deprecated) A general term once used to describe combustible gases, liquids or solids. Now obsolete. See ▶ Flammable.

Inflatable Structures Structures opened or enlarged by input of air and, once enlarged, able to retain the air to maintain the distended position.

Inflow Quench *n* Cooling air for extruded polymer filaments that is directed radically inward across the path of the filaments. The threadline is completely enclosed in a quench cabinet in inflow quenching.

Information Retrieval *n* Recovery of data from a collection for the purpose of obtaining information. Retrieval includes all the procedures used to identify, search, find, and remove specific information or data stored.

Infrared \ˌin-frə-ˈred, -(ˌ)frä-, -fə-\ *adj* (1881) (1) The region of the electromagnetic spectrum including wavelengths from 0.78 to about 300 μm. Infrared radiation serves as a source of heat. (2) Rays which are longer than visible red light rays and are felt as heat. Infrared radiant energy is used to cure coatings.

Infrared Drying (or IR Drying) *n* Method of heating for drying, curing and/or fusing a coating. The radiations involved in this form of heating have wavelengths greater than 780 nm, and they may be derived either from specially designed filament bulbs or special types of gas ovens.

Infrared Heating *n* A heating process used mostly in sheet thermoforming and for drying coatings, employing lamps or heating elements that emit invisible radiation at wavelengths of about 2 μm.

Infrared (or I.) Spectrophotometry \-ˌspek-trō-fə-ˈtä-mə-tər\ *n* An analytical instrumental technique based on selective absorption of infrared radiation by organic and inorganic materials which helps identify them (Stuart, B. H., *Infrared Spectroscopy*, Wiley, New York, 2004). An example of an infrared spectrum of toluene is shown.

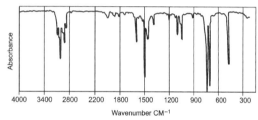

Infrared spectrum of toluene

Infrared Polymerization Index *n* (IRPI) A number representing the degree of cure of phenolic resins, defined as the ratio of absorbances by the sample, corrected for background absorbance, at wavelengths of 12.2 and 9.8 μm (far infrared). The test is often used in conjunction with the ▶ Marquardt Index.

Infrared Pyrometer \-pī-ˈrä-mə-tər\ *n* A narrow- or broad-band instrument that senses the peak wavelength, λ (μm) of IR radiation emanating from a warm surface. By Wien's displacement law, the absolute temperature (K) is given by 2,884/λ.

Infrared Spectroscopy \-spek-ˈträs-kə-pē\ *n* A technique to identify and quantitatively determine many organic substances such as plastics. All chemical compounds have characteristic intramolecular vibratory motions and can absorb incident radiant energy if such energy is sufficient to increase the vibrational motions of the atoms. With most organic molecules, vibrational motions of the substituent groups within the molecules coincide with the electromagnetic frequencies of the infrared region. An infrared spectrophotometer directs IR radiation through a film or layer or solution of the sample and measures and records the relative amount of energy absorbed by the sample as a function of the wavelength or frequency of the radiation. The chart produced is compared with charts of known substances to identify the sample. Peak area is a measure of the percentage of the various compounds present as generated by the Spectrum 100 FTIR (courtesy of PerkinElmer. Inc.).

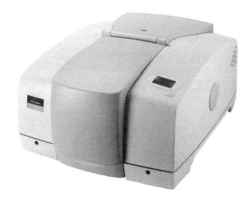

Infrared Thermography \-(▌)thər- ▌mä-grə-fē\ (thermographic NDT) A type of nondestructive testing in which damage development during fatigue testing of composites may be detected without interrupting the test by measuring locally generated infrared radiation. The damage is associated with heat developed through two mechanisms: hysteresis and frictional heating from rubbing at the surface of developing cracks.

Infusible \(▌)in- ▌fyü-zə-bəl\ *adj* (1555) Not capable of melting when heated, as are all cured thermosetting resins and a few special thermoplastics such as ultrahigh-molecular-weight polyethylene, polybenzimidazole, and aramid resins.

Infusorial Earth \(▌)in-fyü- ▌zōr-ē-əl-\ *n* (1882) See ▶ Diatomaceous Silica.

Ingrain \ ▌in- ▌grān\ *adj* (1766) See ▶ Dyeing.

Inherent Flame Resistance *n* As applied to textiles, flame resistance that derives from an essential characteristic of the fiber from which the textile is made.

Inherent Viscosity *n* (logarithmic viscosity number) In measurement of dilute-solution viscosities, the inherent viscosity is the ratio of the natural logarithm of the relative viscosity to the concentration of the polymer in g/dL of solvent: $\ln(\eta_{ref})/c$. See ▶ Dilute-Solution Viscosity.

Inhibition, Inhibitors *n* A substance capable of retarding or stopping an undesired chemical reaction. Inhibitors are used in certain monomers and resins to prolong storage life. When used to retard degradation of plastics by heat and/or light, an inhibitor functions as a ▶ Stabilizer.

Inhibitive Pigment *n* Pigment which assist in the prevention of cedar staining or Redwood staining, mildew, corrosion, yellowing, etc. See ▶ Corrosion-Inhibitive Pigments.

Inhibitor \in- ▌hi-bə-tər\ *n* (ca. 1611) (1) General term for compounds or materials that have the effect of slowing down or stopping an undesired chemical change such as corrosion, oxidation or polymerization, drying, skinning, mildew growth, etc. Also see ▶ Corrosion Inhibitor and ▶ Inhibitive Pigment. (2) A negative catalyst which prevents or retards vulcanization or oxidation. A good inhibitor will retard vulcanization at room temperature to prevent precure but will not retard at normal curing temperatures.

Initial Boiling Point *n* Temperature indicated by the distillation thermometer at the instant the first drop of condensate leaves the condenser tube.

Initial Modulus The ▶ Modulus of Elasticity extrapolated to zero strain from measurements in the low-strain region.

Initial Viscosity *n* A term used in the vinyl plastisol industry to denote the viscosity measured immediately after the plastisol has been mixed. The viscosity normally rises at a declining rate after mixing.

Initiation \i- ▌ni-shē- ▌ā-shən\ *n* (1583) The initial stage of polymerization usually comprising a compound (e.g., peroxide or hydroperoxide), heat or irradiation or combination that produced a free radial or ion that causes monomers to successfully attach (polymerization phase) in a chain reaction; not condensation polymerization which utilizes a catalyst without an initiator.

Initiator \i- ▌ni-shē- ▌āt\ *vt* [LL *initiatus*, pp of *initiare*, fr. L, to induct, fr. *initium*] (ca. 1573) An agent that causes a chemical reaction to commence and that enters into the reaction to become part of the resultant compound. Initiators differ from catalysts in that catalysts do not combine chemically with the reactants. Initiators are used in many polymerization reactions, especially in emulsion polymerizations. Initiators most commonly used in polymerizing monomers and resins having ethenic unsaturation (−C = C−) are the organic peroxides. (Odian, G. C., *Principles of Polymerization*, Wiley, New York, 2004).

Initiator Activator *n* It is usually necessary to add an activator (e.g., $FeSO_4 \cdot 7H_2O$) to attain reasonable rates of polymerization when redox initiators (lauroÿl peroxide and fructose) are used at zero degree temperature or below. (Odian, G. C., *Principles of Polymerization*, Wiley, New York, 2004).

Injection Blow Molding *n* A blow-molding process in which parisons are formed over mandrel by injection molding, after which the mandrels and parisons are transferred to blow molds where the final shape is blown. While the parts are being blown, cooled, and ejected, another set of parisons is being injection-molded. Advantages of the process are that it delivers a completely finished part requiring no neck trimming, etc., closer tolerances are possible, and parison-wall thicknesses can be locally varied as desired.

Injection Mold *n* A mold used in the process of ▶ Injection Molding. The mold usually comprises two main sections held together by a hydraulic or mechanical clamping press, that have sufficient strength and rigidity to contain the high pressure of molten plastic when it is injected. It is provided with channels for mold venting and for circulation of temperature-control media, and with means for product ejection.

Injection Molding *n* (1932) The method of forming objects from granular or powdered plastics, most often thermoplastics, in which the material is fed from a hopper into a screw-type plasticator (or heating cylinder in elderly machines), after which the screw or a ram forces the molten compound into a chilled mold. Pressure is maintained until the mass has hardened sufficiently for removal from the mold. In a variation called *flow molding*, a small additional amount of melt is forced into the mold during cooling of the initial charge to offset shrinkage. Machines employing screws for plastication (as must do now days) are either single- or double-stage. In single-stage machines, plastication and injection are done in the same cylinder, injection pressure being generated by forward motion of the screw while rotation is stopped. This process is called *reciprocating-screw injection molding*. In double-stage machines, the material is plasticated by a constantly rotating screw that delivers the melt through a check valve to an accumulator cylinder, from which it is injected into the mold by a piston. This process is called *screw-and-piston injection molding*. See also ▶ Reaction Injection Molding, ▶ Hot-Tip-Gate Molding, and ▶ Two-Shot Injection Molding.

Injection Molding, Molds *n* Comprises two main sections held together by a hydraulic or mechanical clamping press, that have sufficient strength and rigidity to contain the high pressure of molten plastic when it is injected. It is provided with channels for mold venting and for circulation of temperature-control media, and with means for product ejection.

Injection Molding Nozzles *n* A short, thick-walled, hardened-steel tube that conducts the molten plastic emerging from an injection cylinder into the mold. It usually terminates in a hemispherical tip that closely mates a recess in the mold and is enclosed by band heaters that are controlled by a temperature sensor.

Injection-Molding Pressure *n* (1) The pressure applied to the plastic by the injection screw (or ram) during the injection stroke. (2) The pressure of the melt within the mold cavity just prior to the freezing of the gate.

Injection Nozzle *n* A short, thick-walled, hardened-steel tube that conducts the molten plastic emerging from an injection cylinder into the mold. It usually terminates in a hemispherical tip that closely mates a recess in the mold called a *sprue bushing* and is enclosed by band heaters that are controlled by a temperature sensor.

Injection Ram *n* In two-stage injection molders and elderly machines having no screws but using contact heating, the piston that applies pressure to the plastic during injection. In single-stage screw-injection machines, the screw doubles as the ram during a brief period of stopped rotation.

Injection Stamping *n* A little used modification of injection molding wherein first the plastic melt is injected under relatively low pressure into a mold that is vented during this stage, then, after the cavity is filled, additional clamping pressure is applied to completely close the mold and compress or "stamp" the molded shape. Molds are designed so that even in the venting position no material exudes onto the land areas. Advantages claimed for the process are lower injection time and pressure, and shorter cycle because the injection screw can resume plastication as soon as the mold has been charged.

Injector Molder *n* (1) A person or company that operates injection-molding machines. (2) An injection-molding machine.

Ink \ˈiŋk\ *n* [ME *enke*, fr. OF, fr. LL *encaustum*, fr. neuter of L *encaustus* burned in, fr. Gk *enkaustos*, verbal of *enkaiein* to burn in] (13c) A fluid or viscous substance used for writing or printing (*Printing Ink Handbook*, National Association of Printing Ink Manufacturers, 1976).

Ink Adhesion *n* The tendency of printing ink to stick to the printed surface and not smear or rub off. An informal test of ink adhesion consists of pressing pressure-sensitive tape over the printed area, waiting a few minutes, then stripping it off and observing the a real fraction of ink coming away on the tape (*Printing Ink Manual*, 5th Ed., Leach, R. H., et al., eds., Blueprint, New York, 1993).

Ink Compounds *n* Additives to remedy deficiencies in the ink vehicle or pigment (Apps, E., *Ink Technology for Printers and Students* (3 Vols.), L. Hill, London, 1963).

Ink-Embossed Wallpaper *n* Differs from regular embossed wallpaper in that the ink colors are applied at the time the paper is being embossed.

Ink Fountain See ▶ Fountain.

Inkometer *n* An instrument indicating the tack of an ink in terms of the torque developed by a system of rotating ink-wet rollers. Useful in determining tack differences in series of inks for multicolor wet printing and in the standardization of ink consistency.

Ink Receptivity *n* That property of a sheet of paper (or other material) which causes it to accept and/or absorb ink' (Wolfe, H. J., *Printing and Litho Inks*, MacNair-Dorland, New York, 1957).

Inks, Cover *n* Generally opaque, heavy bodied and have a high pigment concentration.

Inks, Dull *n* Inks that dry with a dull or matte finish.

Inks, Halftone *n* Inks formulated for good reproduction of fine detail such as half-tone dots on coated stock. They generally have high tinctorial strength and are finely dispersed.

Inks, Job *n* Heavy bodied inks formulated to print on uncoated stock using small sheet-fed presses.

Inks, Quick-Setting *n* Inks for letterpress and offset which dry by either filtration, coagulation, selective absorption or often a combination of these with some of the other drying methods. The vehicles are generally special resin-oil combinations which, after the inks have been printed, separate into a solid material which remains on the surface as a dry film and an oily material which penetrates rapidly into the stock. This rapid separation gives the effect of very quick setting or drying.

Inlay \(ˌ)in-ˈlā, ˈin-ˌ\ *n* (1596) Pattern or design formed by inserting cut shapes of contrasting materials into a surface, usually at the same level.

Inlay Printing See ▶ Valley Printing.

In-Mold Decorating *n* The process of applying labels or decorations to plastic articles during the molding operation by which they are formed. Two basic methods, each with many variations, are in use. The first employs a preprinted label of plastic film, paper, or cloth that is positioned in the mold prior to molding. During the molding cycle, the label or its printed image fuses to and becomes an integral part of the article. In the second method, the image is printed directly onto the mold surface with wet or dry ink, or applied to the mold by an offset process. In-mold decorating is done with blow molding, injection molding, and casting.

Inoculum \i-ˈnä-kyə-ləm\ *n* [NL, fr. L *inoculare*] (1902) That portion of the pathogen which is transferred to the plant or plant part for inoculation purposes.

Inorganic \ˌi-(ˌ)nȯr-ˈga-nik\ *adj* (1794) Designation of compounds that generally do not contain carbon. Source: matter other than vegetable or animal. Examples: sulfuric acid and salt. Exceptions are carbon monoxide and carbon dioxide and their derivatives.

Inorganic Coatings *n* Coatings based on silicates or phosphates and usually used pigmented with metallic zinc.

Inorganic Fibers *n* Fibers that are composed of matter other than hydrocarbons and their derivatives and are not of plant or animal origin.

Inorganic Pigments *n* (1) Any pigment derived from naturally occurring minerals or synthesized from inorganic substances. They are always opaque, unlike organic pigments and dyes, and are usually resistant to heat and light. Examples of those used in plastics are titanium dioxide, iron oxides, ultramarines, lead chromates, and cadmium compounds. (2) A class of pigments used in printing ink manufacture consisting of compounds of the various metals. These pigments do not contain carbon such as iron oxide. Example: Chrome Yellow.

Inorganic Polymer *n* Within the scope of the plastics industry, an inorganic polymer may be defined as a polymer without carbon in its main chain and of a degree of polymerization sufficient for the polymer to exhibit considerable mechanical strength, plastic or elastomeric properties, and the ability to be formed by processes used with plastics. Organic-group side chains may be present. The inorganic polymers of greatest commercial importance are the ▶ Silicones. (*Printing Ink Manual*, 5th Ed., Leach, R. H., et al., eds., Blueprint, New York, 1993).

Inpaint *n* To renew damaged areas on paintings or painted surfaces by repainting.

Insect Wax See ▶ Chinese Wax.

Insert \ˈin-ˌsərt\ *n* (ca. 1889) An article of metal or other material that is incorporated in a plastic molding either by pressing the insert into a recess in the finished molding or by placing the insert into the cavity so that it becomes a part of the molding. A common example is a metal bushing, knurled on the outside, threaded on the inside, for attaching the molded article to another article with a screw or bolt.

Insert Bonding *n* Method of incorporating an article of metal or other material in a plastic molding either by pressing the insert into a recess in a finished molding or by placing the insert into the cavity so that it becomes a part of the molding.

Insert Molding *n* The process by which components may be molded into a part. See ▶ Double-Shot Molding.

Inserts An important part of a plastic molding consisting of metal, plastic, or other material that may be molded into position or may be pressed into the molding after the molding is fabricated.

In-Situ Foaming \( )in-ˈsī-( )tü\ *adv, adj* [L, in position] (1740) The technique of depositing a foamable plastic (prior to foaming) into the volume where it is intended that foaming shall occur. An example is the placing of foamable plastics into cavity brickwork to provide insulation. Shortly after being poured, the liquid mix foams and fills the cavity. See also ▶ Cellular Plastic.

Insoluble \( )in-ˈsäl-yə-bəl\ *adj* [ME *insolible*, fr. L *insolubilis*, fr. *in-* + *solvere* to free, dissolve] (14c) Of low solubility; slightly soluble; sparingly soluble.

Inspection \in-ˌspek-shən\ *n* (14c) The process of examining textiles for defects at any stage of manufacturing and finishing.

Instron *n* A tensile or other testing machine made by the Instron Corp of Canton, MA. Their machines have been so omnipresent in plastics testing that "Instron" has taken on a near-generic character.

Instron Tensile Tester *n* A high precision electronic test instrument designed for testing a variety of material under a broad range of test conditions. It is used to measure and chart the load-elongation properties of fibers, yarns, fabrics, webbings, plastics, films, rubber, leather, paper, etc. It may also be used to measure such properties as tear resistance and resistance to compression.

Insulated-Runner Mold See ▶ Hot-Runner Mold.

Insulating Varnish *n* Varnish often used in combination with materials such as mica, fabric, paper, etc., to provide electrical insulation for electrical equipment. These are formulated to have a high resistance to electrical passage.

Insulation Resistance *n* (1) The electrical resistance between two conductors or systems of conductors separated only by an insulating material. The resistance of a particular insulation may be measured by dividing the voltage difference applied to two electrodes in contact with, or embedded in a unit area of the specimen by the current flowing between the electrodes. Tests for thermoplastics include ASTM D 2633, section 10.02. See also ▶ Resistivity. (2) See ▶ Thermal Resistance.

Intaglio \in-ˈtal-( )yō\ *n* [It, fr. *intagliare* to engrave, cut fr. ML *intaliare*, fr. L *in-* + LL *taliare* to cut] (1644) (1) Printing style in which the design is cut into the surface of the cylinder and is thus below the surface. (2) A lustrous, brocade pattern knitted in a tricot fabric.

Integral-Skin Molding *n* A method of producing urethane-foam articles with substantially nonporous integral skins in one operation. Whereas normal urethane foams are inflated by carbon dioxide generated by reaction of isocyanate with excess water in the mixture (see ▶ Urethane Foam), integral-skin foam is expanded by the vapor of a volatilized solvent such as hexane. The mold must be heat conductive so that a layer next to the mold surfaces can be chilled to prevent foaming. The reaction mixture, typically consisting of a polyol, an isocyanate, and the blowing solvent, is introduced rapidly into a closed mold located near the geometric center of the mold. The mold is usually preheated to between 40°C and 65°C, but the main heat needed to foam and cure the mass is provided by the reaction exotherm. After the mass has gelled, the skin should be pierced to equalize pressure and prevent shrinkage. The solvent blowing agent can be removed from the finished article by allowing it to stand at ambient conditions for about a day, or by oven drying for 0.5 h at 140°C. In some instances, integral-skin moldings have replaced composites of vinyl-covered urethane foam articles and cast vinyl skins filled with urethane foam, such as automotive arm rests, crash pads, and instrument-panel covers.

Integrating Sphere *n* A sphere coated inside with a highly reflective, diffuse material and used for the measurement of luminous flux. If the internal surface is perfectly diffuse, the intensity of any part of the sphere is the same. Many instruments used for reflectance measurements utilize such a device for measuring the diffuse or total reflectance from a sample material relative to a reference material.

Integration *n* (**Colorimetry**) The process of summing the products of the three values: the illuminant, the observer response characteristics and the reflectance or transmittance of an object at specified wavelengths in order to obtain the tristimulus values. See ▶ Tristimulus Values.

Intensity \in-ˈten(t)-sə-tē\ *n* (1665) (1) The amount of energy per unit (space, charge, time). (2) The brilliance of a color. (3) The brightness of light.

Intensity of Illumination *n* (Properly called ▶ Illumination). Illumination in *lux* of a screen by a source of illuminating power P at a distance r meters, for normal incidence,

$$I = \frac{P}{r^2}$$

If two sources of illuminating power P_1 and P_2 produce equal illumination on a screen when at distances r_1 and r_2 respectively,

$$\frac{P_1}{r_1^2} = \frac{P_2}{r_2^2} \text{ or } \frac{P_1}{P_2} = \frac{r_1^2}{r_2^2}$$

If I_o is the illumination when the screen is normal to the incident light, then I is the illumination when the screen is at an angle θ. Thus,

$$I = I_o \cos \theta$$

Intensity of Magnetization *n* The rate of transfer of energy across unit areas by the radiation. In all forms of energy transfer by waves (radiation) the intensity I is given by $I = Uv$ where U is the energy density of the wave in the medium, and v is the velocity of propagation of the wave. The energy density U is always proportional to the square of the wave amplitude.

Intensity of Segregation *n* In mixing, the average measure of the deviation in concentration of a component at any point in a mixture from the mean concentration. The ▶ Standard Deviation as determined by sampling is usually used.

Intensity of Sound *n* Depends upon the energy of the wave motion. The intensity is measured by the energy in ergs transmitted per second through 1 cm² of surface. The energy in ergs per cubic centimeter in a sound wave is given by

$$E = 2\pi^2 dn^2 a^2$$

where d is density in gram per cubic centimeter, n is frequency in vibration, per second and a is amplitude in centimeter. the energy reaching the ear in unit time will also be proportional to the velocity of propagation.

Intensive Mixer See ▶ Internal Mixer.

Interaction Parameter *n* Interaction parameter χ in solution theory; Flory-Huggins parameter χ; a measure of the interaction energy ΔE which is a measure of the Gibbs energy, but not of the enthalpy.

Intercoat Adhesion *n* The adhesion between two coats of paint.

Intercoat Contamination *n* Presence of foreign matter between successive coats.

Interdeterminancy Principle *n* (Uncertainty Principle) The postulate that it is impossible to determine simultaneously both the exact position and the exact momentum of an electron. So this aspect of electronics can only be expressed as a probability.

Interface \\ˈin-tər-ˌfās\\ *n* (1882) The common surface separating two different phases, e.g., the resin-glass interface in glass-fiber-reinforced plastics.

Interfacial Angle *n* That angle between two adjoining faces of a crystal.

Interfacial Angle (Polar) *n* That angle between the normals to two adjoining faces of a crystal.

Interfacial Polycondensation *n* Involves polymer formation at or near the interface between two immiscible monomer solutions under very mild reaction conditions.

Interfacial Polymerization *n* A polymerization reaction that occurs at or near the interfacial surfaces of two immiscible solutions. A simple example is the often-performed demonstration of making nylon thread from a beaker containing a lower layer of a solution of sebacyl chloride in carbon tetrachloride and an upper layer of hexamethylene diamine solution in water. A pair of tweezers is gently lowered through the upper layer, closed on the interfacial layer of polymer, and then drawn upward to pull with it a continuous strand of nylon 6/10.

Interfacial Tension *n* (interfacial surface energy) (1) The tension in, or energy of, the interfacial surface between two immiscible liquids. Measurable, like ▶ Surface Tension by capillary rise or with a du Noüy tensiometer. (2) The molecular attractive force between force between unlike molecules at an interface. It may be expressed in dynes per centimeter.

Interfacing See ▶ Interlining.

Interference Color *n* Color which appears in thin films, illuminated with composite light as the result of the reinforcement of some colors (wavelengths) and weakening or elimination of others, depending on the phase differences and amplitudes of the light waves involved. The color produced depends on the angle of incidence, thickness of the film, and the refractive index. Interference color also appears when light strikes narrow slits, the principle of the diffraction grating.

Interference, Constructive *n* The retardation of two light beams by exactly one wavelength or even multiples of one wavelength. The two waves reinforce each other, resulting in brightness.

Interference, Destructive *n* The retardation of two light beams exactly an odd number of half wavelengths, resulting in darkness as the two waves are perfectly out of phase.

Interference Figure *n* The conoscopic pattern of extinction positions of a crystal superimposed on the pattern of interference colors corresponding to the full cone of directions by which the crystal is illuminated, each direction showing its own interference color.

Interference Filter *n* Optical filter made with extremely thin alternate layers of metals and dielectrics so that it passes only very narrow wavelength bands. Such filters are frequently used in abridged spectrophotometers where they serve to isolate discrete wavelengths for measuring, thus effectively replacing a dispersion device in a spectrophotometer. See ▶ Spectrophotometer, Abridged.

Interference Microscopy *n* Optical microscopy in the most common arrangement of which the incident beam is split into two non-parallel oppositely polarized beams. The beams are reflected back from the specimen and its substrate respectively and recombined. The sample beam follows a shorter path length and the beams recombine destructively producing a dark image.

Interference Pigments *n* Materials used as pigments which depend on the interference of light for their color. They are generally made from platelets coated with a thin film of higher refractive index. For example, thin films of titanium dioxide coated on mica flakes are one type manufactured and used today. See ▶ Pearlescent Pigments.

Interferometry \ in-tə(r)-fə- rä-mə-tər\ *n* [ISV] (1897) Any system of measurement based on wave interference between split rays of a light beam, one of which takes a longer or more optically dense path than the other, thus delaying it and generating a phase difference between the two rays. Interferometry has been used to measure a great variety of quantities, e.g., the speed of light, film thicknesses, and chemical concentrations in solutions.

Interfusion \ in-tər- fyüz\ *v* [L *interfusus*, pp of *interfundere* to pour between, fr. *inter-* + *fundere* to pour] (1593) (infusion) A novel method for emplacing hard-surfacing alloys within extruder barrels and, in future, on other parts. Few details have been released (as of late 1992) but it appears the hard-facing material in powder form is distributed around the barrel surface by centrifugal action while induction heating, and perhaps high pressure, are applied to speed the diffusion of that material into the barrel near its inner surface. The process marketers, Inductamentals of Chicago, claim much slower barrel wear than is experienced by cast-in liners. See also ▶ Bimetallic Cylinder.

Interhalogen *n* A compound of two different halogens.

Interlaminar Shear Stress *n* Shear stress between layers of a laminate, an important cause of laminate failure and delamination.

Interlayer *n* An intermediate sheet in a laminate.

Interlining *n* A padding or stiffening fabric used in garment manufacture to provide shape retention. Interlining is sandwiched between layers of fabric.

Interlocked Grain *n* Wood in which the fibers incline in one direction in a number of rings of annual growth, then gradually reverse and incline in an opposite direction in succeeding rings and then reverse again.

Interlock Knit *n* To produce an interlock knit, long and short needles are arranged alternately in both the dial and cylinder; the needles in the dial and cylinder are also positioned in direct alignment. When the long and short needles knit in alternate feeds in both needle housings, a fabric with a type of cross 1 × 1 rib effect is produced.

Interlock Twiner *n* A machine for making three-dimensional braided performs in which yarn bundles are entwined by braiders to form the desired shape.

Intermediate *n* (1650) A compound produced from raw materials that is to be used to synthesize end products. For example, benzene, originally distilled from coal tar and now made from petroleum constituents, and its derivatives, cyclohexane, cyclohexanol, and adipic acid are all intermediates in the manufacture of nylon 6/6.

Intermediate Temperature Setting Adhesive *n* See ▶ Adhesive, Inter-Mediate Temperature Setting.

Intermingling *n* (1) Use of air jets to create turbulence to entangle the filaments of continuous filaments yarns, without forming loops, after extrusion. Provides dimensional stability and cohesion for further processing but is not of itself a texturing process. It is compatible with high-speed spin-drawing and high-speed take-up. When compared with twisting processes, it also permits increased take-up package size. (2) Combining two or more yarns via an intermingling jet. This process can be used to get special effect yarns, i.e., mixing dye variants to get heather effects upon subsequent dyeing.

Intermittent Maximum Service Temperature *n* Maximum temperature at which a material can perform reliably in a short-term application.

Intermittent Pattern *n* A pattern occurring in interrupted sequence.

Intermolecular Polymerization *n* Polymerization which occurs as the result of association between molecules.

Internal Bubble Coating *n* In blown-film production, the circulation of chilled air or carbon dioxide inside the film bubble to substantially reduce the cooling time and thereby increase the production rate. The same practices have been used in blow molding.

Internal Dye Variability *n* The change from point to point in dye uniformity across the diameter and along the length of the individual filaments. Affects appearance of the dyed product and is a function of fiber, dye, dyeing process, and dyebath characteristics.

Internal Lubricant *n* A lubricant that is incorporated into the compound or resin prior to processing, as opposed to one that is applied to the mold or die. Examples of internal lubricant are waxes, fatty acids and their amines, and metallic stearates such as calcium, lead, lithium, magnesium,, and zinc stearates. The lubricants reduce friction and adhesion between polymers and metal surfaces, improve flow characteristics, and enhance knitting at weld surfaces and wetting properties of compounds. They are used primarily in rigid and flexible PRV, high-molecular-weight polyethylene, polystyrene, and acrylonitrile-butadiene-styrene, melamine, and phenolic resins.

Internally Plasticized *n* When a product is synthesized from a reaction involving two or more raw materials, it may be said to be internally plasticized if one of the raw materials is able to confer plasticity or flexibility to it. In other words, the product is plasticized because it is build up from a component which is naturally plastic. For example, an oil-modified alkyd, in which phthalic anhydride and glycerol are combined with drying oil fatty acids, is internally plasticized by reason of the presence of the fatty acid component. If a congo varnish were made by interaction of run congo and linseed oil monoglyceride, it could be described as internally plasticized, in contrast to a varnish made by simple dissolution of run congo in linseed oil, which would be externally plasticized. Also a polyvinyl acetate-acrylic copolymer is internally plasticized while a polyvinyl acetate homopolymer, into which a plasticizer has been stirred, is externally plasticized. See ▶ Internal Plasticizer.

Internal Mandrel Cooling *n* In extrusion of tubing and pipe, the mandrel, which is an extension of the die core, is usually cooled internally so that the hot extrudate, cooled from both surfaces, cools faster than if cooled only on the outside, and with better control of dimensions.

Internal Mix Spray Gun *n* Spray gun in which the fluid and air are combined before they leave the gun.

Internal Mixer *n* A heavy-duty machine in which the materials to be mixed are strenuously worked and fused by one or more rotors designed so that all parts of the charge pass repeatedly through zones of high shear. The shell and rotors may be cored to permit circulation of heat-transfer liquids to control batch temperature. One type, the ▶ Banbury Mixer has long been used in the compounding plastics and rubbers, doing a good job of both *dispersive* and *distributive* mixing. Internal mixers have the inherent advantage of minimizing dust and fume hazards. The disadvantage of most types is that processing is batchwise rather than continuous. See ▶ Banbury Mixer.

Internal Phase *n* In an emulsion, the discontinuous phase. For example, in an oil-in-water emulsion, the oil is the internal phase.

Internal Plasticization *n* Plasticization by means of internally combined groups, such as by copolymerization.

Internal Plasticizer *n* An agent incorporated in, or copolymerized with, a resin during its polymerization to make it softer and more flexible, as opposed to a plasticizer added to the resin during compounding. Common examples occur in oil-modified alkyds, and such materials have considerable practical advantages over externally plasticized systems, where the plasticizer may be lost by evaporation or leached out by water or other agents.

Internal Reflection Spectroscopy *n* A technique for exposing a sample to an infrared beam in spectroscopy, where a thin sample film is placed in direct contact with the reflecting surface of a prism of high refractive index. The output generated is a spectrum similar to the transmission spectrum of the surface layers of the sample.

Internal Stabilizer *n* An agent incorporated in a resin during its polymerization to make it more resistant to high temperatures and other processing and environmental conditions, as opposed to a stabilizer added to the resin during compounding.

Internal Undercut *n* Any restriction that prevents a molded part from being directly removed from its core (force).

International Gray Scale *n* A scale distributed through AATCC that is used as a comparison standard to rate degrees of fading from 5 (negligible or no change) to 1 (severe change). The term is sometimes applied to any scale of quality in which 5 is excellent and 1 is poor.

Interpenetrating Polymer *n* A material with two co-continuous phases or a material with 2 parts intertwined throughout.

Interpenetrating Polymer Network *n* (IPN) A kind of blend formed by swelling a crosslinked polymer with a monomer, than inducing polymerization of the monomer. Another route is to infuse a crosslinking monomer into a crosslinkable polymer. (Zaccaria, V. K., and Utracki, L., *Polymer Blends*, Springer, New York, 2003; Galina, H., et al., *Polymer Networks*, Wiley, New York, 2001).

Interpolymer A type of ▶ Copolymer in which the two monomer units are so intimately distributed in the polymer that the substance is essentially homogeneous in chemical composition. An interpolymer is sometimes called a *true copolymer*.

Inter-Society Color Council *n* (ISCC) A U.S. council made up of delegates from about 30 "member-bodies", i.e., national organizations and professional societies, and of individual members, all of whom have an interest in information bout color and in basic concepts of color. Member-body organizations and individual members represent: (1) creators of color effects in art and design; (2) manufacturers of colored material; (3) producers of color reproduction processes; and (4) fields of science and technology. The work of the Council is carried out in study groups, called "problem subcommittees," and is directed by specialists in the color problems under investigation. News of Council activities and newsworthy items in the world of color are published in the ISCC Newsletter, published bimonthly. Annual Council meetings provide a forum for discussion and resolution of color problems. Specialized interest meetings are held periodically. The ISCC is the official U.S. member of the International Colour Association – Association Internationale de la Couleur.

Interstice \in-ˈtər-stəs\ n [ME, fr. L *interstitium*, fr. *inter-* + *-stit-*, *-stes* standing (as in *superstes* standing over)] (15c) A space between objects; such as between atoms in a crystal.

Intimate Blend n A technique of mixing two or more dissimilar fibers in a very uniform mixture. Usually the stock is mixed before or at the picker.

Intramolecular Polymerization n Polymerization which occurs within an individual molecule. For example, drying oils consist of triglyceride esters, involving up to three unsaturated fatty acids attached to a single glycerol molecule. Under certain conditions, polymerization is believed to occur between neighboring fatty acid chains in a single molecule.

Intrinsic Viscosity \in-ˈtrin-zik-\ n (limiting viscosity number) In measurements of dilute-solution viscosity, intrinsic viscosity is the limit of the reduced and inherent viscosities as the concentration of polymer solute approaches zero. It represents the capacity of the polymer to increase viscosity. Interactions between solvent and polymer molecules give rise to different intrinsic viscosities for a given polymer in different solvents. Intrinsic viscosity is related to polymer molecular weight by the equation $[\eta] = K' \cdot M^a$, where the exponent a lies between 0.5 and 1.0, and, for many systems, between 0.6 and 0.8. *Also known as Limiting Viscosity Number.* See also ▶ Dilute-Solution Viscosity, ▶ Huggins Equation, and ▶ Viscosity-Average Molecular Weight.

Introfaction n The change in fluidity and wetting capability of an impregnating material, produced by addition of an ▶ Introfier.

Introfier n A chemical that will convert a colloidal solution into a molecular one, by improving the solubility of the colloidal material.

Intumesce n To form a voluminous char on ignition. Foaming or swelling when exposed to heat.

Intumescence \ˌin-tu-ˈme-sən(t)s\ n [F, fr. L *intumescere* to swell up, fr. *in-* + *tumescere* to swell] (ca. 1656) The foaming and swelling of a plastic when exposed to high surface temperatures or flames. It has particular reference to ablative urethanes used on rocket nose cones, and to ▶ Intumescent Coating. (*Handbook of Fire Retardant Coatings and Fire Testing Services*, Kidder, R. C., CRC, Boca Raton, FL, 1994).

Intumescent Coatings n A coating that, when exposed to flame or intense heat, decomposes and bubbles into a foam that protects the substrate and prevents the flame from spreading. Such coatings are used, for example, on reinforced-plastics building panels. Examples of such coating materials are magnesium oxychloride cement used on urethane foams, and certain epoxy coatings used on polyester panels. (*Handbook of Fire Retardant Coatings and Fire Testing Services*, Kidder, R. C., CRC, Boca Raton, FL, 1994).

Inventory \ˈin-vən-ˌtȯr-ē\ In injection molding or extrusion, the amount of plastic contained in the heating cylinder or barrel.

Inverse Emulsion Phase Polymerization n The reversal of emulsion polymerization phases; feeding the aqueous phase into a monomer phase (hydrophobic) where the water particles are dispersed in a monomer phase while agitating (and polymerizing monomers) until the phase shifts to an aqueous phase, and the monomer particles becomes dispersed in an aqueous phase; the phases shift when the aqueous component is more than 50% by volume than the monomer phase. (Becher, P., *Emulsions: Theory and Practice*, American Chemical Society, Washington, DC, 2001; *Emulsion Polymerization and Emulsion Polymers*, Lovell, P. A., and El-Aasser, eds., Wiley, New York, 1997).

Inversion Temperature n (1921) A temperature at which free expansion of a real gas produces neither heating nor cooling of the gas.

Investment Casting n (lost-wax process) A metals-casting method in which patterns made from wax or other expendable material are mounted on sprues, then "invested", i.e., covered with a ceramic slurry that sets at room temperature. The set slurry is then heated to melt away the pattern, leaving a mold into which metal is poured

Iodine Number n (Value) A number expressing the percentage (i.e., grams per 100 g) of iodine absorbed by a substance. It is a measure of the proportion of unsaturated linkages present and is usually determined in the analysis of oils and fats. See ▶ Iodine Value.

Iodine Test n Test for the detection of starches and dextrins. It involves the addition of a small amount of a solution of iodine in potassium iodide to an aqueous solution of the suspected substance. Starches give purpose or blue colors according to the degree of hydrolyzation, whereas erythrodextrin gives an orange or reddish-brown coloration.

Iodine Value *n* (iodine number) The number of grams of iodine that 100 g of an unsaturated compound will absorb in a given time under arbitrary conditions. It is used to indicate the residual unsaturation in epoxy hardeners (and many oils); a high value implies a high degree of unsaturation.

Ion \ˈī-ən, ˈī-ˌän\ *n* [Gk, neuter of *iōn*, pp of *ienai* to go] (ca. 1834) An atom, molecule or radical that has become electrically charged by having either gained or lost an electron. When an electron is gained the negatively charged ion is called an *anion*. A positively charged ion is called a *cation*.

Ion Current *n* The migration of ions between the electrodes of an electrochemical cell.

Ion Exchange *n* (1923) A reversible interchange of ions between a solid phase and a liquid phase in which there is no permanent change in the structure of the solid phase. In the leading application, water softening, an ▶ Ion-Exchange Resin extracts "hard," soap-precipitating calcium, magnesium, and iron ions from the water, replacing them with equivalent amounts of soluble sodium ions. Subsequently, the resin loaded with hard ions may be treated with salt solution (regenerated) to bring it back to the original sodium form, ready for reuse.

Ion-Exchange Resins *n* (1943) Any of several small granular or bead-like resins consisting of two principal parts: a resinous matrix serving as a structural portion, and an ion-active group serving as the functional portion. The functional group may be acidic or basic. Complete deionization of water is accomplished by use of both acidic and basic resins, in sequence or in mixed beds. Ion-exchange resins are also used for other chemical processes such as electrodialysis.

Ionic \ī-ˈä-nik\ *adj* [ISV] (1890) Pertaining to an atom, radical, or molecule that is capable of being electrically charged, either negatively (anionic) or positively (cationic) or both (amphoteric), or to a material whose atoms already exist in charged state. (Whitten, K. W., et al., *General Chemistry*, Brookes/Cole, New York, 1994).

Ionic Atmosphere The space around a given ion in solution, occupied largely by counterions.

Ionic Bond *n* (1939) The electrostatic attraction between ions of opposite electrical charge. (Whitten, K. W., et al., *General Chemistry*, Brookes/Cole, New York, 1994).

Ionic Bonding *n* Ionic bonding contributes strength and adhesive to end-use properties by providing a moderate form of crosslinking.

Ionic Initiator *n* A substance providing either carbonium ions (cationic) or carbanions (anionic) that attack the reactive double bonds of vinyl monomers and add on, regenerating the ion species on the propagating chain. (Odian, G. C., *Principles of Polymerization*, Wiley, New York, 2004).

Ionic Polymerization *n* (cationic polymerization, anionic polymerization) A polymerization conducted in the presence of electrically charged ions that become attached to carboxylic groups on carbon atoms in the polymer chain. The carboxylic groups are produced along the polymer chain by copolymerization, providing the anionic portion of the ionic crosslinks. The cationic portion is provided by metallic ions added to the polymerization mixture. The electrostatic forces binding the chains together are much stronger than the covalent bonds between the molecules in conventional polymers. Some polymers produced by cationic polymerization are polyisobutylene, butyl rubber, polyvinyl ethers, and coumarone-indene resins. A typical product of anionic polymerization is Polybutadiene, prepared with an alkali-metal catalyst. (Odian, G. C., *Principles of Polymerization*, Wiley, New York, 2004).

Ionic Polyurethane *n* A urethane resin containing electrical charges in its backbone or side chain. Cationic urethanes (those containing positive changes) can be formed by reacting diisocyanates with diols containing tertiary nitrogen to yield urethanes that are then treated by a quaternization reaction that forms positive charges in the macromolecular backbone or in side chains.

Ionic Solid *n* A solid composed of anions and cations in its lattice.

Ionization *v* [ISV] (1898) (1) Loss of an electron by an atom, molecule, or ion; (2) dissociation of an electrolyte. It also can be written as a chemical change by which ions are formed from a neutral molecule of an inorganic solid, liquid, or gas.

Ionization Chamber *n* (1904) A partially evacuated tube provided with electrodes so that its conductivity due to the ionization of the residual gas reveals the presence of ionizing radiation.

Ionization Energy *n* The energy used to remove an electron from a gaseous, isolated, ground-state atom (or, sometimes, ion). () Also known *as Ionization Potential*.

Ionization Foaming *n* The process of foaming polyethylene by exposing it to ionizing radiation which evolves hydrogen from the molten polymer, causing it to foam.

Ionization Potential *n* The work (expressed in electron volts) required to remove a given electron from its atomic orbit and place it at rest at an infinite distance. It is customary to list values in electron volts (ev.) 1 ev. − 23,053 cal/mol. Also known as *Ionization Energy*.

Ionomer *n* (1) Polymers with repeating ionic groups which tend to form ionic domains that act as physical cross-links. The domains disassociate on heating allowing the material to be processed as a thermoplastic. (Odian, G. C., *Principles of Polymerization*, Wiley, New York, 2004). (2) Thermoplastics containing a relatively small amount of pendant ionized acid groups. Have good flexibility and impact strength in a wide temperature range, puncture and chemical resistance, adhesion, and dielectric properties, but poor weatherability, fire resistance, and thermal stability. Processed by injection, blow and rotational molding, blown film extrusion, and extrusion coating. Used in food packaging, auto bumpers, sporting goods, and foam sheets. (Sepe, M. P., *Dynamic Mechanical Analysis*, Plastics Design Library, Norwich, New York, 1998).

Ionomer Resin *n* A polymer containing interchain ionic bonding. In particular, commercial thermoplastics based on metal salts of copolymers of ethylene and methacrylic acid, produced in the U.S. by the DuPont Co under the trade name Surlyn®. Ionomers are tough and flexible, the many grades ranging in modulus from 14 to 590 MPa. They have been approved for food-contact applications by the FDA, have outstanding resistance to puncture and impact. Applications include sporting goods (most golf-ball covers), footwear, packaging, automotive, and foams. (*Physical Properties of Polymers Handbook*, Mark, J. E., ed., Springer, New York, 1996; *Modern Plastics Encyclopedia*, McGraw-Hill/Modern Plastics, New York).

Ion Pair *n* Pairs of oppositely charged ions held together by columbic attraction without formation of a covalent bond. Experimentally, an ion pair behaves as one unit in determining conductivity, kinetic behavior, osmotic properties, etc. (Whitten, K. W., et al., *General Chemistry*, Brookes/Cole, New York, 2003).

Ion Plating *n* A process for deposition of metals for dielectric films onto plastic substrates with a highly adherent bond. The process is performed in a tank similar to a vacuum-metallizing tank. A negative charge is developed on a metal bias plate located behind the plastic substrate. Next, the plating material is converted to a plasma of positive ions by filament or radio-frequency heating. At this point a phenomenon known as Crooke's dark space appears, enveloping the entire surface of the substrate, and establishing a large potential difference between the ions and the charged plastic surface. This causes the ions of the plating material to strike the plastic with high kinetic energy and to form strong bonds. (*Whittington's Dictionary of Plastics*, Carley, J. F., ed., Technomic Publishing, 1993).

Ion Product *n* (1) The product of the concentrations of hydrogen (hydronium) ions and hydroxide ions in water; (2) the mass-action expression for a solubility equilibrium.

IPA Abbreviation for ▶ Isophthalic Acid.

IPN Abbreviation for ▶ Interpenetrating Polymer Network.

IPS See ▶ Impact Polystyrene.

IR *n* (1) Abbreviation for ▶ Infrared. (2) Abbreviation for ▶ Isoprene Rubber (British Standards Insitution), the *cis*-1,4-type of ▶ Polyisoprene.

Iridescence \ˌir-ə-ˈde-s°n(t)s\ *n* (1804) Color produced from light interference phenomenon. (Serway, R. A., et al., *College Physics*, Thomas, New York, 2005; *Colour Physics for Industry*, 2nd Ed., McDonald, Roderick, Society of Deyes and Colourists, West Yorkshire, UK, 1997) See ▶ Interference Color.

Irish Gum See ▶ Carrageen.

Iron Blue *n* $FeNH_4F3(CN)_6$. Any of various pigments prepared by precipitating ferrous ferrocyanide from a soluble ferrocyanide and ferrous sulfate. (*Kirk-Othmer Encyclopedia of Chemical Technology: Pigments-Powders*, Wiley, New York, 1996; Solomon, D. H., and Hawthorne, D. G., *Chemistry of Pigments and Fillers*, Krieger Publishing, New York, 1991).

Iron Driers *n* Iron salts of naphthenic acid, 2-ethyl hexoic acids or other acids which actively accelerate polymerization at elevated temperatures, but are only feebly active at room temperature. The presence of the very dark ferric ion makes them applicable in dark or tinted finishes. The British call then iron soaps. (Wicks, Z. N., et al., *Organic Coatings Science and Technology*, 2nd Ed., S. P., Wiley-Interscience, New York, 1999).

Iron Mill *n* Paint mill consisting of a corrugated steel disc revolving tightly against a stationary steel shell. As the pigment particles pass between these moving steel parts, the pigment is dispersed into the vehicle.

Iron Oxides *n* (1885) Fe_2O_3. Pigments which are substantially oxides of iron. (1) *Natural*: Ochres, raw and burnt umbers; raw and burnt siennas; red oxides; metallic browns; maroon oxides; and black oxides. Range in Fe_2O_3 content: 20–99%. (2) *Synthetic*: Yellow, red, brown, black oxides. *Red oxide* – Fe_2O_3; *Yellow oxide* – $Fe_2O_3 \cdot H_2O$; *Brown oxide* – $FeO_3 \cdot xFeO$; *Black oxide* - Fe_2O_4. (*Paint/Coatings Dictionary*, Federation of Societies for Coatings Technology, Philadelphia, Blue Bell, PA, 1978).

Iron Oxides, Synthetic *n* (e.g., $Fe_2O_3 \cdot H_2O$) Iron oxides manufactured by a synthetic process. (*Kirk-Othmer Encyclopedia of Chemical Technology: Pigments-Powders*, Wiley, New York, 1996; Solomon, D. H., and

Hawthorne, D. G., *Chemistry of Pigments and Fillers*, Krieger Publishing, New York, 1991).

Iron Phosphate Coating See ▶ Chemical Conversion Coating.

Iron Soaps *n* (British) See ▶ Iron Driers.

Iron Titanate See ▶ Ilmenite.

Irradiance \i-ˈrā-dē-ən(t)s\ *n* (1667) The quotient of the radiant flux incident on an infinitesimal surface element containing the point in question, by the area of that surface element. (Serway, R. A., et al., *College Physics*, Thomas, New York, 2005; Moller, K. D., *Optics*, Springer, New York, 2003).

Irradiation of Polymers *n* (1901) The subjection of a material to high-energy particle radiation for the purpose of producing a desired change in properties or of determining the effects of radiation on the material. Thermosetting resins such as unsaturated polyesters, acrylic-modified polyesters, and acrylic-modified epoxies can be cured rapidly at room temperature and without catalysts by exposure to ionizing radiation. However, if the polymer is overdosed, serious degradation can occur. Radiation sources most widely used in the plastics industry are electron accelerators and radioisotopes such as cobalt-60. (*Degradation and Stabilization of Polymers*, Zaiko, G. E., ed., Nova Science Publishers, New York, 1995).

Irregular Block *n* A block (in a polymer structure) that cannot be described by only one species of constitutional repeating unit in a single sequential arrangement (Odian, G. C., *Principles of Polymerization*, Wiley, New York, 2004).

Irregular Polymer *n* A polymer whose molecules cannot be described by only one species of constitutional unit in a single sequential arrangement (Odian, G. C., *Principles of Polymerization*, Wiley, New York, 2004).

Isanolic Acid *n* Eighteen-carbon fatty acid with acetylene triple bonds at the 9 and 11 positions, and having also an ethylene bond but no hydroxyl substituent. Occurs in isano oil, along with closely related bolekic acid, which is a hydroxy acid. *Also known as Erythrogenic Acid.*

Isano Oil *n* A fatty oil extracted from an African tree of the same name, used as a flame retardant for acrylic resins. When heated to 200°C it polymerizes and may explode. See ▶ Isanolic Acid.

ISCC Abbreviation for ▶ Inter-Society Color Council.

Iso- (1) The strict meaning of this prefix according to chemical nomenclature is "one methyl group on the next-to-last carbon atom, and no other branches". In the plasticizer field, the prefix is used to denote an isomer of a compound, specifically an isomer having a single, simple branching, not limited to methyl, at the end of a straight chain. (2) Equal, same, or constant, as in *isothermal*, at constant temperature. (Morrison, R. T., and Boyd, R. N., *Organic Chemistry*, 6th Ed., Prentice Hall, Englewood Cliffs, NJ, 1992).

ISO *n* Abbreviation for International Organization For Standardization, headquartered in Geneva, Switzerland,. ISO publishes standards in many fields, including hundreds on plastics in Field 170. ISO standards are available from ▶ American National Standards Institute.

ISO 75 An International Organization for Standardization (ISO) standard test method for determination of heat deflection temperature (HDT) and deflection temperature under load (DTUL). HDT is a relative measure of a material's ability to perform for a short time at elevated temperatures while supporting a load. The test measures the effect of temperature on stiffness; a standard test specimen is given a defined surface stress and the temperature is raised at a uniform rate. Alternate test methods for HDT and DTUL are DIN 53461 and ASTM D648.

In both ISO and ASTM standards, a loaded test bar is placed in a silicone oil filled heating bath. The surface stress on the specimen is either: low – for ASTM and ISO both 0.45 MPa; high = for ASTM 1.82 MPa and for ISO 1.80 MPa. The force is allowed to act for 5 min; this waiting period may be omitted when testing materials that show no appreciable creep during the initial 5 min. After 5 min the original bath temperature of 23°C is raised at a uniform rate of 2°C/min.

The deflection of the test bar is continuously observed; the temperature at which the deflection reaches 0.32 mm (ISO) or 0.25 mm (ASTM), is reported as 'deflection temperature under load' or 'heat deflection temperature.' Although not mentioned in either test standard, it has become common practice to use the acronym DTUL for ASTM values and HDT for ISO values. Depending upon the applied surface stress, the letters A or B are added to HDT; HDT/A for a load of 1.80 MPa; HDT/B for a load of 0.45 MPa.

ISO 2039-2 An International Organization for Standardization (ISO) standard test method for determination of indention hardness of plastics by Rockwell tester using Rockwell M, L, and R hardness scales. The hardness number is derived from the net increase in the depth of impression as the load on a ball indenter is increased from a fixed minor load (98.07 N) to a major load and then returned to the minor load. This number consists of the number of scale divisions (each corresponding to 0.002 mm vertical movement of the indentor) and scale symbol. Rockwell scale vary depending on the diameter of the indentor and the major load. For example, scale R corresponds to the ball diameter 12.7 mm and major load 588.4 N Also called ISO 2039-B.

ISO 2039-B See also ▶ ISO 2039-2.

Isoamyl Acetate Rectified ▶ Amyl Acetate.

Isoamyl Butyrate n $C_5H_{11}OOCC_3H_7$. A colorless liquid derived by treating isoamyl alcohol with butyric acid, used as a solvent and as a plasticizer for cellulose acetate.

Isoamyl Salicylate n (amyl Salicylate, orchidae) $C_5H_{11}OOCC_6H_4OH$. A colorless liquid used as a plasticizer.

Isobar \ˈī-sə-ˌbär\ n [ISV is- + -bar (fr. Gk baros weight); akin to Gk barys heavy] (ca. 1864) For chemistry, elements of the same atomic mass but of different atomic numbers. The sum of their nucleons is the same but there are more protons in one than in the other.

Isobaric \ˌī-sə-ˈbär-ik, -ˈbar-\ adj [ISV] (1878) Taking place without change in pressure.

Isobutane \ˌī-sō-ˈbyü-ˌtān\ n [ISV] (1876) (isobutylene, 2-methylpropene) $(CH_3)_2C = CH_2$. A colorless, highly volatile liquid or liquidified gas derived from petroleum, easily polymerized to form polybutene. It has a bp of −6.9°C; a fp of −139°C, a flp of −105°C, and sp gr of 0.6 (20°C). Uses: Production of isooctane, butyl rubber, Polyisobutene resins, *tert*-butanol, methacrylates, copolymer resins with butadiene, and acrylonitrile.

Isobutyl Acetate $(CH_3)_2CHCH_2OOCCH_3$. A colorless liquid with a fruity odor, used as a solvent for cellulosic plastics and lacquers. Its properties are similar to those of butyl acetate, except that its evaporation rate is higher.

Isobutyl Alcohol n $(CH_3)_2CHCH_2OH$. Colorless flammable liquid used as a solvent. Sp gr, 0.806; bp, 108.39. *Also known as 2-Methyl-1-Propanol and Isopropyl Carbinol.*

Isobutylene \ˈbyü-t°l-ˌēn\ n [ISV] (1872) A colorless, very volatile liquid or flammable gas, C_4H_8, used chiefly in the manufacture of butyl rubber.

Isobutyl Isobutyrate n $(CH_3)_2CHCOOCH_2CH(CH_3)_2$. A colorless liquid with a fruity odor and slow evaporation rate, giving resin solutions with good flow and leveling characteristics. It is used as a solvent for nitrocellulose and vinyl resins.

Isocyanate \ˈī-sō-ˈsī-ə-nāt, -nət\ *n* [ISV] (1872) A compound containing the isocyanate group, $-N=C=O$, attached to an organic radical or hydrogen. Isocyanates containing just one $-N=C=O$ group (monoisocyanates) have limited uses in the plastics industry. The term is often used to mean a compound containing two $-N=C=O$ groups (diisocyanate) or several such groups (polyisocyanate). However, in the case of a trimer compound containing three $-N=C=O$ groups in a six-membered ring, the term *isocyanurate* is used. (Morrison, R. T., and Boyd, R. N., *Organic Chemistry*, 6th Ed., Prentice Hall, Englewood Cliffs, NJ, 1992). See also ▶ Diisocyanate.

$$O=C=N^-$$

Isocyanate-Alcohol Reaction *n* Reaction involving a very reactive chemical group, $-N=C=O$ with an alcohol to form a highly crosslinked polymer, importance in polyurethane formation. (Odian, G. C., *Principles of Polymerization*, Wiley, New York, 2004).

Isocyanate Foam See ▶ Urethane Foam.

Isocyanate Generator *n* (hindered isocyanate) A mixture of an isocyanate, a phenol, and a polyester that remains stable at room temperature. When heated to 70°C, the phenol and isocyanate components dissociate and react with the polyester to form a polyurethane.

Isocyanate Plastic *n* A plastic based on polymers made by the polycondensation of organic isocyanates with other compounds. Reaction of isocyanates with hydroxyl-containing compounds produces polyurethanes having the urethane group $-NHC(=O)O-$. Reaction of isocyanates with amine-containing compounds produces polyurea having the urea group $-NHCONH-$ (ISO). (Odian, G. C., *Principles of Polymerization*, Wiley, New York, 2004). See also ▶ Urethane and ▶ Polyurethane Foam.

Isocyanate Resins *n* Resins synthesized from isocyanates ($-N=C=O$) and alcohols ($-OH$). The reactants are joined through the formation of the urethane linkage and hence this field of technology is generally known as urethane chemistry. See ▶ Polyurethanes. (Odian, G. C., *Principles of Polymerization*, Wiley, New York, 2004).

Isocyanates *n* Compounds containing the isocyanate group $-N=C=O$, that are which is very reactive, especially with compounds containing active hydrogen atoms, such as in amines, alcohols, carboxylic acids and water. (*Whittington's Dictionary of Plastics*, Carley, J. F., ed., Technomic Publishing, 1993a, b).

Isocyanurate Foam *n* A foam prepared from an isocyanurate. The unmodified foams have excellent flame resistance but are brittle and of little commercial value. However, isocyanurate foams modified with epoxides, polyimides, or (most commonly) urethane groups and polyols possess flame resistance far superior to that of conventional urethane foams and can be processed into a variety of foam products suitable for insulation. (*Whittington's Dictionary of Plastics*, Carley, J. F., ed., Technomic Publishing, 1993).

Isocyanurate Plastic *n* A plastic based on isocyanate polymers in which trimerization of the isocyanates incorporates six-membered isocyanurate-ring groups in a chain. (Elias, H. G., *Macromolecules Vol. 1–2*, Plenum, New York, 1977).

Isocyaurate *n* A trimer of an isocyanate, formed by the catalytic cyclization of three isocyanate molecular groups into a six-membered ring. (Elias, H. G., *Macromolecules Vol. 1–2*, Plenum, New York, 1977).

Isodecyl Octyl Adipate *n* $C_{10}H_{21}OOCC_4H_8COOC_8H_{17}$. A light-colored, oily liquid used as a plasticizer for vinyls.

Isodecyl Octyl Phthalate *n* $C_{10}H_{21}OOCC_6H_4COOC_8H_{17}$. A primary plasticizer for vinyls, cellulose nitrate, polystyrene, and ethyl cellulose. In vinyls, it performs better than diethylhexyl phthalate (DOP).

Isoelectric Heating Form of heating by means of electrical energy, in which a special form of element is employed, and which is submerged in liquid raw materials. In practice a series of such elements is employed.

Isoelectric Point *n* The pH value at which a substance or system (e.g., a protein solution) is electrically neutral; at this value, electrophoresis does not occur when a direct electric current is applied. (Goldberg, D. E., *Fundamentals of Chemistry*, McGraw-Hill Science/Engineering/Math, New York, 2003).

Isogyres In a uniaxial interference figure, the two black bars that form a cross and represent the pattern of extinction positions of the crystal. In a biaxial interference figure, the two black, curved bars (brushes) that intersect the isochromatic curves (lemniscates) and represent the pattern of extinction positions of the crystal. (Hibbard, M. J., *Mineralogy*, McGraw-Hill, New York; Rhodes, G., *Crystallography Made Crystal Clear: A Guide for Users of Macromolecular Models*, Elsevier Science and Technology Books, New York, 1999).

Isoindolinone *n* Any of a small family of organic pigments, available in bright yellows and reds. They have good lightfastness, heat stability and bleed resistance.

Isolated System *n* A system which can exchange neither matter nor energy in any form with its surroundings. (Whitten, K. W., et al., *General Chemistry*, Brookes/Cole, New York, 2003).

Isomer \ˈī-sə-mər\ *n* [ISV, back-formation fr. *isomeric*] (1866) From the Greek *isos* (the same, equal, alike and *meros* (part or portion), isomers are substances comprising molecules that contain the same number and kinds of atoms, and have the same chemical formula, but that differ in structure, so that they form materials whose properties can differ widely. For example, the gas dimethyl ether, CH_3OCH_3, and the liquid ethyl alcohol. CH_3CH_2OH differ in both their chemical and physical properties, yet both have the empirical formula C_2H_6O. Isomeric polymers are formed by polymerizing isomonomers that link together in different ways. (Morrison, R. T., and Boyd, R. N., *Organic Chemistry*, 6th Ed., Prentice Hall, Englewood Cliffs, NJ, 1992).

Isomeric Polymers *n* Polymers which have essentially the same percentage composition, but differ with regard to their molecular architecture. (Odian, G. C., *Principles of Polymerization*, Wiley, New York, 2004; Elias, H. G., *Macromolecules Vol. 1–2*, Plenum, New York, 1977).

Isomerism *n* (1839) Existence of molecules having the same number and kinds of atoms but in different configurations. (Smith, M. B., and March, J., *Advanced Organic Chemistry*, 5th Ed., Wiley, New York, 2001; Morrison, R. T., and Boyd, R. N., *Organic Chemistry*, 6th Ed., Prentice Hall, Englewood Cliffs, NJ, 1992).

Isomerized Oils *n* Oils which have been isomerized from a molecular rearrangement by several means, such as the catalytic action of alkalis, sulfur dioxide, nickel, etc. Isomerized oils exhibit greater reactivity, more rapid polymerization and greater drying rate than their non-isomerized parent oils. (*Paint: Pigment, Drying Oils, Polymers, Resins, Naval Stores, Cellulosics Esters, and Ink Vehicles, Vol. 3*, American Society for Testing and Material, 2001; *Bailey's Industrial Oil and Fat Products*, Shahidi, F., and Bailey, A. E., eds., Wiley, 2005).

Isomerized Rubber *n* Rubber which has undergone cyclic rearrangement by heating in solution in the presence of a suitable catalyst. (*Whittington's Dictionary of Plastics*, Carley, J. F., ed., Technomic Publishing, 1993). Also referred to as *Cyclized Rubber*.

Isomorphism \ˈī-sə-ˈmór-ˌfi-zəm\ *n* [ISV] (ca. 1828) The existence of two or more chemical compounds with the same molecular formula but having different properties, owing to different arrangement of atoms within the molecule, e.g., ammonium cyanate NH_4CNO, and urea, $CO(NH_2)_2$, are isomers.

Isooctyl Adipate See ▶ Diisooctyl Adipate.

Isooctyl Palmitate *n* $C_8H_{17}OOCC_{15}H_{31}$. A plasticizer for polystyrene and cellulosic plastics.

Ispthalate Polyester *n* an unsaturated polyester based on isophythalic acid.

Isophorone *n* A cyclic, unsaturated ketone with the structure:

It is a powerful solvent for vinyl and cellulosic resins, with moderate power to dissolve nearly all common thermoplastic and (uncured) thermosetting resins.

Isophorone Diisocyanate *n* (IPDI) An isocyanate used in the production of urethane elastomers and foams. It is less volatile than toluene diisocyanate, therefore easier to maintain at low levels in workers' airspace and safer to work with. Its structure is modified from that of ▶ Isophorone, above, in that the oxygen has been replaced by an —N = C = O group, the double bond is gone, and the top carbon atom in the ring has an additional —CH$_2$N = C = O linked to it.

Isophthalic Acid *n* (benzene-1,3-dicarboxylic acid, IPA) C$_6$H$_4$-(COOH)$_2$. Used instead of phthalic anhydride in making unsaturated polyester resins that, when cured, have good stiffness and resistance to heat and chemicals. Molecular weight, 166.13. Crystalline powder which melts at 345–348°C; sublimes without decomposition. (*Merck Index*, 13th Ed., Merck and Company, Whitehouse Station, NJ, 2001).

Isoprene \ˈī-sə-ˌprēn\ *n* [prob. fr. *is-* + *propyl* + *-ene*] (1860) (3-methyl-1,3-butadiene, 2-methyl-1,3-butadiene) CH$_2$ = C-(CH$_3$)CH = CH$_2$. A colorless, volatile liquid derived from propylene or from coal gases or tars, chemically similar to the mer unit of natural rubber. Its polymer of the *cis*-1,4-type of polyisoprene is chemistry's nearest approach to synthesizing the natural product and it has sometimes been called "synthetic natural rubber". It has a molecular weight of 68.06, a mp of −120°C, and a bp of 34°C. (Ash, M. and Ash I., *Encyclopedia of Plastics Polymers, and Resins*. Vols. I–III, Chemical Publishing, New York, 1982–83) Also called *2-Methyl-1,3-Butadiene, β-Methylbivinyl; Hemiterpene.*

Isoprene Rubber *n* The *cis*-1,4-type of ▶ Polyisoprene.

Isopropyl Acetate *n* (CH$_3$)$_2$CHOOCCH$_3$. A colorless, fragrant liquid used as a solvent for cellulose nitrate, ethyl cellulose, polyvinyl acetate, polymethyl methacrylate, polystyrene, and certain phenolic and alkyd resins. It has bp of 90°C, sp gr of 0.930, flp of 33°F, and a vp of 45 mm Hg/20°C.

Isopropyl Alcohol *n* (1872) (2-propanol, dimethyl carbinol) (CH$_3$)$_2$CHOH. A colorless solvent boiling at 82.4°C, moderately polar, which, because of its low toxicity, is enjoying greater use today than formerly in the plastics industry. It has a bp of 82°C, sp gr of 0.791/15°C, a flp of 62°F, and a vp of 34 mm Hg/20°C. Also known as *Propanol-2*.

Isopropylbenzene *n* (isopropylbenzol) Syn: ▶ Cumene.

Isopropyl Butyrate *n* (CH$_3$)$_2$CHCOOC$_3$H$_7$. It has a bp of 128°C and sp gr of 0.879/0°C.

Isopropyl Cellosolve *n* (CH$_3$)$_2$CHOCH$_2$CH$_2$OH. It has a bp of 141°C, sp gr of 0.906/20°C, and a vp of 8 mm Hg/ 30°C.

p,p′-Isopropylidenediphenol See ▶ Bisphenol ▶ A.

Isopropyl Myristate $(CH_3)_2CHOOCC_{13}H_{27}$. A plasticizer for cellulosic resins.

Isopropyl Oleate n $(CH_3)_2CHOOCC_{17}H_{33}$. A plasticizer for cellulose nitrate, ethyl cellulose, and polystyrene, and, with partial compatibility, vinyl resins.

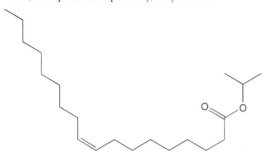

Isopropyl Palmitate n $(CH_3)_2CHOOCC_{15}H_{31}$. A plasticizer for cellulose nitrate and ethyl cellulose.

Isostatic Pressing n The process of molding solid articles from fine powders by enclosing the powder in a closed rubber mold, then immersing the mold in a hydraulic fluid and holding it at a controlled temperature and high pressure for 5 to 30 min. The pressure is equal in all directions and therefore a consistent-molded object is formed. The method though more costly than force-in-mold pressing and much more limited in the shapes producible, provides pieces of more uniform density and strength. (Strong, A. B., *Plastics Materials and Processing*, Prentice Hall, Columbus, Ohio, 2000; *Whittington's Dictionary of Plastics*, Carley, J. F., ed., Technomic Publishing, 1993; *Rosato's Plastics Encyclopedia and Dictionary*, Rosato, D. V., ed., Hanser-Gardner Publications, New York, 1992).

Isotactic \ ▌ī-sə- ▌tak-tik\ *adj* (1955) (1) Denoting a polymer structure in which monomer units attached to a polymer backbone are identical on one side and/or the other side of the backbone. See also ▶ Syndiotactic. (2) Pertaining to a type of polymeric molecular structure containing a sequence of regularly spaced asymmetric atoms arranged in like configuration in a polymer chain (ASTM D 883). Materials containing isotactic molecules may exist in highly crystalline form because of the high degree of order that may be imparted to such structures. (Odian, G. C., *Principles of Polymerization*, Wiley, New York, 2004). See also ▶ Stereospecific.

Isotactic Polymer n A tactic polymer, the base unit of which possesses, as a component of the main chain, a carbon (or similar) atom with two different lateral substituents. These atoms being so arranged that a hypothetical observer advancing along the bonds constituting the main chain finds each of its substituents in the same steric order. NOTE - Here, hydrogen is counted a substituent (IUPAC). Also see ▶ Atactic Polymer, Syndiotactic Polymer, and ▶ Tactic Polymer. (Odian, G. C., *Principles of Polymerization*, Wiley, New York, 2004).

Isotactic Polypropylene n Polypropylene in which each mer unit has the pendant $-CH_3$ group on the same side of the chain backbone. Commercial PPs are about 90% isotactic, conferring high crystallinity and softening range. In contrast, atactic PP is rubbery and weak. (Odian, G. C., *Principles of Polymerization*, Wiley, New York, 2004).

Isotherm \ ▌ī-sə- ▌thərm\ n [F *isotherme*, *adj*] (1859) Constant temperature line used on graphs of climatic conditions or thermodynamic relations, such as pressure-volume relations at constant temperature. (Ready, R. G., *Thermodynamics*, Plenum, New York, 1996).

Isothermal \ ▌ī-sə- ▌thər-məl\ *adj* [F *isotherme*, fr. *is-* + *-therme* fr. Gk *thermos* hot] (1826) When a gas passes through a series of pressure and volume variations without change of temperature, the changes are called isothermal. (Ready, R. G., *Thermodynamics*, Plenum, New York, 1996).

Isotope \ ▌ī-sə- ▌tōp\ n [*is-* + Gk *topos* place] (1913) In chemistry, one of two or more forms of an element ("nuclides") having the same number of protons in the nucleus but differing in mass number because of different numbers of neutrons. Natural elements are usually mixtures of isotopes; thus the observed atomic weights are average values weighted by isotopic relative abundance. (Serway, R. A., et al., *College Physics*, Thomas, New York, 2005).

Isotope Dating A method of determining a period in time using the rate of deterioration of an unstable isotope. (Serway, R. A., et al., *College Physics*, Thomas, New York, 2005). See ▶ Carbon Dating.

Isotope Effect The dependence of a property such as reaction rate on the mass number of an element. (Serway, R. A., et al., *College Physics*, Thomas, New York, 2005).

Isotopes *n* Atoms of an element having different numbers of neutrons in their nuclei, and hence, different mass numbers. Isotopes of a given element have the same number of nuclear protons but differing numbers of neutrons. Naturally occurring chemical elements are usually mixtures of isotopes so that observed (non-integer) atomic weights are average values for the mixture. (Serway, R. A., et al., *College Physics*, Thomas, New York, 2005). See ▶ Atomic Number.

Isotropic *adj* [ISV] (ca. 1860) Substances showing a single refractive index at a given temperature and wavelength no matter what the direction of light through the particle. They show no interference colors between crossed polars. Examples are unstrained glasses, unoriented polymers, and compounds in the cubic system. (Meeten, G. H., *Optical Properties of Polymers*, Springer, New York, 1986).

Isotropic Laminate *n* A laminate in which the strength properties are equal, or approximately equal, in all directions. (Iaac, M. D., and Ishal, O., *Engineering Mechanics of Composite Materials*, Oxford University Press, UK, 2005).

Isotropy *adj* [ISV] (ca. 1860) A materials state in which the material properties are the same in all directions. (Meeten, G. H., *Optical Properties of Polymers*, Springer, New York, 1986).

Itaconic Acid \ˌi-tə-ˈkä-nik-\ *n* [ISV, anagram of *aconitic acid*, $C_3H_3(COOH)_3$, fr. *aconite*] (ca. 1872) (methylenesuccinic acid HOOCC($=CH_2$)CH_2COOH. A white crystalline powder usually obtained by the oxidative fermentation of sucrose or glucose with *Aspergillyus terreus*. It is capable of polymerization alone, or as a comonomer with acrylic acid, acrylonitrile, styrene, methyl methacrylate, and vinylidene chloride. It is used as an additive in acrylic resins to increase their adhesion to cellulose. By polycondensation of itaconic acid with diols, polyesters are obtained that contain methylene side groups. (*Merck Index*, 13th Ed., Merck and Company, Whitehouse Station, NJ, 2001). Also called Methylenebutanedioic Acid and Methylene-Succinic Acid.

IUPAC *n* Abbreviation for International Union of Pure and Applied Chemistry, headquartered in Oxford, England. (*IUPAC Handbook, 2000–2001*, International Union of Pure and Applied Chemistry, 2000; *Progress in Polymer Science and Technology: 2002 IUPAC World Polymer Congress, Beijing, China, July 7–12, 2002*, Wiley, New York).

IVE Abbreviation for ▶ Vinylisobutyl Ether.

Izod See Izon Impact Strength.

Izod Impact See ▶ Izod Impact Energy.

Izod Impact Energy *n* The energy required to break a specimen equal to the difference between the energy in the striking member of the Izod-type impact apparatus at the instant of impact and the energy remaining after complete fracture of the specimen. Also called Izod impact, Izod impact strength, Izod.

Izod Impact Strength *n* A widely used measure of impact strength (www.astm.org) determined by the difference in energy of a swinging pendulum before and after it breaks a notched specimen clamped vertically as a cantilever beam. The pendulum is released from a vertical height of 0.61 m, and the vertical height to which it rises after breaking the specimen is used to calculate the energy expended in that breakage. The notch across the width of the specimen is usually on the side opposite the side impacted, but the reverse setup is provided for and specimens may also be tested unnotched. (Shah, V., *Handbook of Plastics Testing Technology*, Wiley, New York, 1998; *Handbook of Physical Polymer Testing, Vol. 50*, Brown, R., Marcel Dekker, New York, 1999; www.astm,org).